order: in life

EDMUND SAMUEL

ASSOCIATE PROFESSOR OF BIOLOGY
ANTIOCH COLLEGE

PRENTICE-HALL, INC.

ENGLEWOOD CLIFFS, NEW JERSEY

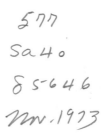

10 9 8 7 6 5 4 3 2 1

ISBN: 0-13-639708-5

Library of Congress Catalog Card Number: 70-177392

Printed in the United States of America

PRENTICE-HALL INTERNATIONAL, INC., *London*
PRENTICE-HALL OF AUSTRALIA, PTY. LTD., *Sydney*
PRENTICE-HALL OF CANADA, LTD., *Toronto*
PRENTICE-HALL OF INDIA PRIVATE LIMITED, *New Delhi*
PRENTICE-HALL OF JAPAN, INC., *Tokyo*

TO YOSHIKO

contents

chapter

i   *the search for order*   1

THE LIFE OF A SCIENTIST   1

SETTING THE STAGE FOR A NEW CONCEPT   3

CONTRIBUTIONS OF PREDECESSOR TO THE CONCEPT—

BELIEF-WEBS   4

THE SYNTHESIS   8

THE INFINITENESS OF THE SCIENTIFIC PROCESS   13

SCIENCE AND OTHER WAYS OF THOUGHT   14

WHAT ABOUT BIOLOGY?   16

ii   *order: in diversity*   18

EARLY SYSTEMS OF CLASSIFICATION   19

JOHN RAY AND NATURAL CLASSIFICATION SCHEMES   22

CAROLUS LINNAEUS AND HIS ARTIFICIAL SYSTEM OF

CLASSIFICATION   23

PLANT CLASSIFICATION SINCE LINNAEUS   28

ANIMAL CLASSIFICATION   28

MODERN SYSTEMATIC METHODS—NUMERICAL ANALYSIS   30

THE SPECIES PROBLEM   33

THOUGHTS   36

iii *order: in fitness* 38

GROWTH OF A SPECIES 40
SPECIFIC SPECIES-SPECIES INTERACTIONS: HOST-PARASITE
PREDATOR-PREY RELATIONSHIPS 45
PREDATOR-PREDATOR COMPETITION: STABILITY AND THE
NICHE 48
THE WHOLE NATURAL AREA 55
THE COMMUNITY CONCEPT 60
THE ECOSYSTEM CONCEPT 62
GROWTH AND STABILITY OF A "COMMUNITY"—
SUCCESSION AND CLIMAXES 65
THOUGHTS 71

iv *order: in space and time* 74

A MEETING AT THE PUB 75
THE YOUNG NATURALIST'S CREDENTIALS 76
NEPTUNISM VERSUS VOLCANISM 78
INTERPRETATIONS OF FOSSILS—PROGRESSIONISM 81
THE VOYAGE BEGINS 84
SIR CHARLES LYELL AND THE "PRINCIPLES OF GEOLOGY" 84
A HYPOTHESIS ON THE FORMATION OF CORAL ISLANDS 86
FOSSIL EVIDENCE FROM SOUTH AMERICA 89
EXTINCTION 92
ENDEMIC SPECIES—THE GALAPAGOS ISLAND ANOMALY 95
AN EXPLANATION 98
THE HISTORY OF THE IDEA—SPECIES MUTABILITY 100
THE SEARCH FOR A MECHANISM 105
CONTRIBUTORS TO THE MECHANISM OF NATURAL SELECTION 108
SOME FINAL QUESTIONS 110
PUBLISHING THE "ORIGIN" 113
THOUGHTS 119

v *order: in form* 124

EARLY CONCEPTS OF THE ORGANISM—ARISTOTLE'S MODEL 125
GALEN'S MODEL OF THE ORGANISM 128

THE RENAISSANCE AND GALEN'S MODEL   133

HARVEY'S MODEL OF THE CIRCULATION SYSTEM   134

VITAL SPIRITS AND AIR   138

THE PLANT AS AN ORGANISM   144

THE GEOMETRY OF THE WHOLE ORGANISM—DA VINCI
AND DÜRER   147

PHYSICAL LAWS OF THE WHOLE ORGANISM—GALILEO   149

MORE ON PHYLLOTAXY—GOETHE   150

D'ARCY THOMPSON—STUDENT OF FORM   152

MORE ON SPIRALS   152

PHYSICAL LAWS AFFECTING SIZE   153

ALLOMETRY   155

THE DETERMINATION OF SHAPE   157

CRYSTALLIZATION AND THE SHAPE OF LIVING ORGANISMS   160

TRANSFORMATIONS OF COORDINATES—DIAGRAM OF FORCES   161

COMPARATIVE ANATOMY AND PHYSIOLOGY—CONTRIBUTIONS
TO EVOLUTION THEORY   165

EARLY CONCEPTS OF THE DEVELOPMENT OF FORM—
PREFORMATION VERSUS EPIGENESIS   168

DYNAMIC MORPHOLOGY EMERGES—WOLFF AND GOETHE   171

MOSAIC VERSUS REGULATIVE DEVELOPMENT—ROUX,
DREISCH, AND SPEMANN   173

DEVELOPMENT AND EVOLUTION   178

THOUGHTS   184

vi   *order: in continuity*   186

EARLY EXPLANATIONS OF HEREDITY   187

DISCOVERING THE CELL   188

THE CELL THEORY   189

PANMERISM   192

WEISMANN'S GERM PLASM   195

STATISTICAL APPROACH TO HEREDITY—GALTON   201

SEGREGATED CHARACTERISTICS—MENDEL   202

THE SYNTHESIS OF HEREDITY FACTORS AND CYTOLOGICAL
OBSERVATIONS—SUTTON   205

THE CHROMOSOMAL BASIS OF HEREDITY   209

THE GENE THEORY OF HEREDITY—CROSSOVER   213

CHARACTERISTICS OF THE GENE   217
REPRODUCTION AND EVOLUTION   219
THE LIFE CYCLE CONCEPT   225
SPECIATION   227
MAN—SOME QUESTIONS ON NATURAL SELECTION AND
SPECIATION   230
THOUGHTS   232

vii *order: in matter and energy*   234

EARLY IDEAS ON THE "SOUL" OF LIFE SUBSTANCE   235
CANDLES AND CLOCKS   236
TRANSFORMATIONS AND ORGANIC MATTER   237
THE CONSERVATION OF MATTER   241
INORGANIC(?) CATALYSIS   243
CELLULAR CATALYSTS   245
AUTOGENERATION   247
AUTOGENERATION AND EVOLUTION   249
ORGANIC MOLECULES—EARLY CONCEPTS   251
THE CONSERVATION OF ENERGY   252
ENERGY AND VITAL PROCESSES   254
SUBCELLULAR ELEMENTS OF LIFE   256
THE MOLECULAR FITNESS OF THE ENVIRONMENT   257
THE IMPROBABILITY OF LIFE TO ARISE FROM INANIMATE
MATTER   259
A MODEL FOR THE ORIGIN OF LIFE   261
SUPPORT FOR THE MODEL   268
THE INEVITABILITY OF LIFE   271
EVOLUTION AND THE ORIGIN OF LIFE   271
THE LEVELS OF ORGANIZATION OF MATTER   273
THOUGHTS   278

viii *order: in composition*   280

PATHWAYS OF MATTER   281
THE FLOW OF ENERGY   286
PROTEIN STRUCTURE   290

ENZYME MECHANISMS   297

PROTEIN COMPLEXES   300

PROTEINS AND GENES   306

PROTEINS AND INTERDEPENDENCY   314

PROTEINS AND REGULATION   317

CELL MODELS AND CONTROVERSY   322

CELL ORIGIN   325

THOUGHTS   329

ix *order: in enlargement*   331

SPATIAL CONTROL IN UNICELLS   333

TEMPORAL CONTROL IN UNICELLS   337

SPATIAL CONTROL IN MULTICELLS   341

TEMPORAL CONTROL IN MULTICELLS   344

THE FINAL DIFFERENTIATED STATE—THE ADULT   349

STABILITY OF THE DIFFERENTIATED STATE   358

THOUGHTS   364

x *order: in awareness*   366

INPUT INFORMATION   368

INPUT WIRING   369

WIRED RESPONSE   372

INFORMATION STORAGE   377

COMPUTERS AND THE BRAIN   384

IDEAS AND THE INNER UNIVERSE   386

THOUGHTS ON THOUGHTS   399

A PERSONAL THOUGHT   395

index   397

preface

First let me tell you what this book is not. It is not an introductory text-book dedicated to reducing the fear or abhorrence of science for the nonscience student. It is not an introductory fact-supplying document for the science major. Nor does it help define or suggest ways that our many physical and social problems may be alleviated by biological means. Rather this book is an attempt to help those who have decided that science is their thing, for some purpose or other, to do that thing a little bit better. Although it is about life processes, it should be valuable to any upper-class science major who has dutifully learned many facts and skills within his discipline. It may even be useful to students of the history or philosophy of science. What it attempts to do is to expose the reader to a variety of views of the forest of knowledge after the trees of facts have been planted in his mind. The assumption is that it is both undesirable and difficult to think about science when one is learning it but that later, in the quietness of one's greater confidence, one is much more in need and able to think about his own ways of thought. And also, if he can find his own best styles of thought, he can perfect these and be that much more successful.

The rationale for using the life sciences as the subject matter for this task is that all styles of scientific thinking for setting up explanations seem to be represented. Biology is rich in multiple and complementary explanations for the same phenomenon. There are explanations as pure as those of mathematics or as empirical as those of psychology. There are the same kinds of limitations and inherent ambiguities in its explanations as there are in those of physics. There are also so many basic unanswered questions. And finally the life sciences are said to be the most rapidly growing.

But this is not a philosophy of science textbook. It deals primarily with the orders that have been discovered in the diverse array of biological matter, the explanations of these orders, and then some analysis of the relationship among these explanations. In order to do this, each chapter is devoted to an analysis of the search for and diverse explanations given for some identifiable order in life. These particular orders are not to be thought of as sacred or definitive. They are this author's choice. Others could have been used and may even be better. They were chosen because they do seem to each make a broad sweep across a great deal of biological material and include a number of concepts. Each tends to have characteristic styles of scientific thought used for their explanations, and they can be related to one another and form a systematic organization. There is, however, an inherent difficulty with this strategy that must be made clear. There is no certain way to categorize the various styles of scientific thinking, and it is misleading to think that one style fits one order to be explained. It should be clear as one reads that diverse styles are used for explaining the same order and that the same style may be useful for different orders. The relative quality of explanations is a typical area of philosophy of science that this author does not feel adequate to discuss. On the other hand, with this understanding, there is hopefully still value in the procedure used. By this device the reader may be able to recognize more easily styles of thought that are most him. And by such a recognition he should be able to realize his limitations and opportunities and improve himself in a realistic and productive manner.

History is used generously to show the progressive patterns of explanations that have been devised for a given order. It applies more adequately in some chapters than others, least ably those involving recently derived explanations. In addition, the sequence of chapters, orders, are organized with orders that were discovered earliest being considered first, with the more recently found and still to be defined orders coming last. This too adds continuity to the book with two major themes running in natural sequence with one another: the first, evolution, and the second, levels of organization of matter and energy. The evolution part consists of a descension, if you would, from macroscopic, local, and more subjective observations to microscopic, universal, and abstract observations and concepts. The second part ascends as simple to complex levels of organization of matter are considered. This descension and ascension pattern passes back and forth through communities, organisms, cells, and subcellular parts, which has the advantage of presenting different styles of thought and explanations for the same size aggregate of living material. This pattern is to achieve one of the main purposes of this book, to encourage pluralistic thinking wherein one actually thinks in terms of the same phenomenon with more than one model or explanation. Whether this book serves that purpose only the reader can say. It seems to me, however, that some book with this purpose in mind is necessary for the students' sake if not for science and truth.

I wish to also acknowledge the generous and sensitive help without which this writing could never have been done. First I must thank my many students who have taught me what is here. I thank Dr. John T. Bonner for helping me formulate, and Dr. James Ebert for encouraging me to write, the ideas as they are expressed here. I would also like to thank my colleagues at Antioch College, Drs. Robert Bieri, Kenneth Hunt, and Martin Murie for their most helpful comments while reading some of the chapters. I am completely indebted to my wife who gave me the support to finally finish this work.

E.S.

ACKNOWLEDGMENTS

The author acknowledges his appreciation for permission given by the following sources for use of quotations or illustrations:

Page 186, from *The Little Prince* by Antoine de Saint-Exupéry, copyright 1943, 1971, Harcourt Brace Jovanovich.

Page 234, from "West Running Brook" from *The Poetry of Robert Frost* edited by Edward Connery Latham. Copyright 1928, 1969 by Holt, Rinehart and Winston, Inc. Copyright 1956 by Robert Frost. Holt, Rinehart and Winston, Inc., and Jonathan Cape Ltd.

Pages 152–154, 158–161, 167 and Figures 5–13, 5–14, and 6–17 from *On Growth and Form* edited by J. T. Bonner, Cambridge University Press.

Figures 6–12 and 6–13 on pages 223 and 224, respectively, from T. Dobzhansky "Changes Induced by Natural Selection in Wild Populations of Drosophila," *Evolution*, Vol. 1, 1–16, 1947.

i

the search for order

One definition of science is that it is the way of thinking that discovers order in nature about us. Discoveries lead to premeditated strategies of searching for more order, defining that order, and supplying explanations for its existence. There have been many kinds of searches for, and explanations given for, the orders found in living phenomena. In this chapter we examine an example of what is considered to be a pure style of scientific thought. We want to understand the human inputs and the social dependence of a scientist's efforts. We can also obtain some insight into the process of science, the imperfections of its explanations, and how the scientific way of thought compares with other ways of thought. It is important, in order to have a good external reference case, that we pick our example from outside biology.

THE LIFE OF A SCIENTIST

Science, like all ways of thought, is a mental process involving the translation of perceptual signals into conceptual products. Thus, if we are to

1

understand science, we must explore the life and thoughts of scientists. For this purpose we shall choose a scientist whose greatness no one disputes and whose explanations of natural phenomena have been held as an ideal for scientifically derived products. Others could well have been used to serve this same end.

The scientist we have chosen was born in England in 1642, shortly after the death of his father, a farmer. After the remarriage of his mother, at the age of 2 he was sent to his aged grandmother who raised him on an isolated farm until he was 14 years old. This setting provided little close parental care, and he was described as a quiet, lonesome, self-inquiring lad who spent much of his time in meditation. During these years he did find some enjoyment in the building of mechanical gadgets. For example, with a great deal of dexterity he built a windmill, a water clock, and a cart that could be self-propelled. In contrast with these accomplishments, however, his performance in formal education was mediocre, and only through a rivalry with a class bully did he manage with hard work to complete his secondary education high in his class. Upon the completion of these studies, his mother asked him to return to her farm to help with the chores. It was soon quite evident, however, that he was not a farmer, preferring rather to read and daydream while the cattle he was tending wandered at will. Knowing little else to do with him, his mother and stepfather arranged for him to enter Trinity College when he was 18 years old. Here he continued as a mediocre student until he became acquainted with Isaac Barrow, a mathematics teacher, who felt that he had unusual mathematical ability and encouraged and guided him. With this late start it is even more remarkable that in the 18 months immediately following his graduation at the age of 23 it has been said that he formulated more significant scientific concepts than any other individual in recorded history. During this period of his life he laid much of the foundation for differential and integral calculus, founded the science of physical optics, and extended and proved the theory of gravitation. Of course, this student was Isaac Newton. Newton is also said to have extended the scientific process far beyond that which it had been and from which it has since been only slightly modified. It has sometimes been said that Newton initiated the "Age of Reason" or, as Alexander Pope wrote, "Nature and nature's laws lay hid in night; God said, let Newton be and all was light!"

If one compares Newton's personality with that of others who have made significant scientific contributions, his would be just a part of a continuous spectrum of characteristics. This overall personality spectrum of the scientist is not very much different from that of the nonscientist. The one significant difference is that a higher percentage of scientists spend unhappy childhoods, being alone much of the time for one reason or another. But, although the incidence of such experience is high in the predestined scientist, it is not necessarily the crucial factor in the development of the scientist. If there is a unique factor common to all scientists, it is not known. It is known that the creation of scientific concepts, as with any creation, requires hard work. Apparently the genius is not a person who naturally discovers or creates things or ideas but one who has unusual abilities to concentrate and is determined.

SETTING THE STAGE FOR A NEW CONCEPT

Many a column could be written about Newton and his works. We shall consider but one of his efforts on gravity. For some reason a great deal has been made of the fact that it all started that day in 1665 when Newton, avoiding the plague sweeping through Trinity by retiring to his parents' farm, happened to see an apple fall from a tree. This story is supported by letters Newton wrote to friends, a close niece, and even Voltaire. In scientific language, such an observation as Newton's is called a *fact*. As such, a fact is any happening or statement absolutely free of any judgments as to its validity, cause, or relationship to other facts. Nowadays when we see an apple fall to the ground, we say that it is due to gravity. But actually, if nothing were known about gravity, we could just as well have said, "apples fall from trees" as we often do for analogous circumstances. This is easily accepted as plain "common sense." Science has mistakenly been defined as a sophisticated form of common sense. Facts are like letters of the alphabet—meaningless until they come together in certain combinations we call *words*. By this analogy, science is the process of describing the arrangements, their resultant meanings, and the restrictions on arrangements of letters necessary during word formation. Long before Newton's time, people had observed apples falling from trees. Newton was the first, however, who was able to fit this observation together with others and arrive at the concept of gravity.

Newton was aware, for example, that the apple was a specific example of a general class of objects that move with increasing speed toward the earth. This generalization had been made by Galileo (1564-1642) a generation before. Galileo had made many experiments with balls rolling down planes as well as with falling objects and found as diagramed in Figure 1–1 that the total distance the object moved, d, was proportional

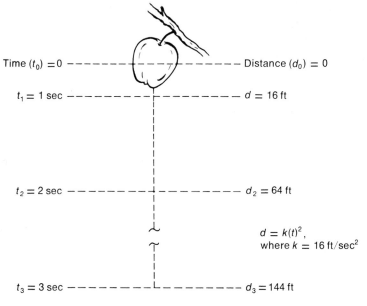

Time $(t_0) = 0$ — — — — — — — — — — — — Distance $(d_0) = 0$

$t_1 = 1$ sec — — — — — — — — — — — — $d = 16$ ft

$t_2 = 2$ sec — — — — — — — — — — — — $d_2 = 64$ ft

$$d = k(t)^2,$$
where $k = 16$ ft/sec^2

$t_3 = 3$ sec — — — — — — — — — — — — $d_3 = 144$ ft

Figure 1-1. Description of apple's fall.

to the square of the time of movement, t; i.e., $d = k(t)^2$. Newton did not have to measure the fall of the apple. It had been accepted by his time that an object such as an apple would behave according to the description above. If this is the case, the value of the observation of the apple in terms of its own motion seems insignificant. The significance of the observation is that it enabled Newton to grasp a more fundamental understanding of motion. In order to see this, we must explore Galileo's understanding of motion, which was the accepted interpretation during this period. And, in order to place Galileo's interpretations into proper perspective, we must also look at the influence of Aristotle, which Galileo felt quite keenly.

CONTRIBUTIONS OF PREDECESSOR TO THE CONCEPT—BELIEF-WEBS

According to Aristotle, motion was the return to balance of the four basic elements: fire, water, air, and earth. It was assumed that there was a constant source of disturbance that perpetuated an unbalanced condition. The "laws" of nature were the pathways the elements took in their return to balance. For instance, fire always went up and earth down (the earliest recorded notion of gravity). Wood was considered to be made of both fire and earth and to have been created by this disturbance. Burning was the act of releasing the two, ashes and fire-smoke, from each other. A rational system, or belief-web, had thus been constructed giving "causes" for observable facts. One of the conclusions of this system was that no void could exist in the world, for if it did this would represent an inconceivable imbalance, and a body moving in this space would reach an infinitely high speed. In other words, even though an imbalance of elements might exist, these elements would still be found everywhere and in a continuously harmonious relationship.

This system of explanations was plausible and adequate for many centuries. About the time of Galileo, however, thoughts about motion were accumulating and certain questions arose that could not be answered by Aristotelian rationale. For example, when a projectile is in motion how could air both move and slow it down? Was there something in the projectile that caused it to move or could something go in and out of it? Such questions as these remained unresolved.

Another aspect of the Aristotelian explanations of that time was the role of the observer. Man's thoughts were included within the observed phenomenon itself since there always had to be a harmony of elements and if man was observing, he being made up of these same elements, was also in harmony. Hence, the reason for the fall of the apple must also directly involve the observer (all men). As difficult as it may be for us to conceive, it was impossible to think of the activities in the physical world without including the activities and purposes of man. Only a small inkling of the pervasiveness of this man-nature involvement can be grasped today through such statements as, "the fall of the apple is the work of an intelligent being." Our subdivision of thinking into the three categories of aesthetics, ethics, and science was impossible then. All thought at that time we would now describe as philosophical.

Although Galileo was not the first or even unique in his variance with this generally accepted rationale, it has been argued that his thinking was facilitated by his passionate antagonism toward any kind of dogma based on human authority. He was aware of the thinking process in which man's body is separated from natural occurrences taking place outside that body. He urged that man see himself as witness to a panorama that was going on with or without his presence. This is not to say that he had no "faith" in the unity of all creation.

Because of this point of view, Galileo was able to emphasize the *description* of phenomena. He isolated individual parts, such as the rolling balls, and accurately measured their movements, thus applying numerical quantification to his observations. He was then able to conceptualize a process, as mentioned earlier, such as that for a falling body, $d = k(t)^2$. By this means Galileo was able to define properties of an object such as inertia as well as occurrences such as those resulting from the effects of gravity on an operational basis. By extending his experiments to many kinds of objects he was able to generalize his descriptions and properties to classes of objects. By refraining from thinking about the "external" causes of phenomena, he effectively introduced, indirectly, properties that could not have been conceived otherwise. Perhaps the reason for the growth and acceptance of this approach is that formulations permit predictability. Knowing the time that an apple starts to fall from the tree, one can then calculate when it will reach the ground. Such predictability allows man to use and feel more at home with nature. In addition, generalizations integrate many occurrences into single abstract explanations. This too is a source of security and encourages greater communication between men since an abstract explanation seems to reduce individual nuances of understanding. "Gravity" explains many things.

Inherent in the descriptive method used by Galileo and presumably all who wish to explain some occurrence is the conceptualization of a *hypothesis*. The hypothesis is a mental construct or "model" that can be expressed in words, diagrams, or symbols. In this way it can be conceived more easily by others. The model is an idealistic replacement of the occurrence itself. It is not the same but represents the real phenomenon to the extent that one understands it. As shown in Figure 1–2, Galileo used

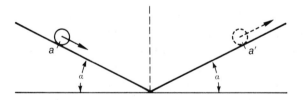

Figure 1-2. Galileo's symmetrical model.

a diagram upon which could be placed representative rolling balls, the earth, and other reference points all located within a plane. Thus mechanical relationships with distances could be indicated and his experimental results could then be symbolized on the diagram. The model so displayed could be analyzed and relationships such as $d = k(t)^2$ deduced. Hence, observations could be transposed to an abstract construct from which in turn unrealized relationships may arise. A relationship, such as the one indicated above, could then be tested by further experimentation with other objects and under different conditions. If the relationship holds after much testing, it is considered a relatively valid explanation and is called a *theory*. If a theory holds, particularly after considering different relationships of its parts as they are involved in other occurrences, over a long period of time, then it might be labeled a *law*.

On first thought, the outline of a scientific method above seems straightforward and simple enough. The emphasis as stated is upon the value of testing through further experimentation. With this feedback of information the validity of a hypothesis or theory is confirmed or denied. Even if denied, the results of testing may suggest another hypothesis that could again be tested, and so forth. Thus, by a selection process, more satisfactory and inclusive explanations are unfolded. The generalization process of including more phenomena within one explanation as testing continues suggests that nature will in time be explained by relatively few laws. The shortsightedness of this assumption is elaborated upon later, but an important flaw in the assumption should be discussed here as we consider hypotheses.

Apparently, in the transfer of information from observations to a model and in the conceiving of relationships between the variables or parts, more is involved than just writing symbols. There appear to be deep-rooted "faiths" as part of the belief-web of the current culture that can shape the process. These show up as those *basic assumptions* that either cannot be tested or are not even realized by the scientist and that influence and, in fact, are necessary in the formation of a hypothesis. Galileo was able to describe occurrences somewhat objectively, but his basic assumptions were still those of Archimedes and Plato. He assumed that nature is geometrically perfect and, more specifically, symmetrical. In one sense this assumption helped him because he felt that nature's symmetry attested to the glory of God and that therefore there was no need for "causes." It hindered the testing of some of his hypotheses, however. He assumed, for example, that a ball rolling down an inclined plane would roll up a similar plane, placed as a mirror image of the first, with symmetrical behavior. Specifically, that the speed of the ball at point a would be identical to its speed at point a^1, the corresponding position on the second plane. Although this assumption seems odd to us, according to Galileo it just had to be, and he did not consider it necessary to perform experiments to verify it.

Basic assumptions are absolutely necessary for making the initial decisions in the process of forming a hypothesis. It is the "faith" in the assumption that permits one to set forth his ideas in the first place. This particular basic assumption, that nature is symmetrical, has continued to influence many hypotheses involving atoms and subatomic particles and was most prominently advocated by Albert Einstein (1879–1955) when he formed his theories of relativity. Hypothesizing in this sense is a purely creative process. As Einstein wrote, "there is, of course, no logical way leading to the establishment of a theory but only groping constructive attempts controlled by careful consideration of factual knowledge."

As basic assumptions play a vital part in the process of hypothesis formation, so do they influence the belief-web upon which all the logic of the day is based. Society builds and depends entirely on the current belief-web. One cannot suddenly alter this web without endangering the rational peace of the period, a safety valve against the whims of an ambitious research worker that might destroy rather than help society. And, even though the creative mind is ever at work, it is also a rational mind, all too capable of rationalizing the current web, perpetuating and building upon it beyond the dreams of its originators. The more applicable the belief-web, the easier this process becomes. It is then perhaps surprising that new belief-webs can occur. They do not necessarily occur because of some social crisis. Even technological changes do not necessarily imply the formation of new hypotheses. New belief-webs seem to arise at any time, unexpected and often not appreciated. When changes in a particular belief-web have taken place, they have occurred gradually with no sharp demarcations. And, there is no reason to suspect that such changes will not continue to occur.

THE SYNTHESIS

Newton's genius seems to pale in the light of what has been said about the thoughts and products of Galileo. What more could be done to describe gravity than to describe the behavior of any object subjected to its influence? A good hypothesis had been established, and it had been tested. The results of these experiments were well-known to Newton. The belief-web partially influenced by Galileo accounted most rationally for the apple's fall. It seems, however, that at the time Newton was watching the apple fall, he was not thinking of the apple per se but rather was deeply contemplating the movements of "heavenly bodies." Somehow, a connection between the two seemingly unrelated phenomena occurred to him. It has been said that his real genius was in asking the question, "Is the invisible 'something' which is pulling the apple to the earth the same as that which is keeping the moon in a circular orbit?" He took the hypothesis involving gravity and extended it beyond Galileo's wildest effort. Until this time, gravity was considered to involve only those objects near the earth's surface, and the role of the earth was not considered except to indicate a direction toward which bodies fell.

The fact that Newton was able to hypothesize a relationship between gravity as exhibited on earth and the orbit of the moon seems even more surprising when one considers that the movements of the "heavenly bodies" were explained quite adequately at that time. Many observers including Ptolemy, Copernicus, Huygens, and Kepler had spent a great deal of effort in trying to establish just which planet or star was moving around which of the others. There were many disagreements throughout the history of these explanations, and yet they all were influenced by the same basic assumption handed down from the ancient Greeks. They all felt that the orbits of the stars and planets must fit into certain symmetrical geometric patterns. Finding a cause for the movement of the moon, for instance, was to show that it moved in a perfect circle. The problem of the orbit of the moon was solved as far as the belief-web of the time was concerned.

Somehow Newton was able to break out of that pattern of logic, bringing the moon down to earth, and considering it like any of Galileo's rolling balls. He then recalled another property of rolling balls that Galileo had described: that a ball would continue to roll in a straight line unless something deflected it from its original course. Because he was able to think of the moon as a moving ball, in contrast to the view of earlier astronomers, he wondered how the moon could be moving in a circular orbit if it had inertia. Having made this assumption, he was then obliged to ask what the something was that continued to deflect the moon from moving in a straight line. Obviously the earth had to be considered since it was in the center of the moon's orbit.

In order to conceive of the question, Newton had to construct a mental mechanical model whereby the apple and the moon were pulled to earth by "something" associated with the earth and each of the two bodies. From such a construct, hypothesis, he arrived at a description of this "something," an abstract entity he called a *force*. Thus Newton had conceived of a totally explicit characteristic, something which could never be seen directly and could only be indicated by some sort of mechanical analogy which had the same influence, such as a spring. Newton, however, did not attempt to describe the mechanical analogy per se.

Once his hypothesis was set up, Newton proceeded to formulate a relationship between the parts. The fact that the fall of the moon toward the earth was equal to the fall of the apple was not enough. It was important to find a relationship that could be tested by applying experimentally derived numbers. He assumed a relationship suggested by Balliades, Halley, Wren, and Hooke that the attraction between two bodies is proportional to the inverse square of the distance between them, $f = kl/d^2$. Thus, a ratio of the force on the apple, f_a, to the force on the moon, f_m, would be equal to $kl/d_a^2/kl/d_m^2$, or d_m^2/d_a^2, where d_a and d_m are the distances from the earth to the apple and to the moon, respectively. He also noted that the apple fell straight down rather than to the side as might be expected since the earth was rotating. Therefore, his d's were assumed to be the distances to the center of the earth. As Voltaire said, "Newton...saw the fruit fall ... which led to a profound meditation on its cause which carried on to all bodies in line which, if extended, will pass closely to the center of the earth."

For his calculations, Newton used the average of the distance of the moon, d_m, from the earth in semidiameters of the earth, 60, as determined by Ptolemy, Copernicus, Huygens, and others. The rate of the revolution of the moon had been estimated to be once every 27 days, 7 hours, and 43 minutes. The distance traveled in one revolution was $2\pi 60R$, where R was the semidiameter of the earth and equal to approximately 20,680,000 feet. Thus, the speed of the moon in orbit was $2\pi 60R/27$ days, 7 hours, 43 minutes or 187,956 feet per minute. As shown in Figure 1–3, it was then possible to calculate the distance the moon was falling per unit time and compare this with the expected distance per unit time an apple fell to see if the same explicit force, gravity, was acting on both sites. Newton's calculation of the apple's fall from the moon data was 13.8 feet per second compared to a known value of 16 feet per second, a difference of about 14%.

Thus, Newton found that his calculations "pretty nearly agreed," and

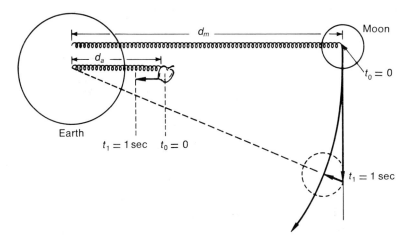

Figure 1-3. Newton's synthesis.

he verified to his own satisfaction that a "force" did exist. Eighteen years later it was discovered that the earth's diameter had been estimated incorrectly and that d_m should have been 69.5R, which accounted for the error in his calculations. Later he was confronted with the more generalized motions of the planets and somehow intuited that the most common orbit was elliptical rather than circular. In order to explain this, he found it necessary to develop both differential and integral calculus, and the results of this work were finally published in what has been called the greatest single feat of analysis in all science, the *Principia* (*Philosophiae Naturalis Principia Mathematica*).

Through this mathematical verification of his hypotheses, Newton gained full confidence in his "laws of motion" and considered them true for all motion. In fact, he thought, as did Democritus, that all phenomena "could be explained by the action of forces repressing either attraction or repulsion depending only upon distance and acting between unchangeable particles," and "the whole burden of philosophy (of nature) seems to consist in this . . . from the phenomena of motion to investigate the forces of nature and then from these forces to demonstrate the other phenomena."

Newtonian synthesis, as the combining of the astronomical and the earthly observations is called, involved much more than simply extending the law of gravity to include the moon. What it proved to Newton, as it has to others who followed him, was that the properties of matter discovered on earth extend to all matter throughout the world. And, more important, it was now possible to have the faith necessary to deal with explicit abstractions such as forces and the mechanical model hypotheses containing them. Newton was now free to define mass, force, and inertia beyond Galileo's descriptions. He could predict and explain all sorts of phenomena such as the mass of the sun or the moon, the perturbations of the moon's orbit, the bulging of the earth at the equator, the earth's axis inscribing a circle, ocean tides, and even the occurrence of comets. What a feeling of security this must have imparted.

While Galileo argued against cause and fought for description, Newton could go on and conceive of the explicit by means of clearer description and more abstract formulations. While Galileo was held somewhat in check by his basic assumption of symmetry, Newton used a different assumption and contributed to the construction of a different belief-web. He felt that any explanation should be simple and uniform throughout nature and, more important, that nature follows the most economical path—an energy concept. When his work was done, Newton turned out to be the greater generalizer. It appears that such a product occurs at random with no assurances beforehand. The successful generalization is selected from a large number of tries. No doubt many a brilliant mind has used a basic assumption that never bore fruit.

Newton, as are all scientists, was also restricted by his basic assumptions. He was never free from the burden of answering to himself and to his contemporaries what the causes were. Although he said, "It is enough that gravity does really exist, and according to laws we have described abundantly serves to account for all the motions of the celestial bodies and of our sea," he nevertheless felt something lacking. He wondered—to himself at least—"why?" "How does the earth know that it has to exert a greater force on the moon than on the apple? Where does the force come from? Does matter generate this force and exert it a distance?" In other words, even Newton, who could deal with the abstract, was unable to completely separate human traits from the physical world. He has been given undue credit for implying that mass and gravity are inherent properties of all matter. In a letter to Richard Bently he said, "You sometimes speak of gravity as essential and inherent to matter. Pray do not ascribe that notion to me; for the cause of gravity is what I do not pretend to know ... it is inconceivable that inanimate brute matter should, without the mediation of something else, which is not material, operate upon and effect other matter without mutual contact." He never used the term "attraction" but rather "bodies gravitating toward one another."

Newton was not free enough to let pure mechanical models with abstract springs and pulleys replace his force. This came later when the industrial revolution of the eighteenth and nineteenth centuries allowed one to think more easily in terms of levers and gears. Newton felt, as had Galileo and Aristotle, that "this most beautiful system ... could only proceed from the ... dominion of an intelligent and powerful Being."

Because of a need to understand the cause of gravity and the desire to withdraw from social contacts because of personal attacks upon his work, Newton departed from his study of physical laws in order to devote all his energies to the study of theology. Newton spent the greater part of his life attempting to find relationships in the Scriptures between phrases, special events, and the times of their occurrence in a manner analogous to that used in deriving his "laws of motion." He used elaborate calculations, but he never satisfied himself with his results. In 1727 he died at the age of 84. The following inscription appears on his grave in Westminster Abbey: "Let mortals rejoice that there has existed such and so great an ornament of the human race."

THE INFINITENESS OF THE SCIENTIFIC PROCESS

From the brief account of the development of the idea of gravity given above, one can begin to understand the methodology and nature of scientific conceptualization. There are certain spectacular incidents in the story, but it nevertheless includes the main characteristics in the development of any scientific idea. One overall characteristic cannot be overlooked. Although there is no way of predicting which basic assumption will be more successful, new hypotheses continue to be built upon old ones with an ever inclusive coverage of the phenomena of nature. With this kind of generalization abstraction (becoming more explicit in time), more quantification and increased predictability develops. Newton's laws were later replaced by the more inclusive Theories of Relativity and the Unified Field Theory proposed by Einstein and others. This does not mean that Newton's laws do not explain what they were meant to explain or that they are no longer useful but rather that they described a limited number of parameters compared with more recent theories and that the predicted relations were valid only within a certain range of these parameters. The history of science has emphasized that no theory or law is sacred, no matter how complete it may seem at the time. Thus, while the various modes of scientific explanation are continually being perfected, no absolute is in prospect or should be expected. There is a temporal certainty with which one must learn to live. And yet there seems to be no end to the creative processes open to man in dealing with natural phenomena, and there is every reason to believe that the satisfaction he gains by such endeavors will continue. A new concept does not crowd out or inhibit the forming of new hypotheses but rather seems to act like a catalyst, opening new possibilities for interrelating facts as evidenced by the ever increasing rate of scientific effort.

SCIENCE AND OTHER WAYS OF THOUGHT

If we compare scientific thought, as illustrated above, with aesthetic or ethical thought, we can find analogies. This is not surprising since each way of thought exists because of the human need to constantly make decisions about basic values; in science it is truth, in aesthetics it is beauty, and in ethics it is goodness. In each there is a creative phase in which a statement (idea or artifact) that is essentially a judgement on the value is put forth by the person, based upon a personal belief, usually an unknown or uncontrollable basic assumption. Examples are Galileo's belief in the absoluteness of symmetry, an artist's belief that the repetition of a given shape is essential for beauty, or a theologian's belief that a certain behavior of man is inherently good. Such statements, as personal judgements, are open to evaluation by others, and thus there are feedback "testing" controls on each statement determining whether or not it becomes a socially acceptable standard of reference. The effectiveness of this testing phase upon the duration of the credibility of a statement varies with each mode of thought. In science, if a single tester disagrees with a hypothesis, it is under suspect and relatively useless unless a substantial number of other testers support it. The result is that very few scientific hypotheses dealing with the same phenomena can exist simultaneously and hence there is a rapid turnover of hypotheses. On the other hand, both aesthetic and ethical statements can remain as active standards, even though surrounded by a sea of dissenters, as long as a reasonable number of testers support the hypothesis. Thus, many of these statements can exist at any one time, they last longer, and there is more room for the individual to choose his standard. Despite these differences, any statement is subject to modification and elimination as testing continues over extended periods of time, and there is a sense of competition between statements. Also, regardless of the kind of value involved, it is usually agreed that all statements are but temporary reference points with no final absolutes possible (an obvious exception is the values of some religious fundamentalists).

Given these similarities among ways of thought, there is some question as to whether the divisions into scientific, aesthetic, and ethical are justified. Such divisions were not made by non-Judeo-Grecian civilizations, and in the Judeo-Grecian civilizations there was extensive overlap among values, intermixing and resulting in a general philosophy. Throughout our history definitive processes have developed in which the meaning and application of the values of truth, beauty, and goodness have become increasingly more separated, as illustrated by Galileo's restriction of truth to descriptions of phenomena and not questions of cause. Each way of thought has become more restricted in application and has created its own vocabulary. As we shall see in many examples involving explanations of living phenomena, there has been rather intense interaction among the various modes of thought, sometimes stimulating and other times inhibiting the creative process within a particular mode. It may have been the net disadvantage of overlap that has resulted in the more exclusive and restrictive use of each mode today, as well as the existence of the three modes. Regardless of the reason, it has become the custom to restrict the application of and the vocabulary used in describing a value to that consistent with its related mode of thought. For example, there is no goodness or beauty per se involved in a description of the force of gravity.

Relatively recently there has been a great deal of thought about whether science, with evolution and time, could or could not create absolutes. Could description become an abstract truth? Some maintain that scientific statements lie outside the domain of values. In the context of this view, value judgments are human artifacts subject to all the peculiar pressures of a culture and therefore subjective, while science strives to be and can be independent of the opinion of man. This position is based upon the assumption that science will eventually be able to describe all the thinking processes of the brain and that any of man's ideas, including values themselves—*if they remain*—may be entirely replaced by scientific descriptions of behavior and impressions now summarized as inducing a sense of value. The difficulty with this view is that although science seeks to explain with abstract, valueless parameters, nevertheless it is a human process involving basic assumptions that themselves are random and artificial, though possibly restricted in degree, and thus must have a value content. Also this view is inconsistent with our basic understanding of the scientific process itself, which depends on the endless selection of a better but necessarily unrealistic model and, incidentally, involves the same mental processes that are presumed to be subject to complete description. Unless these mental processes turn out to involve specific responses for each impression, they must depend to some extent on chance. Under our present state of knowledge it would seem wise to assume that each way of thought has its purpose. Presumably each individual uses all modes even in making common, everyday decisions. In choosing a box for a particular use, for example, while science may accurately define its dimensions, it can hardly state its beauty or its usefulness, if you will permit me the liberty of stretching the value of goodness.

WHAT ABOUT BIOLOGY?

As we now turn to biology, we should realize from the discussion above that there is a long task ahead of us. While there has been much description of living material, there has been a dearth of scientific abstractions on a par with the notion of "force." It is left to the reader to see how many styles of thought are Aristotelian, Galilean, or Newtonian as we proceed. On the other hand it is an open question whether the example used in this chapter may not be misleading us into thinking that all meaningful explanations must involve an extrinsic concept such as gravity. It has become traditional among many physical scientists to feel this way. It is under this assumption that the physical sciences are considered much more advanced and at a higher level. This has often caused the physical scientist to claim that all biology can be explained in terms of physics and chemistry. There is no question that the laws of physics and chemistry are operating with living matter in the same way as with nonliving matter, but the question becomes one of whether the presently constructed laws of physics and chemistry can adequately describe all that occurs in living phenomena. Are there additional laws? And if there are additional laws, are these laws of physics, chemistry, or biology? Or can a distinction be made of this sort? And would these new abstractions necessarily have to be Newtonian—or even related to those used in the "gravity" example? Again I trust the reader will keep this thought in mind as we proceed. Regardless of what we want to call them, I personally believe that biology, as a study of living processes, is ripe for many new abstractions. I believe that the diversity and richness of modes or styles of explanation that exist at present are bound to be stimulating rather than squelching to this task. The present belief-web is directing man's gaze in new directions out of which new doubts of present explanations, new syntheses, and new inspirations are bound to arise. Age and experience do not seem to be necessarily the prerequisites for originality. Hard work will be involved, but with courage, consistency, and time I believe these new abstractions will come forth.

ii

order: in diversity

For thus in an order, than which nothing more beautiful exists in the heavens or in the mind of a wise man, things which are far and widely different become, as it were, one thing.

MATHIS DE L'OBEL (1538–1616)

"Yes, those are dragons all right," said Pooh. "As soon as I saw their beaks I knew."

Now We Are Six, A. A. MILNE

We must begin to examine the orders found in life. It is my judgment that the most "primitive" order is derived from our need to classify a group of related but diverse objects, at least originally, for the purpose of communications. This basic need is demonstrated in our need to name and generalize. One of many examples is found within the language of a particular tribe of Indians in New Guinea. These keen observers have 137 names for different birds whose characteristics have been recognized and categorized into almost identical groups, 138 species, by specialists who have access to birds throughout the world. The subtlety of the significance of this need and its potential reward has been noted by ancients and moderns alike. Isodorus wrote, "If you know not the names, the knowledge of things too is wasted." A recent classifier, Raymond Pearl, has written, "It is the systematist (one who classifies) who has furnished the brick with which the whole structure of biological knowledge has been reared."

In this chapter we shall explore the inherent ambiguities present in any classification scheme and the meaning of their resulting hierarchy and their most basic unit, the species. This search is easily overlooked today as being outmoded, redundant, or too simpleminded. One of the purposes in this chapter is to show that the opposite is true. The diversity, patterns of differences, and many characteristics of species are so complex that no certain or progressive abstractions have as yet been discovered. No area of biology has occupied more of its history. This should not be taken as a sign of the hopelessness of the search. For with the advent of computers, classifications can be tried that have never been tried before and it is not unlikely that some new ideas will emerge in the future.

EARLY SYSTEMS OF CLASSIFICATION

From the need to communicate, presumably organisms have been named and grouped according to their use since the most primitive kinds of hunting or collecting of desirable plants were done by groups of men. They would have had to have identified poisonous from nonpoisonous plants, animals skins or plant leaves that furnish the best clothing or shelter, etc. Such practical criteria for identifying various organisms are known to have become quite extensive and complex by at least 5000 B.C. Many specialized criteria such as specific medicinal use and beasts of burden were added by this time.

It was with the logicians of ancient Greece and China that classification systems were analyzed and developed systematically as well as used. During the fourth century B.C. quite similar ideas were being formulated in these two cultures. Typical of either is the following excerpt from a Mohist (Chinese) scholar:

> The horse and the ox are different, but if someone says that a horse is not an ox because the ox has teeth and the horse has a tail, that will not do. In fact, both have teeth and tails—one has to say that horses are not the same as oxen because the ox has horns and the horse does not. . . . Animals of four legs form a broader group than that of oxen and horses. Everything may be classified in broader and narrower groups. It is like counting fingers; each hand has five, but one can take one hand as one thing.

Herein, in principle, is the nature of classification. Not only can organisms (horses or oxen) be separated from one another by certain unique characteristics, but these organisms can also be grouped by certain common characteristics. The Greeks gave us names for these distinctions: the species (εδοζ), wherein all organisms have the same essence or pattern of attributes; and the genus (γενοζ), which then meant that part of essence shared by a group of distinct species, i.e., an assortment of common characteristics such as teeth, tails, and four legs. Also revealed in this quote is the inherent ambiguity associated with the classification of living organisms. Are four legs a more fundamental or general characteristic than teeth? Teeth could be found in fish without legs. And yet birds have legs without teeth. No one of these characteristics is apparently more significant or general than the others.

Although these logicians often agreed as to the fundamental nature of classification, there was a considerable range of basic assumptions as to the significance of the classifying process and its result. Plato (437–347 B.C.), a firm believer that the real world existed in the mind of man, distrusted conclusions based on direct observations. This so-called a priori type of reasoning caused him to imagine that all life-forms must be able to be divided into two halves on the basis of some one fundamental characteristic, each of these to be divided again and again by other characteristic differences as exampled in Figure 2–1(a). It was this same kind of a priori logic that gave birth to the "Doctrine of Elements" so influential in both Greek and Chinese culture. According to this doctrine all phenomena can be explained in terms of a certain few basic elements, i.e., fire, water, air, and earth in Greece and fire, water, earth, wood, and metal in China. This type of belief-web was the cultural inertia Galileo was struggling against with his study of motion. In fact it has been said that it was the elaboration of such a belief-web in the East that served so successfully as a source of explanation of phenomena that a Galileo-Newtonian-type change of belief-web was not encouraged.

On the other hand there were empiricists of classification such as Aristotle (384–322 B.C.) who derived their ideas from direct observation. This does not mean that they did not use the doctrine of elements for explanations but that they based any questions of nature on direct experiences. Aristotle was concerned with overall shape (morphology) and the reproduction of organisms and used these as bases for his classification. He did not seek any particular meaning in the overall scheme he devised and merely listed the various groups of organisms ranging from simple to

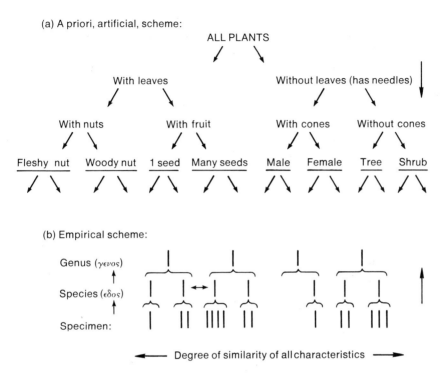

Figure 2-1. Basic classification schemes.

complex, a "ladder of life." Theophrastus (370–287 B.C.) did for plants what Aristotle did for animals, classifying 480 species.

The classification systems of Aristotle and Theophrastus, though often forgotten in the interim, have nevertheless remained influential to some degree up to the present. In fact, until the eighteenth century very little was done to modify them. During this long period more practical systems of classification for plants appeared sporadically and indedently in various parts of the world. Many so-called herbalists, often ministers and doctors, roamed the woods for herbs that would alleviate diseases. Most often these men associated their search with some religious purpose for the betterment of mankind. An extreme example of this was the popular "Doctrine of Signatures" proposed in the sixteenth century by a wandering doctor, Phillipus Aurellus Theophrasus Bombastus of Hohenheim (1493–1541), also known as Paraceleus, who wrote, "I have often declared how by the outward shapes and qualities of things we may know their inward virtues, which God hath put in them for the good of man." According to this "doctrine," long-lived plants could extend the life of man, yellow sap from plants could cure jaundice, kernels of walnuts could comfort the brain, and maidenhair fern could prevent baldness. A similar belief was held by Aztec herbalists. Prior to the eighteenth century, particularly in India, the need for the herbalist to identify his plants resulted in some of the finest morphological descriptions and drawings or woodcuts ever produced. Some herbalists contributed to an understanding of the effect of the soil on the morphological characteristics of replanted herbs. But such systems of classification proved to be transient, perhaps because their use was limited.

The renaissance of classification as a study for its own sake was not the work of any one man or event. With increased explorations and the rather common sentiment that "The study of nature is essentially a religious duty appropriate to mankind as a means to God's greater glory and a deeper understanding of his ways," the number of recorded plant species approached 9000 by the end of the seventeenth century. Something had to be done to order this large mass of material. We shall consider here the work of only a few of those who undertook this task.

JOHN RAY AND NATURAL CLASSIFICATION SCHEMES

It was John Ray (1628–1704), a Latin scholar and classmate of Isaac Barrow, who first suggested that there was a special significance to a hierarchy system resulting from the natural grouping of organisms. With this view of a hierarchy, he was free to or forced to define the species in terms of all the measurable characteristics of the organisms in question. For our Mohist example of the horse, not only the four legs, teeth, tail, and lack of horns will do but an entire description of the horse including its hair length and color, every dimension, and many of its eating and mating habits. This was necessary to yield the most "natural" classification scheme rather than a relationship based upon a few preconceived, "more fundamental" characteristics that was used by all classifiers in previous schemes. As noted in Figure 2–1(b), if one could lump all the characteristics of a species into one dimension and compare this dimension for different species, a natural arrangement of the species would determine the grouping pattern. Ray was also obliged to account for variation, the fact that no two specimens, though obviously quite similar, are identical.

He overcame this confusion by assuming that the variation of any characteristic is a range of expressions of a single potential characteristic. He grew plants under different environmental conditions in order to comprehend the extent of a plant's adaptability. He concluded that there is in any organism an internal effort at consistency, hampered and modified by the many external forces of the environment. Here we have a first suggestion of a basic unit of life, unique yet adaptable, reproducing a potential rather than an absolute replica of itself. Although his ideas were forgotten, they have taken on new meaning in recent times.

While Ray's system was far from natural in that he could only consider a limited number of characteristics, he did manage to classify the plants into 25 genera and his groupings indicated that the type and position of the fruit were decisive and thus fundamental in deciding his genera characteristics. He had less success with animals but concluded that teeth, feet, insect scales, and membranes must be important differences between genera.

A contemporary of Ray's, Joseph P. de Tournefort (1656–1708) increased the hierarchical classification system for plants by concentrating on genera and using only flower parts as definitive characteristics. He ultimately defined 673 genera encompassing 8846 species of plants. He added a higher rank by grouping his genera into 22 larger "*classes.*"

CAROLUS LINNAEUS AND HIS ARTIFICIAL SYSTEM OF CLASSIFICATION

Almost all the basic philosophies and concepts, the contradictions, and the dilemmas of the search for order through systematics can be found in the writings and thoughts of Carolus Linnaeus, a poor man by birth, born in 1707 in the small Swedish town of Rashult. Linnaeus' early life was full of conflicting pressures, for while his parents expected him to be a clergyman, this type of life did not appeal to him. The reasons for this were far from irreligious, for during most of his life he subscribed to the ethical belief-web of his day, creationism. This belief played a major role in helping him formulate his early hypotheses regarding species. It was best stated in his *General Plantarum* of 1737: "There are as many species as the Infinite Being created as diverse forms from the beginning . . . just as there are now no more species than have been from the beginning." After the clergy, medicine was considered the most respectable profession and seemed reasonably acceptable to young Linnaeus. Although he was a good medical student, however, he soon found two avocations that occupied most of his time. One was an insatiable need to read just about every book written and the other was the collection and description of plants. It was said that "he saw plants just as an insect sees them." As a result of his interest in plants he was offered the post of naturalist on a trip to Lapland in 1732, a trip that greatly influenced his continued interest in taxonomy. Linnaeus' enthusiasm and cheerfulness opened up opportunity after opportunity and endeared him to all who met him. He said of himself, "God has suffered me to peep into his secret cabinet." Despite his great devotion to taxonomy, he successfully completed his medical training and proved his worth to his future father-in-law by attending the queen. Thereafter, however, he devoted his life to taxonomy and by the age of 32, had completed 14 botanical works.

In his early work, Linnaeus was primarily concerned with the "delineation and then ranking" of all known plants. Like Ray he thought in terms of the resulting hierarchy of his classification schemes. Unlike Ray, however, he was convinced that a concentration on species was impossible. For one thing he felt that it was pointless to consider the variation within a species and assumed that such variations were mistaken deviations from some discrete form quite similar to the morphotypic one used by the ancients wherein there was thought to be one ideal form of organism for each species. He also opposed Ray's concentration on many characters, asserting, "Color, varying as it does in the same species, is strangely sportive: hence it is of no value as a distinguishing character. Hairiness is a distinguishing character which may very easily become misleading since it often disappears under cultivation. The position of the fruit affords the best distinguishing character. All those plants which agreed in their position of fruiting should be united in one genus." In addition he realized that to consider all these characteristics would only increase the chance of ambiguities as to how the species could be grouped. In this way he supported the use of a few characters, an artificial basis. Like de Tournefort, who shared his skepticism of species variation, he felt that the only possible rank with which to begin delineation was that of the genus. His basic belief in the fixity of species and hence genera helped him establish his famous tenent that "the characters do not make the genus; rather it is the genus that gives the character." To the question, "What gives genus?" he answered, "God."

In order to establish which species belonged to a genus, he used the strict criterion that every species member and no other species must possess a certain type of fundamental characteristic. For his plants this was the position of the fruit and he established 1000 genera on that basis. He then found he could group these genera into "ordos" or orders on the basis of the number of styles (female flower parts). Those with one style were all placed in one ordo, those with two styles into a second ordo, and so forth. Likewise, these orders were grouped into a total of 24 classes based on the number of types of stamens (male flower parts). Linnaeus was quite aware of how arbitrary his characteristics and method of grouping were, but the use of "sexual" parts of the flower and their number not only appeared to him to be fundamental and inclusive but of extreme convenience for the person working in the field to identify a specimen quickly. It was this convenience that made his system far more acceptable than that of any predecessor and gave the study of botany a significant boost.

Linnaeus explored ways to represent his four ranked (four taxa) hierarchy system (taxon). His and those of others are shown in Figure 2–2. In this example, eight ideal specimens are grouped. They are ideal in that their basic defining characteristics are found in simple patterns. For example, characteristic A is not found in specimen 8. If it had, the same diagram could be constructed but would not be completely accurate. Linnaeus had close to an ideal system because of the few arbitrary characteristics he used. In any respect, systems can be designed as shown. Linnaeus developed the branching diagram.

Linnaeus extended his approach to all matter. Although not nearly as complete as his work with plants, he classified 4235 species of animals into 312 genera, 32 orders, and 6 classes (Mammalia, Aves, Amphibia, Pisces, Insecta, and Vermes)—still under the heavy hand of Aristotle. He also was concerned with minerals, for they too existed, he felt, within an interdependent total system. He thus subdivided a grand empire (of matter) into the three kingdoms—animal, plant, and mineral. He felt "The whole world, animate and inanimate, must stand in some natural order as a hierarchy." (*Systema Natura*, 1758).

The real greatness of Linnaeus actually lies in the flexibility he permitted himself in practice and in interpretation once his typologically based system had been completed. He could then relax about exceptions to his defined groups at any level and use different sets of criteria for different species or between different ranks. In his later writings, when he had more time to consider species, he acknowledged, as Ray had, that variation is due to various soil conditions or domestication and that new

IDEAL EXAMPLE SPECIMENS:

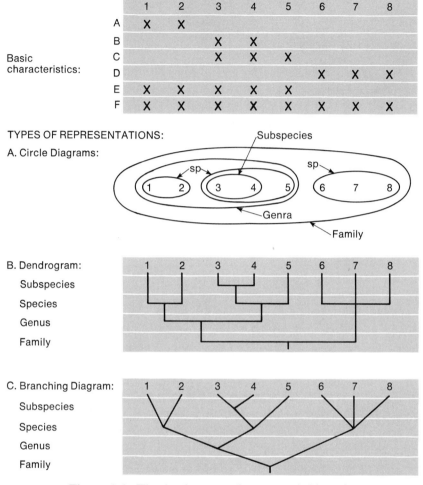

		1	2	3	4	5	6	7	8
	A	X	X						
	B			X	X				
Basic	C			X	X	X			
characteristics:	D						X	X	X
	E	X	X	X	X	X			
	F	X	X	X	X	X	X	X	X

TYPES OF REPRESENTATIONS:

A. Circle Diagrams:

Subspecies

sp

sp

1 2 3 4 5 6 7 8

Genra

Family

B. Dendrogram:

1 2 3 4 5 6 7 8

Subspecies

Species

Genus

Family

C. Branching Diagram:

1 2 3 4 5 6 7 8

Subspecies

Species

Genus

Family

Figure 2-2. The development of a taxonomic hierarchy.

species could arise as the result of hybridization of two different species. Criticisms of Linnaeus have usually been based on his early writings, ignoring this flexibility and growth.

Practicality had first aided and then worried Linnaeus. He had never been content with his arbitrary system and sought against his own design for a more natural one. He attempted to modify his criteria of classification with little success; however, his breadth and imagination did allow him to discard his original monotypic interpretation for a polytypic one. He developed a 12-word polynomial description of every species. This was standardized so that each word was the description of some important characteristic of the organism, the last two being genus and species, in that order. Therefore, although his classification system remained a hierarchy of ranks, his concept of an organism became a broad complex entity. He actually thought of the organism in terms of the 12 words. This breadth is usually lost in any discussion of the legacy of Linnaeus. Most often he is only given credit for setting up a standard method of naming the species, the so-called binomial nomenclature. This method consists of naming the species with Latin or Greek words, the first word being capitalized and indicating genus, while the second is not capitalized and is the species. For example, *Homo sapiens*, man, and *Quercus alba*, white oak. The nomenclature was the last two words of his 12-word description, an abbreviation he used in his notes. This procedure has proved convenient and has remained as a standard that all use.

The dilemmas that Linnaeus faced continue to plague systematists. The need to deal with large numbers of possibly less significant characteristics among vast numbers of species requires some artificiality or selection by the taxonomist even though he may desire to produce the most natural system. Where a system is most useful, it is conceptually dissatisfying: when it is conceptually pleasing, it is impractical, ambiguous, and frustrating. It is small wonder that Linnaeus set the stage for all systematic criteria. As Stearn wrote in 1959, "Whatever Linnaeus failed to do that later people with lesser burdens think he should have done, his definition and classification of these organisms (more than 10,000) stands out as something no one else could do then and no one else has been able to do since."

Finally, we should consider the great influence Linnaeus had on others. His enthusiasm for his subject and his concern for others produced more dedicated students (literally in the thousands) than any other known teacher. Most of these students took extensive collection trips to all parts of the earth, many dying while on their missions. These collections were made with a great spirit of purpose, both for taxonomy and for the man Linnaeus. He reigned supreme and conquered friend and foe with his kindness, often returning to his camp after a trip with an entourage of students, their bounty of new species triumphantly accompanied by great fanfares of French horns, kettledrums, banners, and shouts of "Vivat

Linnaeus!" It is ironical that when he died in 1778, his wife auctioned his manuscripts and collections for only 1000 guineas—fortunately to a botanist James E. Smith.

PLANT CLASSIFICATION SINCE LINNAEUS

Little has happened to the basic practices of the plant taxonomist since the death of Linnaeus. Expansion in the number of known species has necessitated herbaria where specimens can be stored by drying and pressing and later compared. Also, the need to examine the extent of variation in species due to growth conditions and to check fruiting characteristics has led to the establishment of huge gardens where live material can be studied. No garden has yet surpassed the Kew Gardens, begun by George II of England in about 1790. This famous laboratory for systematists was the cause of many an expedition, including the ill-fated voyage of Captain Bligh through the South Pacific. Sir W. J. Hooker (1785–1865) together with his son Joseph D. (1817–1911) and George Bentham (1800–1884) made most use of Kew. An extension of their classification system by O. Tippo (1911–) using a synthesis of the nonvascular plant system of G. M. Smith (1885–) and the vascular plant system of A. J. Eames (1881–) is the most popular taxonomic system today. This system accounts for over 300,000 species of plants. It differs from that of Linnaeus primarily by an increase in the number of declared fundamental characteristics and in a shift back to differences in plant structure and position.

ANIMAL CLASSIFICATION

We are not certain why zoological systematics lagged behind that for plants, both in workers and ideas, until about the beginning of the nineteenth century. Perhaps it was the scarcity of animals considered to be of practical value and those known could be classified adequately by Aristotle's system. It probably was not until enough expeditions had returned from the southern continents with wide assortments of mammals and birds and until more time was permitted for leisurely observing and collecting birds and butterflies that there was any need for animal classification. While interest in plant taxonomy diminished during this period, animal observers sprang up everywhere, and museums and zoos flourished. An indication of the energy devoted to the collection and classification of animals can be appreciated when it is noted that while Linnaeus had described only 4236 species of animals in 1758, by 1859 there were about 130,000 known species, about 500,000 by 1910, and an estimated 1.5 million species today.

Animal systematists have stressed the whole organism for defining groups and have thus for the most part used a natural approach. M. Adanson (1727–1806), a little known animal collector and a contemporary of Linnaeus, denied the use of fundamental characteristics and, although perhaps influenced by Ray, added a unique approach to reach a natural system. "A species consists of individuals with a maximum number of shared characters, a genus with species with a maximum number of shared characteristics, etc." Unlike the criterion of Linnaeus in which at least one characteristic had to be shared by all members of the group, Adanson's system only required that the majority of the characteristics be found in most of the member specimens. This difference is diagrammed in Figure 2–3. Thus he produced a set of criteria that permits greater lumping with

	Linnaean:			Adonsonian:					
Species:	1	2	3	1	2	3	1	2	3
Characteristics:	A	A	B	A	A	B	A	A	B
	B	B	C	E	E	F	B	G	D
	C	D	D	F	G	G	C	H	E
	1 genus			1 genus — majority of characteristics shared			Not same genus		

Figure 2-3. Comparison of Linnaen versus Adansonian methods of classification.

less ambiguity. Also, he felt that there should be at least 65 characteristics considered for each specimen. Since this method requires a great number of measurements and a correlation of these measurements, it received essentially no support until very recently. The introduction of computers as a taxonomic tool has made a neo-Adansonian approach to systematics quite realistic, and it is being actively pursued particularly, as one might expect, by animal taxonomists.

MODERN SYSTEMATIC METHODS— NUMERICAL ANALYSIS

There are several different practical methods for utilizing this empirical Adansonian approach. One popular one is that of numerical analysis. The procedure devised by Sneath and Sokal is worthy of elaboration. They define their method as "the numerical evaluation of the affinity or similarity between taxonomic units and the ordering of these units into taxa (ranks) on the basis of their affinities."

These units may be analyzed at any level but would normally be done on the species level. Each unit is referred to as an operational taxonomic unit, OTU. If 50 different species are being compared, there would be 50 OTU's. At least 40 or 50 characteristics (n) are used in describing each OTU. For ease in programming the computer, the characteristics as applied to any single OTU are answered in terms of a plus or minus. For example, this species is radially symmetrical ($+$) versus nonradially symmetrical ($-$), monocotyledonous ($+$) versus dicotyledonous ($-$), etc. In cases where there are multiple answers such as any quantitative measurement, the plus-minus system is elaborated by adding more pluses or minuses for the single characteristic. The number of stamens could be programmed as $+ - - -$ if one, $+ + - -$ if two, $- + + +$ if five, and so on. If there are no data on a particular characteristic for any OTU, then NC (no comparison) is indicated. A table of OTU's versus characteristics, called a "$t \times n$" table, is constructed. On page 31 is an example of such a table with 4 OTU's (species) and 12 characteristics (n).

Similarities between OTU's for all the characteristics considered can be calculated by an arbitrary numerical index, a coefficient of association, defined as follows:

$$S = \frac{N_s}{N_d} + N_s,$$

where N_s is the total number of $+ +$ combinations existing between two particular OTU's and N_d is the total number of $+ -$ combinations. Thus, for example, between A and B, N_s is 9 and N_d is 1. Where there are $- -$

n	A	B	C	D
1	+	+	−	NC
2	+	+	+	+
3	+	+	+	−
4	−	+	NC	NC
5	+	+	+	+
6	+	+	−	+
7	+	+	−	NC
8	NC	−	+	+
9	+	+	+	+
10	+	+	+	−
11	+	NC	−	NC
12	+	+	+	−

combinations or where NC is present, the characteristics are not compared. Thus S for A and B is calculated as $9/(9+1)$ or 0.90, which is expressed as a percentage, 90%.

A "$t \times t$" table can then be constructed for all combinations of OTU's as follows:

	A	B	C	D
A	100			
B	90	100		
C	60	60	100	
D	57	50	50	100

Once a $t \times t$ table is available, a shuffling of OTU's in order to group them by similarities should reveal patterns for ranking. The example above has too few species to demonstrate this grouping process; thus, a larger sample is shown on page 32. The "cluster analysis" is accomplished by rearranging columns and rows in the $t \times t$ table so that regions of the new table include identical (or approximately identical) S values. This is most easily done if the high S's are grouped as near the cross diagonal as possible.

Construction of a hierarchy is now possible. If the OTU's in this example were specimens, a cluster of similarities with a high value would yield a species. In this example, species can be defined where there is a similarity of 90% between any two of the OTU's. Thus, A, B, E, and I would be specimens belonging to the same species, and H and D, a

	A	B	C	D	E	F	G	H	I	J
A	100									
B	90	100								
C	60	60	100							
D	50	50	50	100						
E	90	90	60	50	100					
F	60	60	80	50	60	100				
G	60	60	80	50	60	80	100			
H	50	50	50	90	50	50	50	100		
I	90	90	60	50	90	60	60	50	100	
J	60	60	70	50	60	70	70	50	60	100

→

	A	B	E	I	C	F	G	J	H	D
A	100									
B	90	100								
E	90	90	100							
I	90	90	90	100						
C	60	60	60	60	100					
F	60	60	60	60	80	100				
G	60	60	60	60	60	80	100			
J	60	60	60	60	70	70	70	100		
H	50	50	50	50	50	50	50	50	100	
D	50	50	50	50	50	50	50	50	90	100

different species; each such group is defined as a *phenon*. Through experience in comparing types of organisms, arbitrary levels of percent similarity could be defined for each taxon. For instance, one might choose 85% as the level of similarity used in separating a species from a subgenus. Continuing the process using this example, C, F, and G with an 80% similarity would then be termed a subgenus. Each of these OTU's, together with J, have a similarity of 70% and so could be grouped as a genus. And, since members of the ABEI group and the CFGJ group each have a 60% similarity, they could be clustered into one subfamily. The overall pattern of relationships can be displayed as a dendrogram, shown in the diagram.

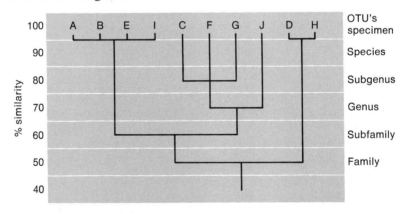

Obviously there is an inherent arbitrariness in determining the level of similarity to be used in defining each rank, and the example above has been oversimplified for clarity. But the point is that when similarities are calculated between OTU's for many characteristics, in practice clusters of similarities do appear about certain values with a fewer number of similarities falling between clusters, thus defining natural ranks.

Critics of these methods shall no doubt continue to exist and to date the majority of taxonomists, especially botanists, agree that an experienced taxonomist with a group of specimens in hand can intuitively arrive at a better relationship of these than can a computer. They argue that the individual is a "natural computer" containing valuable decision-making judgements that are not yet understood well enough to be programmed into a computer. Fortunately, the products of numerical taxonomy are for the most part in agreement with those found by the taxonomist. The computer cannot eliminate ambiguities. But, what it can do is to help standardize the weighting of characteristics or their number to define a given rank. In spite of the criticisms above, these methods represent and permit a change impossible until recently. While they are now generally used as a compromise between the ranked systems described in this chapter, only accounting for more characteristics, they probably will continue to influence basic taxonomic philosophy. The computer could in time suggest new ways to relate diverse organisms other than by a hierarchical system.

THE SPECIES PROBLEM

While the search for a "natural" hierarchy has continued at a rather slow pace during the present century, the definition of the species has been altered considerably. Curiosity as to the meaning of a species grew rapidly once Linnaeus permitted an increasing number of taxonomists to be more content with genera.

The difficulty in defining species, the "species problem," is related to the wide variation in characteristics among individuals within a possible species. The characteristics causing the greatest ambiguity involve morphological factors such as shape, color, and size. Below is a list of some of the factors causing these variations.

1. The two sexes of an organism often have different appearances, termed *sexual dimorphism*. This characteristic is particularly common among trees, insects, fish, and birds as, for example, the king parrot, *Larius roratus*, in which the male is green with an orange bill, and the female is red with a black bill.

2. Immature forms often do not resemble the adult. This difference is marked in many insects and amphibians and is found, to some extent, in all animals and plants. There is also the case of the mosquito, *Anopheles maculipennes*, in which the eggs differ among themselves.
3. In animals having a social hierarchy, such as termites and honeybees, individuals differ in form.
4. The several races of man.
5. Developmental irregularities.
6. There are seasonal and other cyclical variations in form such as trees losing their leaves and animals changing fur and feather color or losing antlers.
7. A number of other environmental factors affect the form of organisms, among them:
   (a) nutrition
   (b) population density
   (c) salt concentration of water
   (d) type of substrate
   (e) water motion and depth
   (f) background color
   (g) disease and injury

Of course, no two specimens are ever identical. If a group of similar organisms are measured for one characteristic such as height, a bell-shaped curve or so-called normal distribution results about an average height measurement. It is no wonder that a typological definition of species was thought necessary by many. This same variation caused others to feel that the concept of species was a rational mistake.

One ray of hope that alleviates the ambiguities caused by morphological variation was the reemphasis of the significance of the reproductive process on the definition of species, first suggested by John Ray. He defined species as "closely related groups of organisms whose parents were similar and who passed their characteristics on to their offspring." This definition in time influenced the development of a completely new point of view concerning species in terms of reproductive ability. Species could then be defined as a group of organisms in which one member could mate (cross) with any other member of the same group and produce like offspring. This is a functional definition independent of morphological characteristics. A yellow oak tree reproduces a yellow oak tree, regardless of variation, and thus is said to belong to the same species. Using this definition, the species is no longer thought of as *an* organism but rather as a population of organisms, and variation per se is no longer a detrimental factor.

In 1910, before this notion of species was widely accepted, there were an estimated 19,000 species of birds. By 1940, this figure had risen to 27,000. With the acceptance of this new definition, however, the number of species has been reestimated to be about 8500. A more dramatic example of the application of this new definition is that the 251 species of oysters described by one investigator are now considered to be but 1 or 2. Oysters are quite sensitive to the nature of the substrate on which they live and the variation in form thus produced no doubt contributed to the description of so many species.

The real power of this definition is that it is based on a very fundamental and unique property of living matter, the ability to reproduce itself, and the fact that by crossbreeding one can test whether or not two organisms are of the same species. Here is an instance where experimentation can be used in supporting a concept.

Although the definition above removed many of the morphological aspects of the "species problem," some ambiguities still remain. For instance, in plants and in some domestic animals, hybridization occurs quite easily, particularly under artificial conditions. Often the ripening of flower parts is timed under natural conditions to eliminate the possibility of self-pollination, but if pollen is applied artificially, such elimination may not be in effect. While the mating preferences of a number of animals normally would prevent hybridization in the wild, in captivity such animals may mate, as do a male lion and female tiger, producing a liger, and a male donkey and a mare, producing a mule. And so an additional experiment to determine the fertility of the offspring must be performed since most hybrids are sterile. A few hybrids, however, particularly among plants, can reproduce themselves.

Given the limitation above, Ernst Mayr's definition of species has been the most widely accepted. "Species are groups of interbreeding *natural* populations which are reproductively isolated from other (such) groups." A *subspecies* is a subgroup of a species that has some distinct morphological, ecological, or behavioral characteristics setting it apart from other members of the same species.

There remain some serious difficulties even with this improved definition of species. One is that if species are to be determined on the basis of whether or not the specimens crossbreed, this would lead to an insurmountable task. Certainly each specimen could not be tested. And it has been estimated that only one five-hundredth of 1 percent of all known species has now been studied genetically, let alone crossed with likely relatives. Given the present rate of discovery—10,000 species per year—it is doubtful that hybridization studies will solve our classification disputes.

Another major difficulty is that not all organisms reproduce sexually. Many reproduce both sexually and asexually. Some organisms like hydras, flatworms, sponges, and earthworms can fragment into pieces from which whole organisms can regenerate. This phenomenon is even more common in plants. In many insects and plants apomictic reproduction occurs in which parthenogenetic organisms develop from eggs that have not been fertilized by sperm. Among microorganisms the common mode of reproduction is asexual by fission or budding. One group, the blue-green algae, reproduce only asexually. Such organisms must be classified on some other basis. For them, morphological characteristics are still used as the most convenient basis, but other attributes are supplementing or replacing it. Examples of these attributes are stain reactions, physiological characteristics, immunological reactions, chemical constituents such as steroids or organic acids, and, more recently, the specific structures of proteins and nucleic acids.

THOUGHTS

Although the classification of organisms is in a primitive state, the style of thinking that is most in evidence during any classification is probably the most abstract of any. The specimens per se are soon lost in a pattern of numbers and parameters and it is the relationship between these latter that becomes the sought after goal. It is the complexity of organisms and their diversity that has, at least so far, kept this style but a necessary method of analysis with little progressive development and essentially no new extrinsic concepts have been conceived. In order to appreciate this loss to biology, we need to but compare some of the products of this style of thought in other disciplines. In mathematics, most of the abstractions and axioms result from a classification of number and geometric entities. This success is due to the inherent regularity and consistency of these parameters. In nuclear physics, the classification of subatomic particles has led to a variety of abstractions, many of which are still incomplete. Nevertheless, symmetries in various properties of these particles have been able to be assumed, and matrix patterns of all the particles have suggested additional abstractions that could develop into a unique description as esoteric as quantum mechanics. But it is probably in chemistry that the result of a classification of purely empirical observations has produced the most amazing sequence of abstractions. I am referring to the Mendeleev chart of the relationship of the various chemical elements derived from their patterns of interaction. From this cyclic array, abstract models of atoms, electrons, and eventually quantum mechanical constructs were generated. We are very far from finding a "table of life-forms" (species), if indeed there are any fixed "atoms-of-life" species known—as yet. We must wonder about this point a great deal.

This search with living organisms has also revealed that classification can be legitimately approached from either a priori or empirical starting points. With organisms, it seems that the extremes of these views may net some results—in fact, one must ask if both may be effective at the same time. With this kind of wide openness permitted to our approaches in classification processes, one must seriously wonder whether any of the derived abstractions are merely human relations or real. We must remember what John Locke wrote: "Genera and species depend on such collections of ideas as men have made, and not on the real nature of things." There is the opposite view, as stated here by Cicero: "The beauty of the world and the orderly arrangement of everything celestial makes us confess that there is an excellent and eternal nature, which ought to be worshipped by all mankind." Could it be that both are right?

iii

order: in fitness

One hill cannot shelter two tigers.

A Chinese proverb

We fat all creatures else to fat us,
and we fat ourselves for maggots.

WILLIAM SHAKESPEARE

In the last chapter we looked at the diversity of the many species per se. We are keenly aware that in so doing we removed these organisms from their natural setting and had to ignore any question of why they were where they were. This latter point suggests an order that must be as ancient in our awareness as the language of classification. Our archaeology, myths, and religions strongly tell us that man had to find explanations for his need for certain plants and animals and for the heat, sunlight and soils that played specific roles in his existence. The seeking of the explanations for this order in fitness of man to organisms, organisms to organisms, and both to their physical environment must have been long and tortuous. We shall but cover the more scientific strategies.

One of the major difficulties with this search is that there are many facts and even some concepts that are well-known, yet not tied together into some whole picture. From one's own experience he is aware of the

38

tremendous impact the physical environment can have on specific organisms, e.g., temperature on leaf fall and fur color; light on flowering, or type of algae that can grow at given depths in the ocean, and mating in many animals; and humidity on germination of seeds and vertical stratification of insect species within a rain forest, etc. Organisms also depend on other organisms in very general ways such as the "food chain" notion where for example a hawk may live on small rodents who in turn live on cereals. In addition we are easily aware of specific organism-organism interrelationships such as the interdependency of certain algae with certain fungi to form lichens, specific parasite-host relationships, and flower-insect dependencies where a specific food need is exchanged for a specific pollination need. Natural history is so much a part of our everyday language that we must agree to some extent with Charles Elton when he wrote about this search and its explanations, "(It) consists in saying what everyone knows in language that nobody understands."

The point is that a variety of views of this subject exists. Some are more productive than others, but even in the most analytical approach, few if any reliable extrinsic concepts have been generated and, even more disconcerting, the wide scatter of facts most often remain unintegrated. The many views make a common theme difficult to find and little synthesis has resulted. One must wonder if there is not a self-restricting factor operating in this search because of its purpose: defining *natural* relationships. For the most part, analysis consists of empirical observations and correlations derived from natural areas as they exist. Little experimentation is possible as this would introduce an artifact and the naturalness of the results would be suspect. In lieu of this device, searchers must look for natural disturbances as a source of change and trust that the response is natural and that the data can be applied to some hypothesis. This is a serious handicap. Man seems to be caught in a dilemma where he is aware of the effect he has upon his own environment, feels that it is often detrimental, and as a result has a strong urge to return his environment to what it was. And yet, he cannot feel free to experiment with his environment in order to understand it and correct his own effect. How curious it is that this most urgent need as felt by many progressive people is frustrated by a basically conservative attitude they must also believe in.

In this chapter we must omit many facts and points of view and concentrate on examples of various styles of thought used and the concepts they produced. Little of the long history of natural history can help us and only some history is used. Analysis of the growth of a single species, mainly man, is considered first since it came early and could be applied to other organisms. General relationships of a given organism with others and their physical environment follow. We are then ready to look at a complex natural area from a purely analytical descriptive view of species abundance, distribution, and species-species interaction. Some special kinds of interaction such as parasitism, predation, and competition are analyzed in some depth to reveal the significance of several products of this analysis—Gause's law and the niche. Then various methods of defining a natural area, mainly as a community and/or an ecosystem, permit many dynamic qualities such a growth of the area, stability, species diversity and exchange, succession, and optimum energy utilization to be revealed. Although these latter more recent concepts are incomplete, they are most exciting and suggest that many more concepts and syntheses are due.

GROWTH OF A SPECIES

In this section alternate explanations for the patterns of population growth are surveyed historically. These early ones were based on assumptions with little data, and even when some data was available, the relative effect of different possible regulatory factors remained ambiguous.

One of the earliest considerations of the interaction of organisms was that involving populations of humans. Plato left a lasting influence when he wrote that man unconsciously sets an upper limit on the populations of cities by encouraging abortion, infanticide, celibacy, and other methods. This was not generally questioned until the sixteenth century when it was suggested that the amount of land or food available and disasters such as famines, wars, and epidemics were more important than self-regulation in controlling population. Giovanni Botero (1540–1617) stated in 1588 that all populations strive to increase in geometric progressions, checked in this growth only by external resistance factors. This assumption formed the basis for estimating the age of the earth. One calculation, based upon an estimate that the earth's population has doubled every 64 years, concluded that Adam and Eve and the earth were 5610 years old. A different explanation as to the number of organisms present at a given time was suggested by William Derham (1657–1735) who concluded that "the whole surface of our globe can afford room and support only to such a number of all sorts of creatures. And if by their doubling, trebling or any other multiplication of their kind they should increase to double or treble that number they must starve or devour one another." He assumed that the exact number of all organisms, not only man, was divinely maintained by the altering of their life-span. This conclusion was based on the fact that according to the Scriptures man lived about 900 years during the time of Noah. With the shorter life-span today more individuals could be born at any one time and yet the total number would remain constant.

By the eighteenth century, an extension of the explanation put forth by Boltero became widely accepted, especially through the writing of the prominent biologist and contemporary of Linnaeus, Georges L. L. de Buffon (1707–1788), who wrote, "An unbound fertility (reproductive potential) of every species" is counterbalanced "by the innumerable causes of destruction (environmental resistance) which are perpetually reducing the produce of that fecundity—so as to preserve nearly the same number of individuals in each species" (*Historie Naturelle*, 1756–1758). De Buffon considered that this balancing force could consist of physical environmental factors and that these acted upon all organisms, including man, in like manner. He even allowed for flexible oscillations of the balancing point so that fecundity one year is balanced by sterility the next. But, his most unique contribution was to conclude that biotic control of population size, especially through predation, was more important than control by physical factors. Predation not only benefited the predator but also the prey: "If prodigious numbers of them (herring) were not destroyed (by fishing), what would be the effects of their prodigious multiplication? By them alone would the whole surface of the sea be covered. But their number would soon prove a nuisance; they would corrupt and destroy each other. For want of sufficient nourishment their fecundity would diminish; by contagion and famine they would be equally destroyed; the number of their own species would not increase, but the number of those that feed upon them would be diminished. As this remark is alike applicable to any other species, so it is necessary they should prey upon each other."

In addition to the writings of de Buffon, interest in the regulation of growth was given a strong boost by the writings of Thomas Robert Malthus (1766–1834), particularly *An Essay on the Principle of Population as It Affects the Future Improvement of Society*, written in 1798, which proved to be one of the most influential essays in literature. While he only repeated what had been said over 200 years before, that man's population is following a geometrical pattern of increase, while the amount of food increases arithmetically, he said it at a time when England was in danger of being overpopulated and, as a result, great controversies resulted. He was a forceful writer who dramatically emphasized the hopeless state of society by arguing that certainly man needs food and yet that the sexual passions of man will and must remain. He felt that man would inevitably outstrip the earth's food supply and that "misery (would be) an absolute necessary consequence (and) vice ... a highly probable consequence. ..." He envisioned a violent competition and struggle for existence. His arguments and conclusions are of a social nature containing many value judgements that cannot be proved or disproved. If all his predictions did not come true in his time, the issue has nevertheless continued to be of major importance and new predictions of an even worse nature are being put forward today.

One thing which Malthus reemphasized from earlier works and which was not obviously considered by de Buffon was the role of intraspecies competition among the predators as a control on their population size. Malthus emphasized the harmful effects such competition would have on society—especially on the moral state. Such an emphasis has produced an endless controversy on the evils of competition. This connotation of competition has influenced its assigned role on the interactions of organisms of all sorts. It has been difficult to convince one that competition can be indirect, such as two cows eating a common grass or, even less obvious, two oak trees competing for a common supply of water or sunlight. Yet most competition is of this nature.

There is another point that even de Buffon is guilty of when he said, "by contagion and famine they would be equally destroyed." If the food supply required by a stock of cows was an acre of grass and there was only three-fourths of an acre of grass available, according to this notion, all the cows would die. This is not the case, however. Within the species there is variation in the ability to obtain, utilize, and even require food. No moral judgements can be ascribed to those cows with greater or lesser abilities. There is inequality and, in the case of the food shortage, some will die; however, most will remain. In this light, it could be argued that inequalities have served to preserve the species. We may then wonder if this rule can be applied to man.

Nevertheless, de Buffon did set the stage, and the details of the dynamic interactions of species were sure to follow. He encouraged the development of a mathematical formulation of the growth of populations. This was first done by Quetelet (1796–1874) in 1835 when he likened growth to the speed of a falling body as described by Galileo. That is, the change in population ΔN, during a period of time Δt is proportional to the population N at that time, or $\Delta N/\Delta t = rN$, where r is a growth constant representing the growth potential of the population. This is analogous to the falling body problem where N would be distance; $\Delta N/\Delta t$, speed; and r, equivalent to a gravitational constant. With this analogy in mind, Quetelet stated and P. F. Verhulst later formulated (1838) that the resistance to growth would be proportional to "the square of the rapidity with which the population tends to grow." Or, in terms of the analogy, there would be a resistance to the free fall due to the media through which it is falling proportional to the square of its speed. Thus the actual growth would be $\Delta N/\Delta t = rN - \phi N^2$, where ϕ is a constant of resistance to growth. Verhulst was inclined to feel that the power of N in this last term may well vary from one kind of population to another. Nevertheless, the squared power term has been retained, and the Verhulst equation has been named the logistic equation and rewritten as $\Delta N/\Delta t = rN(1 - N/K)$, where K equals r/ϕ and represents the population at equilibrium between the potential growth and the resistance factors. This can best be indicated by the curves in Figure 3–1. In this plot of population N versus time t, curve

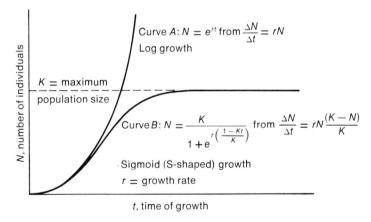

Figure 3-1. Population growth.

A represents the growth characteristics where there is no resistance to growth; thus, N increases geometrically. Curve B includes a resistant factor K. Notice that if N is small with respect to K, the increase in population is determined only by the growth potential constant r. But as Botero first mentioned, factors opposing this growth appear as the population size increases. As N approaches the value K in size, the growth approaches zero. Growth of the population will cease and the population will remain at a constant size. The result is an S-shaped curve.

This formulation implies that resistance to growth is related to the size of the population, but it does not reveal its source—physical or biological. It has not been until recently that controlled experiments were carried out to test the hypothesis of growth or its controlling factors. S-shaped curves have been found for many microorganisms, insects, and mammals. In typical experiments the time required for the population to reach a constant steady state has been about 18 hours for yeast, 40 days for fruit flies, and about 75 years for sheep. In these experiments, various investigators were able to show that the lack of nutrients could yield the S-shaped curve. Other factors such as metabolic products secreted into the surrounding medium are also believed to affect the rate of growth of microorganisms. These experiments and formulations involve only a single species and are under simple conditions; they do not necessarily speak to a natural, complex "woods." They do, however, form a basis for understanding growth and the importance of resistance factors as well as the possibility, at least, that food alone could account for this resistance.

The de Buffon-Malthus influence also stimulated a further analysis of the immediate environment that may effect the organism's growth, the so-called microenvironment. This was especially thought of in terms of the microelements about which much was being learned at this time. Among those most influential in focusing attention on the physical micro-environment was Justus Liebig (1803–1873) whose major concern was to define the chemical conditions essential to life. The results of his many detailed investigations and experiments were summarised in 1840: "A beautiful connection exists between the organic and inorganic kingdoms of nature. Inorganic matter affords food to plants, and they, on the other hand, yield the means of subsistence to animals. An animal requires for its development and for the sustenance of its vital functions a certain class of substances which can only be generated by organic beings possessed of life. (This) primary nutrient must be derived from plants. ... Hence one great end of vegetative life is to generate matter adapted for the nutrition of animals out of inorganic substances, which are not fit for this purpose."

In other words, there is a flow of matter from organism to organism that is by its necessity for their existence a binding force in keeping them together. Another important conclusion is that the physical environment is not just an external impingement upon the species but at least some parts of it become the species.

SPECIFIC SPECIES-SPECIES INTERACTIONS: HOST-PARASITE, PREDATOR-PREY RELATIONSHIPS

The concern for a balance of numbers of organisms in a predator-prey relationship, as stimulated by this tradition above, caused further theoretical and some experimental studies of special cases. A particular case was of major practical concern, that of the parasite malaria. It was Sir Ronald Ross who found in 1911 that malaria grew as an S-shaped curve under ideal conditions. Knowing this, W. R. Thompson in 1922 and Alfred Lotka in 1925 modified the logistic equations of growth as follows:

$$\frac{\Delta N_1}{\Delta t} = rN_1 - kN_1N_2,$$

$$\frac{\Delta N_2}{\Delta t} = KN_1N_2 - d_2N_2,$$

where N_1 is the number of individuals in the host (insects or man) population; r is the growth constant of the host; N_2 is the parasite population; k is the coefficient of invasion of host by parasite, a function of contact between the two; K is the growth constant of the malarial parasite, a function of successful egg laying and hatching; and d_2 is the death rate of the parasites. Lotka found solutions to these equations as constant oscillations of both the host and parasite just out of phase with one another with a period of $2\pi/\sqrt{rd_2}$, as shown in Figure 3–2. There was also a special

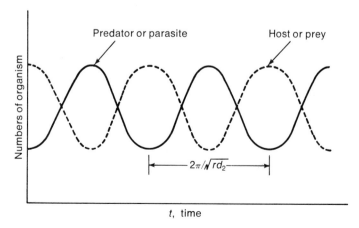

Figure 3-2. Host-parasite oscillations.

case, depending on the values of the variables, for the oscillations to damp out with a constant ratio of host to parasite remaining. This latter case is not considered a probable condition and, besides, the existence of oscillations was supported by empirical host-parasite observations. Although this may apply to many host-parasite relationships, malaria can be eliminated if the host (mosquito) is reduced in number without being entirely removed. Lotka was able to find an answer with his questions that satisfied the malaria case by extending the number of terms in the series of its solution. In this case, the oscillations are unstable and the parasite can be eliminated as indicated in Figure 3–3.

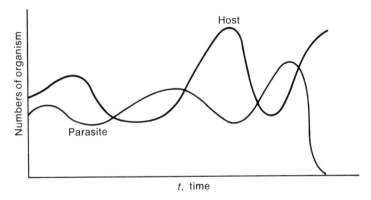

Figure 3-3. Parasite elimination if oscillations are unstable (e.g., malaria).

Lotka also expanded his analysis to include predator-prey relationships. The same equations as the host-parasite one above apply if it is assumed that the prey have an excess of food independent of their population changes. Also, in this case, r is the growth constant of the prey; N_1 is the number of prey; k is the predation constant, a function of contact made between predator and prey species; K is a constant related to growth of the predator due to successful assimilation of the prey; N_2 is the population of predators; and d_2 is the death rate of the predator.

The same equations should yield the same solutions, which would be for most cases an oscillatory one. Lotka argued against the chance elimination of the prey, possible for the parasite, on the basis that this must occur just prior to an extremely high prey-predator ratio, which he maintained was impossible due to the size of the predator which must be for the most part larger than the prey, just opposite for that of the normal host-parasite size ratio. The large size of the predator would require that it assimilate many prey in order to survive, and it could not increase in number greater than the prey for any extended period of time or the predators themselves would suffer severely. Lotka's conclusion was that the prey could never be eliminated, at least by predation.

Such oscillations, or fluctuations as they are more commonly known, have long been observed in nature. One of the first thorough investigations of these was carried out by Charles Elton and reported in 1924. He studied the arctic fox-Canadian lemmings in northern Canada and the red fox-snowshoe rabbits in southern Canada. Significant amounts of data were made available from skins collected by the Hudson Bay Company. Elton wrote, "Any abundance in fur returns one year necessarily implies abundant food supply in the preceding year"—as the young grow in numbers, there would be a lack of food that also makes them easier to trap. Elton calculated the period for the fox-lemmings to be about 3.6 years. In spite of this irregular time, he still thought that the fluctuations in the lemmings and rabbits were due to some climatic change. He noted that the snowshoe rabbits had a period closer to 10 to 11 years, which agreed with sunspot activity cycles. Lotka would have disagreed with a need for an external factor and would have concluded that the fluctuations are the common mode of balance between all predators and their prey.

Lotka's conclusions have still not been confirmed to everyone's satisfaction. Many controlled experiments in the laboratory have been carried out and the results are ambiguous. Most often the relationship is unstable, and the prey are eliminated, often without any oscillations.

There have, however, been found several interesting ways to stabilize a simple predator-prey system. One has been to space the prey as heterogeneous clumps. In this case, the predators build up at one clump by devouring the prey. As the prey at this point are almost eliminated, the predators also decrease in number until some of them find another "area of discovery." Here a second burst occurs, etc., producing stable oscillations over extended periods of time because the depleted clumps of prey have time to recover and serve their role again. An example of this kind of stable oscillation has been found in nature where knapweed, which is used by gallfry for its egg-laying site, is found in scattered clumps each about 0.25 square mile in area.

Another artificially produced stable system has been demonstrated in the laboratory by adding the prey in pulses. An example of this was observed for daphnia grown in an aquarium. A most interesting result of this experiment was that the fluctuations of daphnia occurred with a constant period of about 14 to 18 days independent of the regularity or period of adding the prey. In this case the cycle is believed to be set by maturation of the reproductive potential of the daphnia each generation. It is interesting to note that either an internal or external factor could possibly stabilize the oscillating systems. We might wonder if oscillations themselves are necessary for stability. There does not seem to have to be fluctuations of members within a food-web. Where does artificiality cease and natural occurrence begin with such laboratory experiments?

If the system is made complicated by adding additional interacting species, perhaps more can be said. Lotka concluded from solutions to his equations involving two prey subject to predation by the same predator that this three-member system is probably unstable and that one of the prey species would be eliminated, thus returning to the stable oscillating doublet.

PREDATOR-PREDATOR COMPETITION: STABILITY AND THE NICHE

It was A. J. Nicholson who first brought in the notion in 1933 that interspecies competition could possibly stabilize predator-prey relations. He assumed that it did so by "automatically regulating the severity of its (predator-prey) actions to the requirements of each (species)," otherwise it was all too random and thus wasteful. He also said, "For the steady state (balance of nature) to exist each species must possess some advantages over all the other species with respect to some one, or group, of control factors to which it is subject." He assumed the possibility that predators could shift their prey such as is known for hawks, who shift from rodents to frogs at different seasons, resulting in a more stable system. Or, several predators may act on different parts of the prey cycle producing stability as is known for two species of wasps who prey on the bean weavil. One of these species acts at the maximum density point and the other at the minimum point. Combining these two examples, we see the value of having a flexible system where the species fill specific functional roles of a food-web and are interchangeable, thus producing a more stable whole. Nicholson proposed that competition is effective by keeping itself at a minimum with the two wasps keeping each other in their specific trophic roles. In this sense competition between these two predators brings about a more stable system.

As well as by competition between species, external physical conditions could force some interchange of species. In a like manner, with age there are shifts in predator-prey relationships of a food-web.

The first detailed analysis of competition was carried out in 1934–1935 by G. F. Gause. In the tradition of Lotka, he had managed to obtain a stable oscillating predator-prey system in the laboratory using paramecium and yeast. He found this was possible if the growth constant r was such that the yeast doubled each day and the death rate of paramecium d_2 was maintained at 0.45. This required the daily addition of yeast and the constant removal of paramecium. He then extended his experiments and theoretical analysis to include a consideration of interspecies competition. Gause did not favor Lotka's equations but rather considered that the addition of another term to the logistic equation to account for competition would be more logical. Thus the two competing species would have growth characteristics as follows:

$$\frac{\Delta n_1}{\Delta t} = r_1 N_1 \left(\frac{K_1 - N_1 - \alpha N_2}{K_1} \right)$$

$$\frac{\Delta N_2}{\Delta t} = r_2 N_2 \left(\frac{K_2 - N_2 - \beta N_1}{K_2} \right),$$

with α and β the coefficients of competition for the two species. Thus the rate of growth would depend on a logarithmic increase indicated by the first term $(r_1 N_1)$, reduced by any self-crowding or other self-limiting effects indicated by the second term $(- N_1^2 r_1 / K_1)$ and also reduced by any detrimental effect due to competition with the other species as indicated by the third term $(- \alpha N_1 N_1^2 r / K_1)$.

The solutions of these equations fall into three categories as follows:

1. When $\alpha > K_1 / K_2$ and $\beta > K_2 / K_1$, either N_1 or N_2 species are the sole survivors depending on the initial conditions. According to plot A in Figure 3–4, if the initial conditions are such that the system would start above the o–s line, then N_2 would be the surviving species; if started below this line, only N_1 would remain.

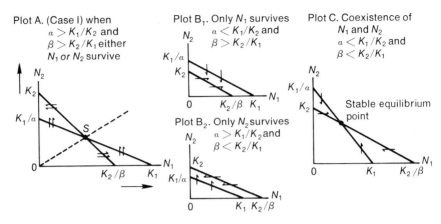

Figure 3-4. Plots of Gause's growth characteristics under competition.

2. When $\alpha > K_1/K_2$ and $\beta < K_2/K_1$, then N_2 is the sole survivor as indicated in plot B_1; while if $\alpha < K_1/K_2$ and $\beta > K_2/K_1$, then N_1 is the survivor as indicated in plot B_2.
3. When $\alpha < K_1/K_2$ and $\beta < K_2/K_1$, the two species will shift to a stable ratio in number and be able to coexist as shown in plot C.

Gause followed the growth of *Paramecium caudatum* and *Paramecium aurelia*, both separately and mixed, competing for a common yeast food. Both species grown separately follow S-shaped curves with *P. aurelia* the faster. Of interest is the finding that when mixed, both species grow somewhat unaffected by each other for the first 5 days, while there is an excess of food. But when the food becomes limiting, competition develops and *P. aurelia* begins to replace *P. caudatum* until finally the latter is entirely absent! See Figure 3–5.

From the experiments above and many others, Gause concluded "As species of the same genus usually have, though by no means invariably, much similarity in habits, constitution, and always in structure, the struggle will generally be more severe between them, if they come into competition with each other, than between the species of distinct genera." This suggested to Gause that each species has its own "space" in a natural area and that when these "spaces" overlap due to some competition, the amount of "space" actually occupied by the species is usually less than that which can be tolerated by the species.

These "spaces" are now known as *niches* and their meaning has had

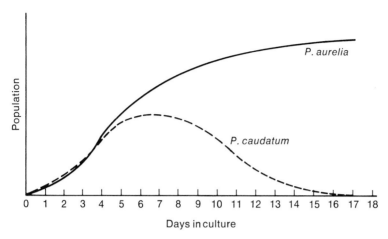

Figure 3-5. Experimental results with *Paramecium* (*caudatum* versus *aurelia*).

to be defined and redefined but confusion or disagreement still exists. A rather abstract definition of a niche is the sum composite of the *ranges* of all environmental factors, biotic as well as physical, that can allow the species associated with it to exist *throughout* its life. This definition can be imagined by laying out an influential factor such as temperature along one dimension, another such as oxygen concentration along a second, and for example size of berries along the third dimension. Then if the range of the physical factors that can be tolerated is marked off, and the size of the berries that can be ingested (for instance for a bird species), these limits would define a volume (see Figure 3–6). Now if we could imagine adding another dimension for every factor that influences the species and add a dimension of time for the life cycle of the species, the resulting hypervolume would be the niche. The niche includes the idea of what the species does—where it fits into the food-web—what eats it, and what it eats. If a species is a population "title," so the niche is its "profession." We might ask the question, what is man's niche? Is his niche greater than all other organisms? Does man expand his niche when he touches down on moon?

Defined in this way, the niche, according to the "Gausian principle," can only support one species—or another way to put it is that two species could only occupy the same niche for a short period of time. One of the species must be eliminated by competition. Some have even attempted

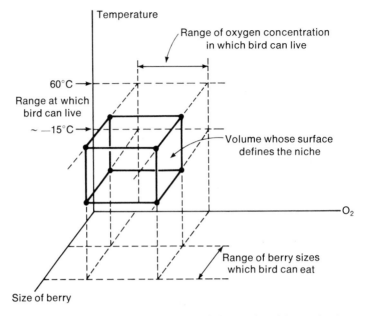

Figure 3-6. Model of the partial definition of a niche: only three parameters.

to define the species according to the description of its niche. This is difficult and it becomes a bit circular to do so.

Gause also found by experiments that if he made his system more complicated by adding a nonselective predator that removed individuals from both species of paramecium depending on their density, he could get them to coexist. This is just contrary to Lotka's theoretical conclusion and was counter to Gause's own exclusion principle. Gause's contention was, however, that under this condition neither of the two species was occupying all of its potential niche. Also even if there was some overlap of niches as a result of their being similar species, the species were restricted by competition to separate parts of the overlapping regions. Their numbers were kept low by their common predator, permitting the relinquishing of some of their respective niches. If this were the case, it is easy to see the importance of food-webs versus food chains for stabilizing the size of the species populations. In addition, more species could be accommodated; thus stability and diversity would be associated. Similar stabilization has been found in laboratory cultures of hydra species through the presence of a nonselective predator and also with insects by the addition of a parasite. In cases where there were any oscillations, the predator or parasite would dampen these fluctuations out altogether.

There is a major complication which may be operating wherever predation is thought to exist which could invalidate or make incomplete the theoretical treatments and conclusions above. This is that the prey species may well die by means other than predation. For instance, Niko Tingergen calculated on the basis of hatch size and frequency that sparrow hawks would have had to kill 5569 coal tits from June 16 to September 15 (1938) in order to maintain a constant population of tits within a given area of land that he had been observing. Yet only 138 were killed. Likewise of 2906 great tits, only 526 were killed by the hawks. As hawks are the only predators for these birds, some other cause of death had to be found. Disease, accidents, or perhaps in some cases even natural death were the apparent causes. Although these causes are no doubt a controlling factor of the prey population, this is not the type of control we would require for a self-regulating system. In this case, there is definitely a sloppy interrelation between predator and prey. This is also the case in a study of rodents preyed on by owls. One could find some increase in predation when the prey was extra large, but it was not directly related to the prey increase and far below that required to account for the death rate. Another very common example of this is demonstrated where fish or game studies have been made. It is often possible to catch three times as many fish and yet maintain the same balance! If the fish are not removed by man, apparently there is that much more of a chance for epidemics, or accidents. In other words, predation could be a regulator and stabilizer, but often it is not and in fact many believe it to be a by-product; some of the wastage of overreproduction is taken care of by this means.

There are yet other cases. Often there is a sudden obvious loss of life of many animals with no predators and with an excess of food. As these occurred when there had previously been a high density of animals, it was proposed by John J. Christian in 1950 that perhaps this crowded condition permitted or produced a greater number of antagonistic interactions between individual organisms. According to this notion the interactions would overstimulate the adrenal gland, which in turn could inhibit the reproductive rate and lower the resistance to disease. He found such was the case in experiments on mice and later with elaborate details for deer. He also wrote in 1964, "Furthermore we believe that environmental factors in most instances probably act through these mechanisms by increasing competition." A similar result has been found now in eagles and grouse and is a suspected cause for the lemmings' fluctuations. One of the additional curious findings is that it is exactly at the point where the populations become detrimental to their own environment that the inherent negative regulation takes place and relief pursues.

And so while predation and other interactions may help in some cases, competition of either an intra- or interspecific type is the main factor now believed to be involved in "the balance of nature." Both of these are density dependent and could act as the original "growth resistance" term in the logistic equations. It is now believed that if the response is direct and thus rapid, the growth curve will level off at a stable and steady value, while if there is a time delay in the effect of the competition due to the number of steps involved such as with the adrenal mechanism, then oscillations of a stable type can result.

The niche concept does not necessarily suffer from this decrease in importance in predation. Robert MacArthur set up a theoretical study in 1957 to test the Gausian principle. He contended that if a species utilizes the environment in a way to exclude other species from its particular niche, the environment is like a stick of given fixed length. If single points are selected along the stick and the stick is broken into two pieces at this point, then one part of the stick is like a given species and the remaining part is like an excluded species. By adding up the distribution of lengths of stick found by a large number of separate selections of a point and breaking the stick once for each throw, a distribution of lengths equivalent to the abundance of different species expected in a natural area would result. A different distribution equivalent to that expected when species are independent of one another with randomly overlapping niches is obtained when pairs of points are selected at random along the stick and the abundance of species is related to the distance between the two points. A plot of these two theoretical conditions as abundance of organisms per species versus $\log_e$ of number of species of any one size in descending order of size is shown in Figure 3–7. Studies on natural populations, with both birds in Peru and snails in Hawaii, agree quite closely with the nonoverlapping curve. This adds some support for the one species-one niche hypothesis.

The concept of niche has added a new dimension to a meaning of the species. In fact some systematists use the niche as a classification characteristic. The niche-species relationship reminds us that a species has a very complex life style and must fit with all sorts of other species it may compete with, be eaten by, eat, etc., throughout its life. As one thinks

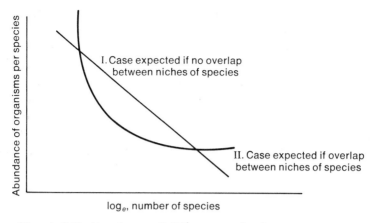

Figure 3-7. Abundance of different species in a natural area.

about this back toward the niche, one must realize that our abstract definition of the niche is a simplification of possible more complex realities. It can be quite difficult to define the niche of man, for example, especially if his entire culture must be taken into account. When man is on the moon, is he still in his niche? Does he or any other organism fit their niche as well as they might?

THE WHOLE NATURAL AREA

The above emphasized the development of the restrictions and fitness of a species with other species with which there was an obvious direct relationship. The depth of this development was made possible by the small number of interacting species that allowed a theoretical treatment and experimental verifications of the resulting descriptions. Of equal concern but of considerable more difficulty to express is the order of the whole natural area. In other words, is there a natural arrangement and balance of numbers of the members of the species found in a natural area? Also, do certain combinations of species fit best to form the makeup of a natural area? This concern that an order of the whole must exist is implicit in most of the early explanations, especially emphasized by de Buffon and Liebig. With a further realization of their ideas, and the extension of the food chain to the food-web, all became aware that in order to really understand the factors that regulate the size of any one species, more than its direct predator-prey relationship would have to be considered. Indeed, every organism of every species in the natural area must have some effect on any given organism or species. It is impossible to use the logistic equations for a complete description of this integrated whole natural area for several reasons. For one, additional terms would have to be added to the point that the mathematics would be too complicated to handle. And even if this mathematics could be managed, there would be a compounding of possible imperfections in construction, meaning, and determination of the terms and constants of the resulting equations. We are reminded of the ambiguity of applying these equations to natural or experimental data as especially made clear by the difference between the Lotka and Gause equations. And finally, the logistic equations, as they are presently constructed, do not take into account the circular or feedback nature of the many interacting members of the natural area. Obviously different strategies and styles of thought must be applied to the whole area.

There have been a variety of strategies used to describe the structure of a natural area. We shall consider only several of these here, for as interesting as they are for comparing styles of thought, most of them have been unproductive.

An important characteristic of the structure of a natural area is the arrangement of the individual organisms of a given species found in that area. This turns out to be one of the most difficult of all characteristics to express. One can describe a regular arrangement such as the trees in an orchard, but other arrangements require a statistical method with all their inherent limitations. A popular statistical method is one that uses the number of actual versus expected organisms present in fixed square areas (quadrats) into which the natural area has been divided. It is known, for example, by Poisson distribution calculations, that if the organisms are distributed completely at random, then 48.8% of the quadrats should contain no organisms (of the one species); 35.1% should contain only one organism; 12.7%, two organisms; and only 3.8% of the quadrats should contain three or more organisms. This method is demonstrated for a study of the distribution of yellow oaks in Figure 3-8. Note that in this case a higher percentage of the quadrats (67%) contain no trees (yellow oaks), less than expected (17%) contain one tree and also less than expected (3%) contain two trees, while more quadrats contain three trees (6%), four trees (5%), and five trees (2%) than expected by random distribution. There is evidently a nonrandom clumping of yellow oaks. If there had been an overregular nonrandom arrangement of these trees, the observed

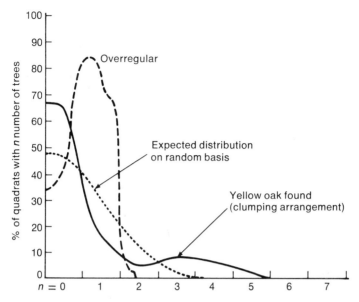

Figure 3-8. Frequency distribution of yellow oaks (an example).

number of quadrats without trees would be lower and the number of quadrats with one tree higher than expected with random distribution. Obviously these are only crude methods of describing arrangement, but they attest to our present limitations in describing patterns mathematically.

Nevertheless, in spite of this limitation, analyses of different natural areas have revealed rather consistent classes of nonrandomness among the species of plants represented. This is done by simplifying the quadrat method above and stating only the total percentage of quadrats containing one or more trees of the given species. For the yellow oak case above, this would be 33%. Then the different species are defined into classes depending on this percentage figure for quadrats with one or more represented organisms: class A species occurring in only 0 to 20%, class B occurring in 21 to 40%, class C occurring in 41 to 60%, class D occurring in 61 to 80%, and class E found in most (from 81 to 100%) of the quadrats. The yellow oak example would be a class B species. Frequency profiles of these classes for various types of natural areas are shown in Figure 3–9. When there is a heterogeneous assortment of species found in clumps, there will be many class A species. When the area contains a few species that would then most likely be homogeneously arranged, these species

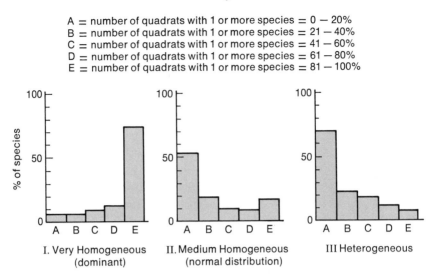

A = number of quadrats with 1 or more species = 0 − 20%
B = number of quadrats with 1 or more species = 21 − 40%
C = number of quadrats with 1 or more species = 41 − 60%
D = number of quadrats with 1 or more species = 61 − 80%
E = number of quadrats with 1 or more species = 81 − 100%

I. Very Homogeneous (dominant)

II. Medium Homogeneous (normal distribution)

III Heterogeneous

Figure 3-9. Class frequency profiles.

would fall in class E. When most natural areas in temperate climates are surveyed, an intermediate "Raunkaier" profile is found. These contain close to 53% of the class A, 14% of the class B, 9% of the class C, 8% class D, and 16% of the class E species. Such empirical findings have even suggested that a "law of frequencies" must exist wherein the frequency of occurrence of species classes shall be such that $A > B > C > D < E$. However, no known mechanisms or hypotheses have been offered to account for such a "law." The finding does support the contention that most species are nonrandomly distributed. It is also true, however, that most of the species in class E are not only homogeneous in distribution but are also the most abundant in the area. Such results, easily observed in most areas, have led to that area being named according to these "dominant" species, e.g., oak-hickory, spruce-hemlock, etc.

As distribution class could be associated with some kind of specific species-species interaction, finer analyses of associated species has been devised. For example, if 40 quadrats of the total number observed contained species A, and 30 of these also contain species B, then the "association index" of species A and B would be 30/40 or 0.75. This is a simple and therefore practical determination but it is not necessarily complete, for it does not take into account the possible wide distribution of B and thus accidental occurrence with A. A more informative definition is the "coefficient of association" or dc/ab, where d is the total number of quadrats observed; a, the number of these that contain only species A; b, the number of these that contain only species B; and c is the number of quadrats containing both A and B. The difficulty with this definition is that it is arbitrary and there remains the uncertainty of whether the association is a strong one or not. By far the best way to analyze for associations is again statistically comparing observed associations per quadrat with those expected assuming a random distribution of the species. In this way significant associations or repulsions can be determined. This may even reveal that the presence of a third species has an effect on an affinity or repulsion of two other species. This method gives some reliable descriptions of species-species interactions but cannot reveal the nature of the interaction. Also it cannot reveal those subtle effects that would be too small to show up as statistically significant associations.

Another interesting strategy, different from the quadrat approach and yet revealing certain characteristics of natural area structure, is that of indicating species diversity by counting the additional species found within an area as the diameter of the area increases. If the number of species S present versus their containing area A or their total number of specimens N is plotted as S versus $\log_e A$ or $\log_e N$, it has been found for most natural areas that a straight line results as shown in Figure 3–10. This empirical result can thus be formulated as $S = C'' + \alpha \log_e A$ or $= C' + \alpha \log_e N$, where C'' and C' are constants and α is the slope of the straight line. Since

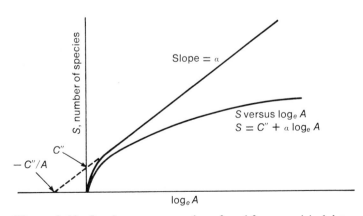

Figure 3-10. Species versus area plots: found from empirical data.

the greater the slope, the greater the number of different species exist in the same area, α is directly related to species diversity of the area. Since it is often so that density of organisms is constant in the area being considered, then N can be equated to A by the relation $N = DA$, where D is the density in terms of number of specimens per area. Thus the two kinds of plots are identical: $S = C' + \alpha \log_e DA$ or $= C' + \alpha \log_e D + \alpha \log_e A$, and since all the terms are constant, C'' can be set equal to $C' + \alpha \log_e D$. In any respect, since the area is usually quite large, the intercept constant can be ignored and S can be written as equal to $\alpha \log_e A$. Here is a case where a nice formulation can be constructed but the significance of the intercepts or $\log_e$ relationship is not known and no particular extrinsic abstractions are obvious.

THE COMMUNITY CONCEPT

One of the difficulties with any natural area study is the defining of the natural area. Where does the area begin and where does it end? Without boundaries our task of describing the area is impossible and the sense of an integrated whole is lost, for there are obvious shifts in patterns of species as one views natural scenes across wide expanses. There are abrupt breaks in the physical environment, such as land-ocean, land-lake, desert-woodland, etc., that must influence the composition of species on the two sides. But, is this the only way that composition is determined, i.e., by physical means? Studies of species interaction have suggested to many that there must be a biological glue holding certain species into specific combinations. But, again, if this is the case, where and how are the boundaries determined?

One of the most intriguing concepts toward answering this question was first put by Karl Mobius in about 1877. He considered that there were both intra- and interspecies interactions that set up balancing forces to maintain a "community" of organisms. There was considered to be a specific combination of species to produce the proper balance of forces. Mobius saw the community as analogous to an organism with self-regulatory, self-correcting, and self-propagating properties. It would have such a composition of species that "If favorable temperatures makes one species more fruitful, it will, at the same time increase the fertility of all the other ... but since there is neither room nor food enough in such a place for the maturing of all (the) germs, the sum of individuals in the community soon returns to its former mean." He felt that the community was "the highest measure of life which can be produced or maintained there." This unit had permanence "by means of transmission" of the sum total of its germs (seeds) of all its species, and it could resist "all assaults." Although there were considered to be too many different kinds of communities, depending on their composition, he also assumed that one could be transformed into another if "one of the external conditions of life should deviate for a long time from its ordinary mean." The pattern of species complement would shift accordingly.

This interesting concept stimulated several methods of comparing communities in order to determine if the species composition of areas having similar physical environmental conditions are significantly similar. The assumption is that a best fit composition of species will occupy an area depending on the physical environment. One popular method, that of determining the "coefficient of communities," uses the species diversity-area relationship described above. If we let $S_a = \alpha \log_e A$ be the number of species expected in community A, and $S_b = \alpha \log_e B$ the number of species expected in community B, then the total number of different species expected in the two communities would be $T = \alpha \log_e (A + B)$. The expected number of species to be in common between the two communities would be $S_a + S_b - T$ since the species in common will be duplicated by adding S_a to S_b. It is also convenient to use the equations $T - S_a$, which is equal to $\alpha \log_e [(A + B)/A]$, and $T - S_b$, which is equal to $\alpha \log_e [(A + B)/B]$.

An example of such a calculation is worth mentioning. Guernsey and Alderney are islands off the coast of Britain; the former has 804 species in 24 square miles and the latter has 519 species in 3 square miles. From this data, $T - S_a$ is $T - 804 = \alpha \log_e (^{27}/_{24})$, and $T - S_b$ is $T - 519 = \alpha \log_e (^{27}/_3)$. Combining these two equations, T is found equal to 820 species. Therefore, the expected number of species to be found in common is $(804 + 519) - 820$ or 503. The actual number of species in common is 480. This would seem to suggest that the two islands are quite similar, but it is difficult to estimate whether this is a reasonable closeness or not. Other community-community studies where the two are under different physical environments usually produce less similar figures, but there is an arbitrariness about this procedure that causes concern. It must also be remembered that the empirically derived equations used here may well not apply with all natural areas, especially if they are small.

Community boundaries have been sought by looking for breaks in the slope of the species diversity-area plots. This would seem like a fairly good criterion for a boundary. The results obtained by using this method remain ambiguous, however. No breaks appear where one might expect them, or just the opposite. One major source of inaccuracy, which has rarely been taken into account in using this procedure, is that microorganisms that could well play a significant role in community structure are not included. Microorganism composition of the soil is difficult to ascertain and with its uncertain role in boundary formation the question of the validity of the community concept must remain unanswered. Although this is the case, the community concept serves as a focus for framing questions of natural areas, and other styles of analysis and models as presented below will keep it in mind.

THE ECOSYSTEM CONCEPT

A different model of a natural area (or community), but one that concentrates on the dynamics rather than the structure of whole, is that of the ecosystem. This is really a description, resulting from a further development of the ideas of Liebig and food-webs, involving the circulation, transformation, and accumulation of energy and matter through the living things in a defined area. Thus the natural area is considered a system in which the flow of energy, especially, is measured at the various trophic levels. A natural area, as is any living system, is an open system in that energy must flow into and out of this system in order for it to exist.

Sunlight is the form of the input energy where some of it is converted into chemical energy within the organic molecules of plants. These molecules are continuously broken down releasing the energy, some of which is used for forming other parts of the plant. Also some of these molecules are consumed by animals, and part of the energy released in the animals is used for the formation of their own organic molecules. These latter molecules are also continuously breaking down with some of their energy going to form new molecules of the same animal. Again some of these latter molecules may be consumed by a second animal, etc., there thus being a constant loss of the initial sunlight energy and yet some of the energy is flowing up the trophic levels. In addition to this loss of energy by the selection of only a part of the prey to the predator, there is some loss at each trophic level due to microorganisms decomposing the organisms that are diseased or die. These decomposers rapidly break down their ingested organic molecules and some of the energy released goes into forming their own organic molecules as found in those organisms of the next higher trophic level. The decomposers are not usually consumed by another organism, however, and thus there is no sequence of trophic levels involved.

The flow of the system is due to the fact that every time an organic molecule is broken down or formed only a part of the energy involved in the transfer is retained as chemical energy. For, according to the second law of thermodynamics, in every chemical reaction some of the energy must go into a heat form rather than remain in the chemical form. The heat form of energy cannot be used directly in the formation of molecules; thus it is a nonutilizable form. Eventually all the energy that enters the ecosystem leaves it in the form of heat. This is, indeed, why there is a flow.

The organisms at any one level exist because energy in a chemical form accumulates as a pool. That is, energy is flowing in and out of the organism as water may pour into and leak out of a bucket. If the flow is constant, there is a "steady state" of living material present. The ecosystem can best be understood by flow diagrams with compartments for the various trophic levels and decomposers as shown in Figure 3–11.

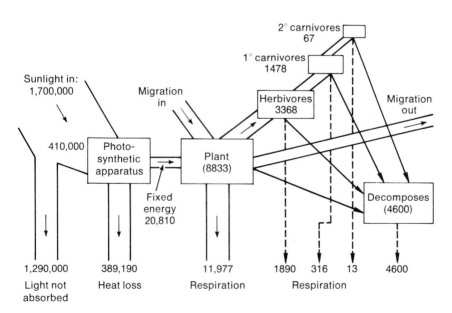

Figure 3-11. An ecosystem: data from an actual lake study (values in kilocal/m^2/year).

Note that very little, only about 1%, of the sunlight energy becomes fixed into plant material. The most significant relationship is the so-called efficiency of production, the ratio of energy actually available for the next higher level divided by the fixed energy (steady state) of the level. These efficiencies are shown for the particular case diagramed. It is of most interest that this efficiency tends to remain constant from one level transition to the next and regardless of the type of ecosystem being studied. This value is usually around 10%, although some studies in the ocean may reach as low as 5%.

This is a tremendous loss of energy and accounts for the smaller amount of fixed energy content (and usually biomass or numbers of organisms) at higher levels, thus forming a pyramid of energy on a trophic diagram. This shape dictates that there can only be a limited number of trophic levels, so many carnivore levels. The low efficiencies also tell us that if one had a choice, based on energy availability alone, he should be an herbivore. Analysis of various ecosystems has shown that those with the most complex food-web and thus large number of interwoven niches have the highest efficiencies and hence flow of energy and can support higher trophic levels more easily. Food-webs permit more parts of an organism to be consumed by many predators and with the greater amount of inter-species competition existing presumably the most efficient combination of species will be selected. Thus we can see that an ecosystem description and measurements suggest how the structure and composition of a natural area (or community) are limited and possibly determined.

One other empirically derived relationship that may prove a useful index of the efficiency of a natural area is the "turnover," the number of times the energy present in the steady state condition of the community is replaced by inflowing energy. This is equal to the net production rate P divided by the biomass B over a given period of time. For the study given above an equivalent mass of the production rate is 6390 grams per square meter per year, which maintains a biomass of 809 grams per square meter. Thus the turnover P/B is equal to about eight times per year. It should be kept in mind, however, if turnover is an index of efficiency, it should, as Eugene Odum wrote, "be measured by the number of times the community turns over *without* change in its composition." For instance, areas of small organisms that persist for a year with a daily turnover are considered as stable as an area of large organisms that lasts 300 years turning over only once per year. With this in mind, the lower the turnover number, the more stable and diverse the species should be since a greater biomass is being maintained with a smaller flow of energy; e.g., there would be a more efficient utilization of energy. A lower limit to turnover in addition to inherent inefficiencies may be the need for a rapid inefficient use of energy and matter in order to keep the microelements recycling adequately and also perhaps to maintain rapid responsiveness to changes or damages to the community. One way in which energy and matter flow can be speeded up is by having a constant leakage to the rapid acting decomposers that can return microelements to the soil.

GROWTH AND STABILITY OF A "COMMUNITY"— SUCCESSION AND CLIMAXES

Taken as one of the best supports for the community concept is the ability of a natural area to recover from disturbances, thus demonstrating responsiveness and stability. An indication of this ability was first sensed by Henry Cowles in 1899 when he was studying the zonation associated with the sand dunes of Lake Michigan. Cowles described the three zones, hydrophytic, zerophytic, and mesophytic, existing in that order from the shore inland. What was more important was his conclusion that these represented stages in transition from sand to a moist forest. He noticed that the sand which was washed and blown ashore was held in place by grasses which moved in from offshore locations. There was a natural reclaiming of the lake by the shore. And what was more, with time the grasses were replaced by other plants more characteristic of first the xerophytic and then the mesophytic zones. This phenomenon, now referred to as "succession," was further studied at Lake Michigan over the next 40 years with the results supporting Cowles' conclusions. As shown in Figure 3–12, the first stage consists of an invasion of "pioneer" flora and fauna onto a soiless base. The Lake Michigan pioneers, which require only about 3 to 5 years to become established, are grasses; willow, cherry, and cottonwood trees; and some tiger beetles, spiders, and grasshoppers. In spite of the fact that each association is gradually replaced with continual mixing by the next succeeding association, each is distinguished by dominant species forms—for the pioneer group, the cottonwoods. The second stage, taking 25 to 100 years, is dominated by the jackpines with other spiders, sand locusts, wasps, and ants being found. A third stage can be recognized as dry forest dominated by black oak and containing other grasshoppers, antlions, wireworms, and snails.

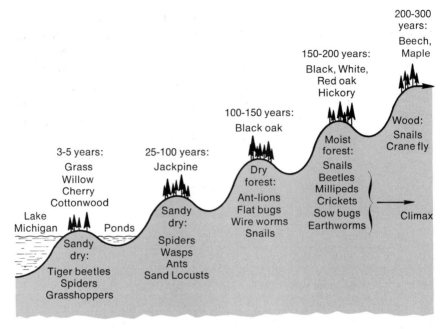

Figure 3-12. Succession: Cowles' study of dunes at Lake Michigan.

This in turn is replaced by a moist black, white, and red oak and hickory-dominated stage with snails, beetles, millipeds, crickets, sowbugs, wood roaches, and earthworms and requires at least 150 years. Finally there is a beech- and maple-dominated forest with some oak and hickory, most of the fauna in stage four, plus wood snails and crane fly larvae. It has been estimated that this latter stage would take at least 200 to 300 years to be established from the time of the pioneer invasion. This stage also potentially extends over that entire region of the country.

The study has also confirmed that there is continuous community-environment interaction. The grass modifies the sand distribution, which in turn helps support more sand. The decaying grass supplies some nutrients to the sand and offers some shade for seedlings to get started. This too is a modification of the environment. The cottonwoods add more shade and decayed matter to the sand, producing a soil which can support microorganisms which can in turn add other nutrients necessary for other plants to take hold. With succession the environment becomes more compatible for more kinds of organisms to exist but not necessarily the early types.

Many generalities can be made from this and similar studies. The different stages of succession, called "seres," follow a sequence similar to the land stratification, starting from the colder zone at a pole toward the warmer one. Tolerance of the organism to the total environment of the seres is a limiting factor on the kinds of organisms present. Factors other than temperature are involved. Light-demanding plants precede shade tolerant ones as shade becomes abundant and influences the type of seedling that can grow. Generally annuals precede perennials, and shrubs are found before these. Simple assimilation systems precede the more complex root systems dependent on humus. This in turn will influence the size, weight, and coverage of tree that the roots can support. The size of the fauna will also usually increase. And of most importance the density, diversity, number of trophic levels, and the rate of energy flow per area through their ecosystem increases with succession. The food-web becomes more complex with more interspecies competitive interactions and a tighter niche-niche relationship results.

Of special interest is the last sere in this succession. It is called the climax of the succession and appears to be the final stage in these changes, opposing changes by resisting the invasion of new organisms and yet supporting the reproduction of its own kind. It is thought of as being in a dynamic equilibrium condition with itself. This is supported by the finding that different sequences of seres result when the initial substratum is different but that the same climax ensues and remains if the climate is the same. It would seem that the climax is primarily determined by temperature and humidity. The stratifications of life zones are the different climaxes that the different temperatures can support.

There has been debate about whether indeed there is such a thing as convergent succession to the same climax and whether there is such a thing as a climax itself. Those opposed to the concept point to the exceptions where not all the seres expected show up or for some reason a stand of trees exists for long periods, and yet these "edaphic" climaxes are not of the climatic climax type. Because of the exceptions to the sere sequence, opposers to the climax concept claim that pioneers are merely those organisms with the greatest dispersal mechanisms, and any sere is dependent on chance availability of the organisms. Other observers, however, contend that there are good reasons for all these exceptions. A notable example was the finding by pollen analysis that a pine or oak-hickory climax existed for many years in New England but that it has been replaced relatively recently by a beech- and maple-dominant type. As the climate has not changed significantly over this period of time, it was thought to be an edaphic climax. Since this conclusion it has been learned that Indians present in precolonial days burned the deciduous trees to produce greater open spaces for hunting deer. Here is a case where man was maintaining a "subclimax." There appear to be natural causes such as local blights and galls, fires, floods and so forth that could have the same effect.

In addition, there is too much regularity during succession to believe that it depends on chance dispersion. Man has leveled whole fields in some cases to bare rock and has been able to find expected successive seres. In one long study of a field with new soil, the first year produced horse-weed, by 2 years there were aster and bloomsedge, and by 5 years there were pine seedlings. In 15 years the herbs were eliminated by the pine, and by 26 years there was an understory of gum, red maple, and dogwood. The oak-hickory understory was starting at 50 years and was estimated to climax by 200 years.

Others did much to support the notion of succession by testing the susceptibility of different seres to the introduction of new species. This work has led to the definite conclusion that the closer to climax, the less likely a foreign species will take. If a community is disturbed by fire, flood, man, or other sources, a new species has a reasonable chance of becoming established. It will not remain if the community recovers by succession. Apparently the new species is not as efficient as the normal sere members. If the area remains disturbed, the new species may continue to thrive there. If there are large disturbances such as huge forest fires, the whole area may be reduced to a primitive successional level and a series of successions will follow. The soil conditions resulting from a fire are different from bedrock and the consequential sequence of seres may be different from earlier ones. Even so there is a convergence of the various succession seres types back to the same climax.

Likewise in the laboratory, a very reproducible experiment is simply to place hay infusion into a jar with water. A count of the organisms over a period of time invariably shows first Monads, then Colpoda, then Hypotrichs, then Parameciam, then Vorticella, and finally Amebas as dominants all within a few days. Amebas remain as a climax until adverse conditions of the culture cause their deterioration.

It was Fredric Clements who in 1916 emphasized the likeness of succession to growth and of climax to the mature state of a community in Mobius' sense of the word. Clements' analysis convinced him that succession was universal and predictable. E. Odum and others since have analyzed the maturation of communities in terms of rate of production P and rate of respiration R. He considered that all pioneer stages of succession are autotrophic and therefore the $P/R > 1$. On the other hand if we consider any of the heterotrophic levels, they will have a $P/R < 1$ since they do not need to synthesize as much organic matter. Although we usually think of succession as starting from a soilless condition with pioneer plants, it is possible to have local conditions of high decomposition such as compost piles or hay infusions, and in these cases the P/R is definitely less than unity. Odum feels that regardless of the initial condition the total ecosystem that is unequal in either production or respiration will have to change to balance these two against each other and this process of establishing a balance is the maturation or successful process. He feels that the requirements of energy balance will tend to make the P/R equal to unity, and this would be the steady state equilibrium climax. Odum is not concerned with the steps. Any sere that is closer to the balanced state will replace a less mature one. According to this hypothesis the force produced by an unbalance in P versus R causes the restoration. Figure 3–13 represents a graphic description of Odum's analysis.

Another analysis of succession and climax has been that of Ramon Margalef who in 1963, using simple laboratory models, demonstrated that succession could take place in a predictable fashion with an increase in species diversity and a decrease in P/B, the turnover. Here we see a strong relationship of the optimum ecosystem conditions being a natural consequence of succession. Competition between successive forms favors the more efficient system, and the diverse food-web climax is the ultimate.

In spite of all that has been said above, we still have not confirmed the existence of a unit of life matter we can unequivocally define as a community. Clements was completely convinced that the property of succession and climax resolved the issue in favor of the affirmative. He felt that the community was quite analogous to an organism. "The developmental study of vegetation necessarily rests upon the assumption that the unit or climax formation arises, grows, matures, and dies. Its response to the habitat is shown in processes or functions and in structures which are the record as well as the result of these functions. Furthermore, each climax formation is able to reproduce itself repeating with essential fidelity the stages of its development." But, without denying the existence of succession or climax, it can still be argued that there is nothing that demonstrates the integrated unit of the community concept. Food-webs

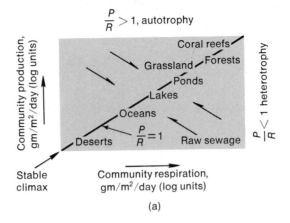

$$\frac{P}{R} > 1,\ \text{autotrophy}$$

Community production, gm/m²/day (log units)

Coral reefs
Grassland — Forests
Ponds
Lakes
Oceans
Deserts $\frac{P}{R} = 1$ Raw sewage

$\frac{P}{R} < 1$ heterotrophy

Stable
climax

Community respiration,
gm/m²/day (log units)

(a)

$$\frac{P}{B} = \frac{\text{Production}}{\text{Biomass}} = \frac{\text{Energy fixed}}{\text{Resulting biomass}} = \text{Turnover rate}$$

$\frac{P}{B}$ should decrease with maturity and succession
with increased complexity (number of
branchings and species) of food web

(b)

Figure 3-13. Models of forces acting to produce succession and stability.

can certainly form whether there are boundaries or not, for there could be continual flow of energy and matter from area to area. Thus we are uncertain still whether it is legitimate to use the community concept in a strict sense.

We spoke earlier of the dilemma a conservationist as a researcher must feel. On the other hand we may put to him that, with our increase in population with its increase in waste and thus pollution, the combination is very rapidly changing his environment, probably a great deal faster than he could do as a researcher. There are those who advocate living simply and stopping all technology and scientific study on the grounds that the latter two are the real culprits in our rush to destroy our environment. We must seriously wonder about this. Conservationists have learned that they can be overconservative. The strict prevention of forest fires has often resulted in greater, more destructive fires when they do occur and the loss of propagation of certain plants that depend on bursts of fire to disperse seeds. Also the highly emulated Indians did a great deal of destruction. They eliminated at least 50 species of mammals from North America and their practice of burning off trees to grow crops over wide expanses has caused severe erosion problems that are increasingly becoming unmanageable in parts of Mexico. We forget that technology has actually helped much of the land yield a higher abundance of crops without self-destruction. We have also forgotten that man is really a great deal healthier today than Indians of the past, although some would deny this. Instead more and more ecologists would say that we cannot go back and that we need technology and science more today than ever to reduce waste and pollution by fulfilling the biogeochemical cycle of matter including all wastes. Man has temporarily interfered with this cycle, but there is faith that he can find the ways to complete the needs by pursuing this search even more. This is a faith and the reader must decide where he stands on it. As many older natural areas have been disrupted, it may be necessary to even create new artificial ones. Man may have to play God and feel that his artificial ones could even be better and more efficient than the natural ones. This would require a great deal of faith. One would have to add a clear additional meaning to those lines by Rachel Carson in her last book, *The Sense of Wonder*: "Those who contemplate the beauty of the earth find reserves of strength that will endure as long as life lasts. There is a symbolic as well as actual beauty in the migration of the birds, the ebb and flow of the tides, the folded bud ready for spring. There is something infinitely healing in the repeated refrains of nature (including man, his thoughts, and his deeds—author's addition)... the assurance that dawn comes after night, and spring after winter."

THOUGHTS

The section on growth reminds us of how easy it is to find a formal description of a phenomenon such as growth that fits some empirical observations and to extend it by modifying old or adding new terms with further observations. This style of thought becomes centered around the equations and their solutions, these being required to fit the observations. In this case with the development of these equations to include the interaction of species, ambiguity resulted, e.g., Lotka versus Gause. The interactions of this sort are too complex, being many in kind with too many unknown or unquantified factors that influence the interactions. Thus it is difficult to generalize or derive any explicit concepts. Gause's law is about the strongest concept that comes through—through because it is a hypothesis that existed before the equations were applied to it. The niche per se does not fit these equations as well because of the difficulty in either defining it or including all its characteristics. The strategy of using an equation as a key description that can be modified to fit more of the data and generating new concepts has been very successful and one need only look at the results when it is applied to physical mechanics, astronomy, and especially quantum mechanics. Ecological equivalents have been attempted and theoretical equations as complicated as any of those of physics have been produced but their meaning, accuracy, and applicability leave something to be desired, at least as yet.

The latter half of the chapter surveys a number of strategies for looking at the whole of nature. A number of empirically derived descriptions and correlative data are available but the whole is even more complex than the parts and there is an obvious shortage of formal descriptions for dealing with gestalt-type observations. The method of correlating data between quadrat areas by direct or various statistical treatments is necessarily limited. The styles are good but again they do not apply to pattern as well as desired and the whole is very complex. The interesting order of the fitness that must exist in the whole is suggested by the discovery of the "law of frequencies" and other correlative data, but more must certainly exist. A good naturalist senses pattern, feels its significance, can ignore exceptions or deviations, and yet cannot describe it formalistically. This is not to say that exciting descriptions of some aspect of the whole is missing. The species diversity-area logarithmic relationship that is in the empirical tradition of Galileo is most exciting even though its meaning remains obscure and a synthesis with other parameters is needed. The ecosystem model of describing a natural area or community is very valuable for its clarity. Flow models are used in all areas of science where there is flow of some given entity, energy in the ecosystem. They do not necessarily add new extrinsic concepts but their clarity permits easy elaborations and generalizations. They also allow easy reassortment of the components of the model that can suggest new dynamic interrelationships. Interesting concepts such as community, succession, and climax, although supported by some observations, remind us of how close we still are to our metaphysics of harmony, balance, or symmetry with these hypotheses.

There needs to be a great deal of synthesis between all of these concepts and apparent suborders. Especially lacking is some synthesis between the results of the species-species mathematical analysis and those concepts dealing with the whole, i.e., the parts with the whole. Gestalt observations and sensitivity are rarely found in individuals with abilities for the exacting mathematics required with complex equations. Nevertheless, concepts are moving fast and the opportunities are great to break this problem and contribute to solutions of our social problems.

iv

order: in space and time

Poor man, he just stands and stares at a yellow flower for minutes at a time. He would be far better off with something to do.

Darwin's gardener

Descended from apes! My dear, let us hope that it is not true, but if it is let us pray that it will not become generally known.

The wife of the Bishop of Worcester

The search for order in space came after man was both able to travel great distances and collect observations from various places. A comparison of these data plus a synthesis of observations of geological formations at any one site resulted in the discovery of an order. There was no apparent need to explain this kind of order until it was discovered in contrast to the orders of diversity and fitness. Indeed, some explanations of these latter two orders were probably necessary for the discovery of order in space and time. Certainly some of the same data were necessary. This situation allows us to do an in depth analysis of the role social conditions and metaphysical factors have on, the effect other acceptable concepts have on, the effect the personal characteristics of the searcher have on, the role direct observations and their progressive synthesis have on, and the role that interpretations and acceptancy have on the development of a

74

key biological concept, mainly evolution and its mechanisms of action. In order to fulfill the above, the search will concentrate on the particular period of history, the early part of the nineteenth century, in which the concept was made clear to most people. Throwbacks are used to elaborate the considerations above, but very little recent data need be included. They add little new to explaining the order in space and time, and much of this will be known by the reader. There are other reasons to concentrate on this one period. It gives us a chance to explain more fully the limitations on, the opportunities of, and the incompleteness of the explanations by any one searcher, even Darwin. It is also important to take a deep look at evolution since it is often considered obvious or redundant now that it has been accepted by so many. Nothing is further from the case. It is a very complex notion that generalizes about a great number of observations. In terms of great spaces and long times, evolution cannot be proved by scientists of the purists' schools. But more important, it is being challenged today as being incomplete in the same way that Newtonian physics was challenged and superseded by Einsteinian physics. We must continue to think about evolution and one of the best ways is to examine critically the factors that contributed to its development.

A MEETING AT THE PUB

The time was September in 1831; the place, a small pub in London; the occasion, a meeting between a young though experienced Royal Navy ship's captain and a still younger inexperienced naturalist. The 26-year-old sea captain, Robert FitzRoy, had been charged with the task of circumnavigating the globe, extending chronological and longitudinal measurements, and completing a survey of South American coastlines. He had been asked to take the naturalist along. He was unhappy with this request since his ship, the 10-gun brig HMS *Beagle*, was so small that he would have to share his own cabin with the naturalist; two of his friends had already turned down his invitation. Although or perhaps because our captain is an aristocrat (the grandson of a duke and the nephew of an earl), he has a fierce quick temper and it was put on edge at the sight of the young candidate. For you see, Captain FitzRoy was also well learned in the skill of phrenology, and one look at the young man's face, having a weak nose and forehead, convinced him that this youth was certainly in lack of character and endurance to stand the journey.

THE YOUNG NATURALIST'S CREDENTIALS

If FitzRoy had known the qualifications of the 22-year-old so-called naturalist, he would have rightly been infuriated. The lad was also of aristocratic inheritance, his grandfather was a famous physician and the founder of the "Wedgewood Pottery," and his father was another well thought of physician. His mother died when he was but 8 years old, but he was well taken care of by tutors and loving sisters. His first brush with natural history came as it did to all the sons of the aristocracy—through tutored visits to collections of all sorts of things from coins and minerals to shells and insects. He always enjoyed these excursions if for no other reason than to put another new name on an already long list of names.

In spite of a good memory, he disliked and did poorly with his typically classical schooling, which was heavy with Greek and Latin. He did have a brief flare for chemistry of sorts, encouraged by his older brother who liked to do experiments at home. For these "nonacademic" activities his teachers and friends chided him and called him "Gas." But this interest did not last long and his skills developed in quite a different direction with certain esteem by his aristocratic peers—mainly in the art of hunting and fishing.

As the years went by he had little but the interest above so he was encouraged to follow the footsteps of his father and brother and try medicine at Edinburgh University. He was not, however, the student his brother was and his lack of self-discipline was heightened by an increasing awareness of his wealthy father, his grand allowance, and his expected inheritance. Also observation of several operations convinced him that there must be a less gruesome way to exist in society. But during these 2 years at Edinburgh he did manage to learn a little more about natural history. There were several lectures he found tolerable, and he met several field naturalists who took him on collecting trips. It was the informality of the outdoors and the social life associated with it that interested him the most, for he found ways of combining some of these excursions with his main interests, talking, riding, fishing, and that skill of all skills, hunting. These activities were made especially easy for him by his sports-loving uncle who introduced him to many influential friends including naturalists while on hunts together.

Even our young playboy began to realize, however, that he needed some title or position in society. When his father suggested the clergy as a possible discipline he could manage, he agreed most heartily. The position of a parson in those days was most pleasant with security, respect, and, above all things, time for the hunt. And so with concerted effort and the necessary special tutors, he attacked Cambridge in spite of a continued dislike for the classical subjects he had to master. He also disliked and did poorly in mathematics. But in spite of the pain, his memory served him well enough to pass the examination. And if he was successful in his course work, he was even more so in his avocation. He extended his social graces of riding and hunting to gambling and drinking and enlarged his friendships considerably. There were further excursions and additional naturalists, among them a professor of botany, John Henslow. Fortunately, Henslow had the facility of explaining his subject with such clarity and enthusiasm that our young clergyman could not help but develop a curiosity about the subject and in the informal setting of a ride or walk the two exchanged answer for question at great length. The book-hating young clergyman was initiated into a more scientific framework of thought about natural history and he was fascinated. He later said that Henslow had a greater influence on his interest in science than any other person. Henslow took a liking to the young man, finding him an excellent conversationalist and listener and even invited him to social gatherings where discussions went on for long hours with some of the leading natural historians of the day. With time the young man actually began to read as well as talk about natural history. Among the books that influenced him the most was Alexander von Humbolt's *Personal Narrative*. This book fired in him a desire to travel and to learn directly about natural history. It was Henslow who encouraged him to read about and consider geology as that branch of natural history that would be most advanced and most rewarding in the coming years.

It was also at this time that he was introduced to one of the leading geologists of the day, Professor Adam Sedgwick of Cambridge University, and after some discussion a field trip to Wales was arranged. Sedgwick was another gifted teacher and our clergyman found no difficulty in barraging the learned man with question after question. It was while on this field trip that Sedgwick demonstrated by example how much more valuable direct observations in the field were if they could be correlated than the vague generalizations that filled many of the books of the day. He stressed that too much of the book would prevent one from finding exceptions to already outworn dogmas of the time. As one could imagine, the young apprentice was delighted to reinforce his belief in the value of an informal undisciplined approach to natural history.

After this impression above it is understandable why the most memorable event of the trip was when the great Sedgwick violated his own teachings and amazed his student friend by stating that a tropical fossil that they happened upon in the middle of England must have been dropped there by someone. Otherwise it would be a great misfortune to geology and overthrow all known about that area.

It was at about this time that Henslow learned of the FitzRoy venture. He encouraged the young clergyman-turned-naturalist to apply. But, incensed with his son's dabbling at useless outings, the youth's father strongly refused to allow him to go. It was his riding companion and uncle, a man respected by the father, who turned the tide of dissent and made the pub meeting possible. There was also a glowing letter of praise for the young candidate sent to Captain FitzRoy by Henslow.

NEPTUNISM VERSUS VOLCANISM

Thus although his face was not suited for the job, the letters of introduction forced FitzRoy to at least consider this pub companion. Besides, there was no one else to consider. Fortunately for the young man, their aristocratic pleasures ran close, and with further lubrication by stout and the able flow of words by his companion, FitzRoy was encouraged to overlook phrenology. But, it was the satisfaction with answers given to questions exceedingly important and sensitive to FitzRoy that decided the case. For FitzRoy had been greatly disturbed by debates between leading naturalists over the interpretations of certain geological findings and their possible ramifications for basic religious beliefs. FitzRoy stood clearly on one side of this argument, and his naturalist was not only to be in agreement with him on this point but hopefully to devote a considerable amount of energy finding new evidence in unexplored areas to bolster their position.

This deep-seated debate was primarily over the origin and meaning of fossils. Fossils had been described as the remnants of once living forms off and on since the time of Xenophenes (576–480 B.C.), but there the matter rested. That is, until about the sixteenth century when new movements among some theologians were encouraging those who explored scientific methods to use their findings as a demonstration of the existence of God. One way to do this was to seek evidence in support of the origin of the earth as described in Genesis. As most scholars were mixtures of philosophers, theologists, and natural historians, these trends went quite smoothly.

Gottfried W. von Leibnitz (1646–1717) did reaffirm with clear evidence that fossils were remnants of organisms that were no longer in existence and since some sort of sedimentation had probably covered these, they had existed in the past. Benoit de Maillet (1656–1738) put forth a further correlating principle that fossils were probably the result of water coverage that later evaporated leaving a covering sedimentation layer. But it was de Buffon (1707–1788) in his *Theorie de La Terre* (1749) who went the farthest in setting up a hypothesis of 7 ages of the earth to correspond with the 7 days of Genesis. These were as follows:

1. The earth was thrown off from the sun as a result of a sun-comet collision, and this hot mass took the shape of the present earth by cooling while it rotated—all taking about 3000 years.
2. The earth continued to cool and congealed into a solid body—this took about 32,000 years.
3. Enveloping vapors cooled and formed the oceans—taking 25,000 years.
4. The water subsided by disappearing through crevices, exposing land, leaving marine fossils at various heights, and vegetation on the surface—taking about 10,000 years.
5. Land animals appeared—taking 5000 years.
6. The ocean continued to sink dividing western and eastern hemispheres and forming the islands—taking 5000 years.
7. Man appeared—this happened 5000 years ago.

This would thus total 85,000 years for the age of the earth. Buffon was quite aware that there were far too few fossils known at that time to support his hypothesis, and he emphasized the need to study fossils. He noted, "After me others will come (who will solve this problem)."

Although there was disagreement as to the age of the earth, it was generally conceded to be short, 6000 years being the most popular guess. And there was general agreement at that time that subsiding water could account for the presence of marine fossils being found on the tops of mountains. There was also general agreement that each group of organisms —marine, land, and mammal—had been created once and remained as it was when created. This fundamentalist belief was the "creationism" that had influenced Linnaeus.

The finding that fossils and rock types exist in layers as one digs down into the surface of the earth required a further elaboration of the theory above. This was done most successfully by Abraham G. Werner (1749–1807), Professor of Mineralogy at Freiberg, Saxony. Werner and his school of thought, "neptunism," explained this stratification as the result of a sedimentation of different rock types precipitating out in a specific sequence. Thus there were "primitive" rocks, such as granite, followed later by slate. Fish were thought to be created at this time. Then water began to recede rapidly into the earth, forcing the dead fish and slate into a common layer. The lowered water exposed landforms upon which mammals were created. Further recession of water caused the erosion of limestone first, and then clay and sand, which sedimented out in that order along with the remains of dead mammals. Finally volcanic action produced mountains and lava flows that formed a top layer at about the time man was created.

But it was against neptunism that a new school of thought emerged to cause a major disagreement among naturalists and the one with which FitzRoy was concerned. This disagreement resulted from differences of opinion as to the origin of the strata of rock. Objections were raised as to how sedimentation could account for the innumerable number of sharp breaks and angles found in the strata and in the variations in the order of rocks other than those prescribed by the theory that can be found. Also there was criticism as to the excessive amount of water that would have been required to hold all these rocks in suspension and then somehow disappear. The opposing school was led by James Hutton (1726–1797), Professor of Geology at Edinburgh, who put forth his ideas in his *The Theory of the Earth* (1795) in which he proposed that the earth's crust consists of two kinds of rocks, an igneous type forming a base upon which an aqueous type rests except where upheavals reversed this order. Weathering and erosion are constantly depositing sands and pebbles upon ocean beds that are then converted into solids by heat. He could think of no logical way that water could do this. Heat is also continuously uplifting landforms after they have formed on the ocean bottoms, hence the volcanic activity, mountains, and odd angles of strata. Hutton derived his "volcanism" theory from a study of volcanoes. He found that they were not all erupting at one time or as far as he could ascertain ever did. There seemed to be a continuous irratic series of eruptions.

As Hutton thought about it, he could indeed find no reason not to assume that all present landforms had resulted from the same dynamic process since the earth has existed. Hence a consequence of his thinking was that there was no need to consider that the earth's surface had experienced a series of abrupt changes as had all previous theories. And further, since there has been no observable changes in landforms within recorded history, the age of the earth must be considerably older than was then believed to be the case. He suggested that the same natural laws are acting today as when the earth formed, a principle known as "uniformitarianism." Hutton based his ideas as much as possible on known changes in geological formations, while Werner based his on detailed mineralogical studies of rocks. It is understandable how the two schools arrived at different conclusions.

The neptunists argued that there was no basis to assume that rocks could be formed from sand and pebbles by heat especially while under the water of the ocean. There were also reactions against volcanism on the basis that it did not fit with the description prescribed by Genesis. It should be noted, however, that separate individuals from both sides accused the other of not following Genesis.

INTERPRETATION OF FOSSILS— PROGRESSIONISM

There was no settling of this dispute by any experimental means since the neptunists were speculating on something that had happened long ago, and the ongoing process of the volcanists was too slow. Time was again required for new evidence. Again fossils were the culprits. The main contributions were made by two contemporaries, an obscure drainage engineer by the name of William Smith (1769–1839) and the great professor of zoology in France, Georges Cuvier (1769–1832). By the early nineteenth century enough fossils had been investigated that Smith was able to begin to characterize the successive strata by their specific fossil type and their "facies" or mineralogical characteristics. This was actually an extension of what Werner had tried to do, but Smith was able to find particular species of fossils in particular strata and no others, and he proved the continuity of certain groups of strata by fossil means where the mineral type was lacking. He thus supplied a tool (fossil identification) to study rock formations, and paleontology could begin. Fossils could from hereon be used to form a chronological basis for determining the relative age of landforms. Smith used this tool to decide where canals and tunnels could be most easily dug. Also Smith's studies revealed that fossils from lower strata are fewer and have less resemblance to present forms than those from upper strata. These results were undisputed facts put forth between 1813 and 1815.

Cuvier presented most of his ideas in his *Discours sur les Revolutions de la Surface du Globe* (1817–1827). He had done even more extensive studies of fossils than Smith had and he revealed that there was an apparent succession of organic forms that go in sudden jumps from strata to strata with fish, amphibians, reptiles, and then mammals existing in that order as one approaches the earth's surface. No human fossils were found by him or anyone else at this time. He could also conclude that the older the strata, the more extinct species were found.

It was the sudden break in kind of organism found between strata and the comparatively short period of time estimated to be the age of the earth within which all of these changes had to occur that led Cuvier to the only logical conclusion that sudden "cataclysms" had occurred to produce the break between each strata. He assumed as a carry-over from the neptunists that these cataclysms were due to transient floods suddenly covering the land. As he wrote, "the cataclysms that produced them (sea changes) were sudden—(as evidenced by) great quadrupeds, encased in ice and preserved with their skin, hair, and flesh, down to our time. . . . The animals were killed, therefore, at the same instant when glacial conditions overwhelmed the countries they inhabited. . . . The dislocations, shiftings, and overturnings of the older strata leave no doubt that sudden and violent causes produced the formations we observe, and similarly the violence of the movements which the seas went through is still attested by the accumulation of debris and of rounded pebbles which in many places lie between solid beds of rocks. Numberless living things were victims of such catastrophes . . . leaving only a few relics which the naturalist can scarcely recognize." On the basis of river, sand, and peat deposits, he estimated that the last cataclysm occurred about 6000 years ago, a fact that was well received by neptunists.

Cuvier himself did not attempt to extend any religious significance to these results or conclusions, but in England the very popular lecturers Professor William Buckland (1784–1856) at Oxford and Professor Adam Sedgwick (1785–1873) at Cambridge pounced upon the work of Cuvier with glee. It was Buckland who created the "catastrophist's synthesis." He wrote, "In all these we find such undeniable proofs of a nicely balanced adaptation of means and ends, of wise foresight and benevolent intention and infinite power, that he must be blind indeed who refuses to recognize in them proofs of the most exalted attributes of the Creator." Buckland went on to insist that Cuvier's evidence proved that Noah's flood existed and that this would have been discovered independently of the Book of Genesis. The great book was right after all this fuss. Buckland's own evidence came from cave explorations where the complex heaps of bones attested, he felt, to the products of a sudden deluge. He made calculations of the mud and silt deposits in these caves and estimated their cover to have taken 5000 to 6000 years to have accumulated.

Sedgwick wrote, "new tribes of beings were called into existence, not merely as the progeny of those that had appeared before them, but as new and living proofs of creative interference; and though formed on the same plan, and bearing the same marks of wise contrivance, often times unlike those creatures which preceded them, as if they had been matured in a different portion of the universe and cast upon the earth by the collision of another planet." The lectures and writings of Buckland and Sedgwick paved the way for the necessary modification of the views of the "creationists." It was now only necessary to say that instead of all the organisms of the earth being created at one time and existing today as they did, there was a series of creations and catastrophies producing at each step a more complex and perfect form until finally the ultimate, man, was created as he is today. This belief became known as "progressionism." These ideas were generally accepted and volcanism with its corollary, uniformitarianism, was quietly forgotten.

It would seem that Captain FitzRoy would have been delighted by this final turn of events since he was a devoted fundamentalist, but he was still not satisfied. For one thing he was not very happy with the gradual changes that the fossil evidence had forced upon the creationists—it caused doubts as to whether the battle between neptunists and volcanists was really over and where any simple fundamentalist would find himself from one moment to the next. There were also natural conservative reactions to a poorly understood French revolution among the aristocracy of Britain. Both the government and the church were ill at ease. And what was worse, a new book was just beginning to be talked about that was said to be threatening the whole basis of catastrophism. This book was *Principles of Geology* by Charles Lyell (1797–1875).

THE VOYAGE BEGINS

It was with great delight that FitzRoy found that the new candidate was not only a safe aristocrat, but a trained clergyman who had thus accepted the beliefs of the Church of England. And most acceptable of all he was a protege of Professor Sedgwick and agreed quite easily with the argument of the catastrophists. This was easy for a young man anxious to travel and disinterested in any deep philosophical theories. The more he heard about the voyage, the more he considered secretly the sport and delight of a llama or some other new animal hunt.

An accumulation of such answers to doubts that he might have had, plus an additional positive letter about the young candidate, convinced FitzRoy to offer him the position. There were some delays in the departure. The ship had to be repaired and there was even some confusion at the last minute as to whether the government would advance the funds needed to make the entire journey or only those to reach South America and back. These matters were finally settled as originally planned and the bold HMS *Beagle* set sail from Devonport in late October 1831. But as luck would have it, a violent storm came up and nearly sank the ship. The ship's new naturalist proved to be especially prone to seasickness and both he and the captain began to consider whether or not the trip was really meant for him. They returned to port, all deciding after a pause to try again, only to find the storm worse than before. The second return to port left the poor youth feeling that the whole thing was some gigantic nightmare and he became very depressed. He even felt severe pains in his chest and thought that he must have a heart disease. But with some encouragement from FitzRoy and the crew he decided to stick with the ship. And so it was that young Charles Darwin (1809–1882) set off on the twenty-seventh of December (1831) for a trip that would take 4 years, 9 months, and 2 days and would not only change him but much of the world's understanding of life.

SIR CHARLES LYELL AND THE "PRINCIPLES OF GEOLOGY"

Darwin set his mind on making good use of his time on the voyage. While making the Atlantic crossing, he read Lyell's book, which he brought along on the recommendation of Henslow and Sedgwick and to help placate FitzRoy. If he were to understand the debates of his newly claimed interest, geology, and know what the opposition was claiming, this book must be read. And so it was that Darwin came to know Lyell.

Sir Charles Lyell had quite a similar background to that of Darwin, having well-to-do parents, leaving time to enjoy the luxury of natural history as only an aristocrat could afford. It was under the spell of Dean Buckland while at Oxford in 1816 that Lyell set his goal, an understanding of geology. And like Darwin he was greatly influenced by travel, in his case on the Continent, where he never tired of observing the structures of rivers, mountains, volcanoes, and even glaciers.

In Lyell's early years he was convinced by Buckland and others of the arguments that led to a catastrophist's interpretation of fossils and earth changes. But as his observation accumulated, conflicts appeared. The catastrophies were thought to be sudden volcanic eruptions or inundating floods that swept the earth of all living inhabitants at one time and laid down deposits of rock in the process. His observations suggested that if the sedimentation deposits were produced by such floods, the floods were not uniform in time or number from one region of the continent to the next. There could not have been a uniform holocaust. Also rivers were found to cut valleys through lower strata or lava flows, attesting to their recent formation and hardly an instrument to have laid down the most recent deposit of corpses and stones as claimed. Also, how could a river be of a meandering type if the torrents were uniform and sudden? There were also increasing doubts as to the time required to form these deposits and rocks. Could this really happen so suddenly? It was the accumulation of considerable evidence by 1826–1827 that volcanic deposits were laid down in a continuous fashion and that volcanoes in different regions of the earth were not erupting simultaneously, the same but less documented evidence that turned Hutton against the neptunists, that decidedly convinced Lyell that the catastrophists' theory was wrong. And it was from this point on that he struck out to overthrow the popular notion—an effort finally summarized in his *Principles of Geology* in 1830.

Lyell used the evidence listed above and replaced catastrophism with his own understanding of the history of the earth. He concluded that the changes to the earth's surface occurred in slow cyclic fashion, each resulting from a local volcano, flood, river, glacier, or earthquake. He wrote extensively including many facts on the effects of these disturbances upon landforms. He also added detailed descriptions of the Tertiary period, characterizing its subdivisions into Pliocene, Miocene, and Eocene epochs. What is probably most remembered about this volume was the way in which Lyell methodically convinced himself of the forgotten thesis of Hutton, namely "that all former changes of the organic and inorganic creation are referrible to one uninterrupted succession of physical events, governed by the *laws now in operation* . . . the principles of science must always remain unsettled so long as no fixed opinions are entertained on this fundamental question." It was this restated definition of uniformitarianism that caused the greatest disturbance to the creationists as well as the catastrophists. Also for Lyell the earth was a great deal older than previously thought. No wonder Buckland, Sedgwick, and Captain FitzRoy were concerned. Lyell was one of the first geologists to think historically and to feel that geology was "(that science) which investigates the successive changes that have taken place in organic and inorganic kingdoms of nature."

The *Principles of Geology* became one of young Darwin's favorite companions. He wrote, "The great merit of the Principles was that it altered the whole tone of one's mind and, therefore, that when seeing a thing never seen by Lyell, one yet saw it partially through his eyes." Lyell's hypotheses seemed so daring and straightforward that the young, fresh and as yet unsolidified mind of Darwin found in them the necessary mold with which his own observations could be compared.

A HYPOTHESIS ON THE
FORMATION OF CORAL ISLANDS

Although Lyell's book may have helped, it alone could not have generated the enthusiasm, energy, and skill with which Darwin demonstrated almost immediately in his collecting of facts and observations. No detail was too small to escape his eye even when riding high in the poop deck. This lanky six-footer soon stood out among the members of this ship as one who never shirked from duty, though constantly sick while on the sea. In addition he was always peering, collecting, and examining most everything about him. He was well liked by everyone and was soon called "the flycatcher" or "the philosopher" by the crew. Why a man with such little training became one of the "greatest field naturalists that ever lived" can only be conjectured. He himself wrote, "The excitement from the novelty of objects (revealed especially while traveling), and the chance of success, stimulated him to increased activity. Moreover, as a number of isolated facts soon become uninteresting, the habit of comparison leads to generalization." Or as his granddaughter wrote, "The love of close observation of natural fact, and his need for a theory to explain everything he saw, form the closely woven tissue which constituted his genius." Regardless of the reason, Darwin's decision to write a book on geology prompted him to write down a prodigious amount of material. When the journey had ended, he had managed to fill up 18 daily notebooks with scribblings and details that in turn served as a source for two sets of notes, one a notebook with about 2000 pages and the other his diary.

It is most difficult to categorize the observations of Darwin, as he applied any one fact to different purposes of thought and never developed one idea during only one time of the voyage. He could store vast amounts of data in his mind and somehow easily shuffle them back and forth forming different correlations at will. But his early observations were those associated with islands, the Grand Canary island and the Cape de Verd Archipelago. He soon concerned himself with the volcanic origin of these islands, their shapes, weathering, organic composition, and existing biotas. It was obvious that these islands were of volcanic origin from their lava flows and he wrote, "The volcanic nature of these oceanic islands is evidently an extension of that law, and the effect of those same causes, whether chemical or mechanical, from which it results that a vast majority of the volcanoes now in action stand either near sea-coasts or as islands in the midst of the sea." He distinguished oceanic islands as those of volcanic origin, the volcanoes rising from the ocean bottoms until their erupting peaks pierced the sea's surface. Other islands were formerly part of a continent and hence called continental islands. They became separated by a channel of water as the sea level shifted upward with respect to the land.

On the oceanic islands he found shells and possible coral beds and began to wonder how coral reefs came into being long before he ever saw them. These structures found in the South Pacific and Indian Oceans had been a curiosity to geologists since their discovery. Darwin could not test his ideas until later on the trip but he reasoned that volcanic islands must be sinking as the ocean floor beneath them subsides, raising the sea level with respect to the island. The point being that if the coral animal became attached to the islands and then the sea level changed upward covering these corals, new ones could grow on top of their former ancestors, which must die due to a lack of oxygen. If this continued, the coral would form a reef standing off from the island with the initial rim of coral formed defining a high point of growth. And as the island became submerged its new shores would recede inward as shown in Figure 4–1.

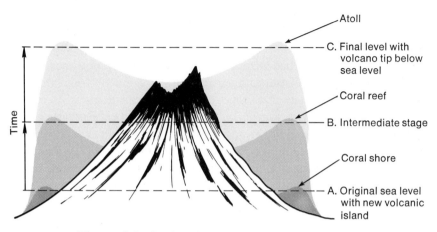

Figure 4-1. Coral reef formation: Darwin's hypothesis.

This could explain the rather common pattern of coral reefs found about some islands. With further submergence the island would disappear altogether leaving atolls. Hence all coral formations would represent different stages of island submergence with respect to sea level. Submergence of the islands was the only way that Darwin could relate all the different forms of corals. Darwin based his reasoning on the assumption that coral could not grow very far beneath the ocean level and thus could not have started on submerged craters and then grow up to meet the water surface, a common theory at the time. He found to his delight that there were only dead coral skeletons at depths below 20 to 30 fathoms. There was also no other explanation for the evenness of heights of the coral reefs regardless of the mountain heights on the islands. Like the atoll, coral beds had sides that were always much steeper away from than toward their atoll's center. He completed the picture by noting that even barrier reefs were similar to atolls, and therefore if continents such as Australia are sinking, they too will leave a fringe of coral.

There was no reason to doubt that the rate of coral formation could keep up with the apparent rate of island submergence. All the facts fit his hypothesis so well that his only conclusion was that indeed the changes of sea level must be gradual and continuous. He wrote, "This conclusion (that the change was slow) is probably the most important one which can be deduced from the study of coral formations; ... we may thus like unto a geologist who had lived his ten thousand years and kept a record of the passing changes, gain some insight into the great system by which the surface of this globe has been broken up, and water interchanged." Thus it is that Darwin on his own found some support for Lyell's general thesis, that land changes are slow and gradual. He was also forced to agree with what Lyell wrote; namely, "The amount of subsidence of land in the Pacific must have exceeded that of elevation, from the area of land being very small relatively to the agents there tending to form it, namely the growth of coral and volcanic action."

Darwin would have been famous for his study of coral formation alone. Sedgwick, hearing of some of Darwin's observations and conclusions while still in South America, wrote, "If God spares his life he will have a great name among the naturalists of Europe." It is interesting to note that Darwin's theory could not be fully tested until it was possible to drill down through some of these thick coral reefs to see whether or not volcanic rock was beneath them. When this became possible, in the 1950's, volcanic rock was found and Darwin's theory was upheld. But, now the subsidence is not believed to be due to a change in the ocean floor but rather due to the weight of the coral compressing the volcano's cone.

FOSSIL EVIDENCE FROM SOUTH AMERICA

As coral was an index of sea-level changes around islands, so Darwin considered marine fossil shell deposits a convenient index of ocean versus land mass changes involving continents. About the Patagonia regions he wrote, "There the tertiary formations appear to have accumulated in bays, here along hundreds of miles of coast we have one great deposit, including many tertiary shells, all apparently extinct. When we consider that all these pebbles, countless as the grains of sand in the desert, have been derived from the slow-falling of masses of rock on the old coast-lines and banks of rivers; and that these fragments have been dashed into smaller pieces, and that each of them has since been slowly rolled, rounded, and far transported, the mind is stupified in thinking over the long, absolutely necessary, lapse of years." He also found that there were often terraces with the shell deposits of presently existing species. About this he wrote, "This fact appears to me highly remarkable; for the explanation generally given by geologists, of the absence in any district of stratified fossiliferous deposits, of a given period, namely, that the surface then existed as dry land, is not here applicable; for we know from the shells strewed on the surface and embedded in loose sand or mould, that the land for thousands of miles along both coasts (of southern South America) has lately been submerged. The explanation, no doubt, must be sought in the fact that the whole southern part of the continent has been for a long time slowly rising." Lyell had offered the same explanation for the occurrence of the terraces.

Darwin also agreed with Lyell as to the cause of this land uplifting, mainly from his own experience in an earthquake while in Chili. "The most remarkable effect of this earthquake was the permanent elevation (2 or 3 ft) of the land; it would probably be far more correct to speak of it as the cause." He combined statistics of the occurrence of earthquakes and of volcanoes erupting on nearby islands along the Chilean coast and concluded, "From the intimate and complicated manner in which the elevatory and eruptive forces were shown to be connected during this train of phenomena, we may confidently come to the conclusion, that the forces which slowly and by little starts uplift continents and those which at successive periods pour forth volcanic matter from open orifices, are identical." He was not at all surprised to find shell fossils as high as 14,000 feet in the Andes. Thus it was that Darwin, using his own observations, arrived at the same basic conclusions as Lyell had. Lyell had made the suggestions and Darwin objectively convinced himself of their truth. The earth was very old, and the landmass changes were slow and continual. From this time on Darwin was a confirmed uniformitarianist.

It was also along the Argentine coast that Darwin happened upon the fossil remains of some land mammals no longer in existence. There was a gigantic horse-like creature, a megatherium, others like rhinoceros, several as large as elephants but unique, and still others like armadillos— one having a long neck like a camel. Judging by the teeth, they were all gnawers and hence vegetarians. Darwin noted that they could not have lived on the present flora. In fact, one had claws like the present-day sloth but so large that the trees must have been larger than any existing ones. He wrote, "There are extinct species of all the thirty-two genera, excepting four; of the local terrestrial quadrupeds the extinct species are much more numerous than those now living; there are fossil ant-eaters, armadillos, tapirs, peccaries, guanacos, opossums, and numerous South American gnawers and monkeys, and other animals. This wonderful relationship in the same continent between the dead and the living, will, I do not doubt, hereafter throw more light on the appearance of organic beings on our earth and their disappearance from it, than any other class of facts." The fact was too coincidental to ignore—that these particular fossils were closely related to present existing animals, both of which were highly unique to South America. If there had been catastrophies, and replacement, they had not occurred universally to the same organisms.

An area of concern generated by the quadruped fossils was that of the geographic distribution of mammals. Darwin had been quite aware of the problem of trying to account for the diversity of kinds of organisms found in different regions of the world. He wrote, "Neither the similarity or dissimilarity of the inhabitants of various regions can be wholly accounted for by climate or other physical conditions. . . . There is hardly a climate or condition in the Old World which cannot be paralleled in the New . . . at least as closely as the same species generally required . . . not withstanding the general parallelism in the conditions of the Old and New Worlds, how widely different are their living productions." Considering the distribution of these living forms with his fossil finds, he wrote, "The existence in South America of a fossil horse, of the mastadon, possibly of an elephant, and of a hollow-horned ruminant . . . are highly interesting facts with respect to the geographic distribution of animals. At the present time, if we divide America, not by the Isthmus of Panama, but by the southern part of Mexico in latitude 20 degrees, where the great table-land presents an obstacle to the migration of species, by affecting the climate, and by forming, with the exception of some valleys and of a fringe of low land on the coast, a broad barrier; we shall then have the two zoological provinces of North and South America strongly contrasted with each other. South America is characterized by possessing many peculiar gnawers, a family of monkeys, the llama, peccari, tapir, opossums, and especially several genera of Edentata, the order which includes the sloth, ant-eaters, and armadillos. North America is characterized by numerous peculiar gnawers, and by four genera (the ox, sheep, goat, and antelope) of hollow-horned ruminants; of which great division South America is not known to possess a single species. (As shown by the fossils) South America possessed a mastadon, horse, hollow-horned ruminant and the same three genera of Edentata. (Fossils from North America over the same period contained essentially the same mammals.) Hence it is evident that North and South America, on having within a late geological period these several genera in common, were much more closely related in the character of their terrestrial inhabitants than they now are. The more I reflect on this case, the more interesting it appears; I know of no other instance where we can almost mark the period and manner of the splitting up of one great region into well characterized zoological provinces. The geologist, who is fully impressed with the vast oscillations of level which have affected the earth's crust within late periods, will not have fear to speculate on the recent elevation of the Mexican platform, or, more probably, on the recent submergence of the land in the West Indian Archipelago, as the cause of the present separation of North and South America. (The character of these fossil mammals) seem to indicate that (the two great divisions were) formerly united. . . ." Darwin goes on to indicate that in this early period the fauna of North America was much like that of Eurasia, as indicated by fossils in these regions over the same period being of the same kind, and that the Old and New Worlds were probably joined across the Bering Straits. He concluded that the elephants, mastadons, horse, and hollow-horned ruminants probably migrated from the Old World into North America and then even into South America before those two continents became separated as they presently are. And he went on, "They (probably) mingled with the forms characteristic of that southern continent, and have since become extinct."

EXTINCTION

The fossil evidence also caused him to turn to the question of extinction —one that was foremost in the minds of catastrophists and Lyell. He wrote, "The great number, if not all, of these extinct quadrupeds lived at a late period, and were contemporaries of most of the existing sea-shells. Since they lived, no very great change in the form of the land can have taken place. What then, exterminated so many species and whole genera?" It was clear by the fossil finds that there were more actual kinds of organisms (more species) living then than now. There must have been a great deal of extinction.

Darwin went on, "The mind at first is irresistibly hurried into the belief of some great catastrophy". He thought about the descriptions he had heard from the inhabitants of the La Plata region about the great droughts. Vegetation failed, brooks dried up, and the whole country "assumed the appearance of a dusty high road." "I was informed by an eye-witness that the cattle in herds of thousands rushed into the Parana (river), and being exhausted by hunger they were unable to crawl up the muddy banks, and thus were drowned. Without doubt several hundred thousand animals thus perished in the river; . . . Subsequently to the drought ... a very rainy season followed, which caused great floods. Hence it is almost certain that some thousands of the skeletons were buried by the deposits of the very next year. What would be the opinion of a geologist, viewing such an enormous collection of bones, of all kinds of animals and of all ages, thus embedded in one thicky earth mass ? Would he not attribute it to a flood having swept over the surface of the land, rather than to the common order of things ?"

Also, "No one will even imagine that a drought, even far severer than those which cause such losses in the provinces of La Plata, could destroy every individual of every species from southern Patagonia to Bering's Straits." Again, he found no evidence that any kind of abrupt change in climate could or did occur universally. He threw out the possibility of a temperature change, glaciers, or man as the cause of their extinction by similar reasoning.

Finally Darwin wrote, "In cases where we can trace the extinction of a species through man, either wholly or in one limited district, we know that it becomes rarer and rarer, and is then lost. The evidence of rarity preceding extinction is (even) more striking in the successive tertiary strata." Thus it was that he again followed Lyell's argument and was forced to conclude by his own evidence that his mentors were wrong to believe that catastrophies as they had described them were the causes of extinction.

Darwin began to wonder if some change in the natural environment could have been responsible for the extinction of the quadrupeds. He had been developing his ideas when the second volume of Lyell's *Principles of Geology* reached him in South America in 1832. In this volume Lyell was again answering the catastrophists and creationists, but in terms of the apparent progression of organic forms. Influenced by Linnaeus' earlier writings, he agreed that each species was probably created in pairs, multiplied, and eventually occupied "an appointed space on the globe." But by his more up-to-date knowledge of geographic distribution he modified Linnaeus' hypothesis to assume there had been a number of "foci of creations" from which the organism spread upon the environment. He could not accept the catastrophists' idea that separate kinds of organisms existed at successive times. Rather, he concluded that all the organisms that had ever existed except man had been created at one time and coexisted until some became extinct. Because of the total lack of any human fossils he assumed that man was a special creation. He did not consider this a break in the uniformity of natural laws—it was an instantaneous single event with no relationship with the ongoing natural laws.

Lyell's explanation for extinction was that "every species which has spread itself from a small point over a wide area must have marked his progress by the dimunition or the entire extirpation of some other, and must maintain its ground by a successful struggle against the encroachments of other plants and animals. . . . (Also) the successive destruction of species must be part of the regular and constant order of nature," and must result from the continual geological changes that create different environments. Either the local forms move out or, if they cannot, they become extinct either directly by the harsher environment or by an invading organism that was destined to be in that environment.

Lyell was implying that there were unique adaptations that each organism (species) had that fitted particular environments. Again Darwin could agree with this from his own observations. He wrote, "A fish not only swims in water but obtains oxygen from the water that it needs by means of its gills. A mammal uses lungs specially adapted for its oxygen and coordinates the use of these lungs with the help of its ribs. A woodpecker has two of its toes turned backwards on each foot and stiff tail-feathers for clinging to the bark of trees. Is there any structure of an organism which does not fit its whole mode of existence within its environment? Yet these are the things which stick out in our mind as we compare the closely related species. . . . Each species has its niche due to some modification of a basic similar body plan." He also agreed with Lyell when the latter wrote, "(There is) a capacity in all species to accommodate themselves to a certain extent, to a change of external circumstances, this extent varying greatly according to the species." But there were limitations to the accommodation and Lyell considered that adaptations were both necessary for the organisms existence as well as a cause for their extinction when they could not adjust to their environment.

In addition, Darwin was drawn through his observations of the apparent gradation between species—the fact that so many were quite similar to others in structure and behavior and yet were clearly different species. Was there any significance to having so many groups of similar species? Another curious generalization was that the closer two species were in characteristics, the closer they usually were in distance to one another. He also noted that a species may have a hybrid of adaptations from two other species. For example he wrote, "One snake (a *Trigonocephlus*, or *Cophias*), from the size of the poison channel in its fang, must be very deadly. Cuvier, in opposition to some other naturalists, makes this a subgenus of the rattlesnake, and intermediate between it and the viper. In confirmation of this fact I observed a fact which appears to me very curious and instructive, as showing how every character, even though it may be in some degree independent of structure, has a tendency to vary by slow degrees. The extremity of the tail of this snake is terminated by a point, which is very slightly enlarged; and as the animal glides along, it constantly vibrates the last inch; and this part striking the dry grass and brush wood, produces a rattling noise, which can be distinctly heard at the distance of six feet. As often as the animal was irritated or surprised, its tail was shaken; and the vibrations were extremely rapid. Even as long as the body retained its irritability, a tendency to this habitual movement was evident. This *Trigonocephalus* has, therefore, in some respects the structure of a viper, with the habits of a rattlesnake; the noise, however, being produced by a simpler device."

In spite of the many areas of agreement in theory, Darwin could not find any direct evidence to support Lyell's explanation of the creation or extinction. He also found it difficult to believe that many of the present living forms had existed and yet had no fossil record while the extinct ones did. Also there must have been a very large number of different kinds of organisms existing earlier since his own fossil evidence was quite telling as to the huge number of organisms that had become extinct. But, the most disconcerting thing was simply that the number of fossils that Darwin had found were far too few in number or kind to say anything definite about these past events. He felt constantly hampered by this lack.

ENDEMIC SPECIES—THE
GALAPAGOS ISLAND ANOMALY

It was late in the voyage, after both coasts of South America had been examined and various trips inland had been managed, that as they departed from this continent westward they came to the Galapagos Islands lying about 650 miles off the coast of Ecuador on the equator (see Figure 4–2). This group of 10 islands were apparently of geologically recent volcanic origin, adorned with "at least two thousand craters." For the most part, these were ugly islands, with twisted, dull-colored, and foul-smelling vegetation scattered sparsely over irregular lava fields. There seemed to be little life present until a search was conducted. It was not long, however, before these islands added facts that were quite new to Darwin. He found

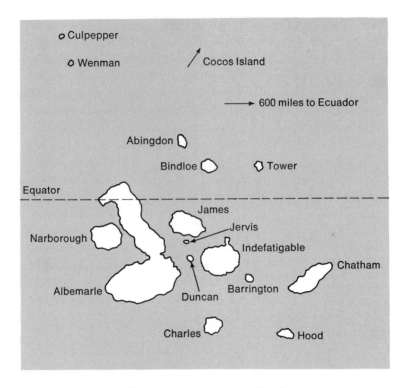

Figure 4-2. Galapagos Islands.

large and unusual tortoises, which were unlike any others known. As he continued to investigate, it was not only the reptiles but also most of the birds, insects, plants, and even the seashells that were unique to this group of islands. In all he found 5 reptiles, all endemic; 26 kinds of land birds, all but one endemic (a lark-like finch common in North America)—this included 1 hawk, 2 owls, 1 wren, 3 flycatchers, 1 dove, 1 swallow, 3 mocking thrushes, and the rest finches; 47 out of 90 seashells and 100 out of 185 flowering plants collected were endemic; and finally 2 mammals, a mouse on one island and a rat on another island, both of common types. Curiously enough, both of these islands had been visited or were even then inhabited by buccaneers or colonizers. Thus the mammals may well have been introduced.

In addition he wrote, "I have not (mentioned) ... by far the most remarkable feature in the natural history of this archipelago; it is that the different islands to a considerable extent are inhabited by a different set of beings. My attention was first called to this fact by the Vice-Governor, Mr. Lawson, declaring that tortoises differed from the different islands, and that he could with certainty tell from which island any one was brought (see Table 4–1 for plant deposition). ... I never dreamed that

TABLE 4–1

GALAPAGOS FLORA

Island	Total number of species	Number of species in other parts of the world	Number of species confined to Galapagos	Number of species confined to this island	Number of species confined to Galapagos but found on more than one island
James	71	33	38	30	8
Albemarle	46	18	26	22	4
Chatham	32	16	16	12	4
Charles	68	39(29)[a]	29	21	8

[a] 10 believed to be imported by colonizers.

islands about fifty or sixty miles apart, and most of them in sight of each other, formed of precisely the same rocks, placed under a quite similar climate, rising to a nearly equal height, would have been differently tenanted. Their apparently recent volcanic origin renders it highly unlikely that they were ever united. ... (It is that) these species having the same general habits, occupying analogous situations, and obviously filling the same place in the natural economy of this archipelago, that strikes me with wonder." He added, "The most curious fact is the continuous gradation in the size of the beaks in the different species of *Geospiza*, from one as large as that of a hawkfish to that of a chaffinch. ... Seeing the gradation and diversity of structure in one small intimately related group of birds, one might really fancy that from an original paucity of birds in this archipelago, one species had been taken and modified for different ends."

He also noted that most of the nonendemic species were common in the Americas. But what was more striking was that the endemic species in essentially all cases could be said to be similar to some presently living American group. He added, "I have said that the Galapagos Archipelago might be a satellite attached to America, but it should rather be called a group of satellites, physically similar, organically distinct, yet intimately related to each other, and all related in a marked, though much lesser degree, to the great American continent." He thought of other oceanic islands and noted, "The islands of the Cape de Verd group resemble, in all their physical conditions, far more closely the Galapagos Islands than these latter resemble the coast of America; yet the aboriginal inhabitants of the two groups are totally unalike; those on the Cape de Verd Island bearing the impress of Africa, as the inhabitants of the Galapagos Archipelago are stamped with that of America."

AN EXPLANATION

Knowing these facts and placing oneself in Darwin's position in time, we may well ask what our own explanation might be for such an unusual phenomenon as the Galapagos Islands. Further studies of this group of islands and other similar oceanic ones have only reinforced the same kind of data. It should be noted that continental islands are never found with the same characteristics as oceanic ones. They rarely have endemic species, and there are many more species of plants and animals. In any respect, we must pause and really wonder about these data before we convince ourselves with the explanation that Darwin gave. For if it is wrong, as any hypothesis can be, or if it is incomplete, which probably every hypothesis is, we may accept his thoughts too easily and miss some greater truth.

But as it went, Darwin wrote in his notebook sometime in September of 1835 the following: "When I recollect the fact, that from the form of the body, shape of scale and general size, the Spaniards can at once pronounce from which island any tortoise may have been brought: When I see these islands in sight of each other and possessed of but a scanty stock of animals, tenanted by these birds but slightly differing in structure and filling the same place in Nature, I must suspect they are only varieties . . . . If there is the slightest foundation for these remarks, the zoology of Archipelagoes will be well worth examining: For such facts would *undermine the stability of species.*"

Such a conclusion could not be easily accepted. After all, Lyell had discussed this possibility in his second volume and had been most insistent upon its absurdity. Lyell had written, "The reader will immediately perceive that amidst the vicissitudes of the earth's surface, species cannot be immortal, but must perish one after the other, like the individuals which compose them. There is no possibility of escaping this conclusion without resorting to some hypothesis as violent as that of Lamarck." He is referring to a theory that suggested that the species may be modified into other species by the environment. He went on, "It is idle to dispute about the abstract possibility of the conversion of one species into another, when there are known causes (environmental changes) so much more active in their nature which must always intervene and prevent the actual accomplishment of such conversions." How could an organism that was accustomed to fresh water have the time to become a salt-tolerant species if it was exposed to saltwater?

Lyell was also inhibited by the similarity between the modification of species from one form to another and the step-like changes in organisms that the progressionists dwelt upon. If the species are modified, or if this change occurs in steps, how unnatural it seemed that it would result in some ultimate such as man, as the progressionists claimed. Lyell could not disassociate the two ideas, and he could not accept catastrophism. Besides, there was no need or evidence for such a drastic assumption as species mutability.

These were strong arguments but Darwin could not escape from considering the idea. After all, if the Galapagos Islands were historically recent, as even Lyell would be quick to agree, where did the present living organisms come from? They could not have been created or if they were, it was very recent and a most unusual pattern of forms. The catastrophists' and progressionists' explanation would certainly not fit; the present forms were not like those elsewhere; and why would these unique creatures be placed there, especially without man? Lyell would agree that some of the organisms might have migrated there from the South America coast. But if this were the case, why were they not the same as those on the mainland or elsewhere? This did not fit Lyell's explanation. Again Lyell would have to say that the endemic ones were created on the island, but within recent times? No. Thus only contradictions remained. Not all Lyell's explanations for the creation or extinction of species could be correct.

So it was that for the first time Darwin had an uphill struggle against the beliefs of his own mentor, Lyell. He was on his own. And by this time in his life Darwin had become a very humble person who constantly doubted and checked his own ideas. He wanted to be very certain.

THE HISTORY OF THE IDEA—
SPECIES MUTABILITY

Although it seems clear that the evidence and logic above directly forced Darwin to conceive of the mutability of species, there is little doubt that he had heard of the idea in some general way before this time. The question is, what did it mean to him before Galapagos Islands? For instance, he later wrote that he "never happened to come across a single naturalist who seemed to doubt about the permanence of species." Yet, there is no question that he had heard of Lamarck's theory of changing species. Lyell devoted a considerable amount of time in his second volume explaining and denying the theory. There were also the writings of his own grandfather, which he read against the wishes of his parents, which were constantly promoting the idea of transmutating species. And surely, Sedgwick or others had mentioned it in their many discussions. But in order to clarify, as well as we can, the meaning these past references to the idea held for Darwin, its history must be unfolded. This will also help us to appreciate Darwin's actual contribution to the idea.

The ancient Greeks had no qualms about the mutability of species, for they thought in terms of the general transformation of organic forms. A commonly accepted notion was that all organisms arose from primordial slime and then "evolved" through stages including fish and eventually formed man. It was generally assumed that it was the environment that produced the changes but no mechanism was suggested. Throughout most of our written history some modification of this idea has remained. Francis Bacon (1561–1626) agreed and claimed that flying fish and bats were intermediates in the process. Rene Descartes (1596–1650) felt both inorganic and organic forms could progressively change. Gottfried von Leibnitz assumed that habitats not only could change animal species but that they must exist in all gradations of forms if only they could be found. John Ray and later Linnaeus thought that new species could arise by hybridization. It is clear that there was no well-developed concept of species mutability throughout this period.

A thread of some continuum seems to have started with the writings of Pierre Louis M. de Maupertius (1698–1759), who dwelt at some length on the subject in his *Systeme de la Nature* (1753). He gave examples of what he considered changes in which the most perfectly adapted varieties would successfully replace other varieties and with successive generations produce a "different species." De Buffon, though influenced by de

Maupertius, modified this mechanism considerably. Instead of the change coming about through a normally occurring variation, he felt that environment had a direct affect on the organisms. These altered characteristics were then inherited by the next generation. He restricted this affect to one of degeneration and wrote, "The ass may be of the family of the horse, and that one might differ from another only by the degeneration from a common ancestor—we might be driven to admit that the ape was of the family of man, that he is but a degenerate man, and that he and man have had a common ancestor."

It was, however, Darwin's grandfather, Erasmus Darwin (1731–1802), who first popularized the idea of the transmutation of species. He was a brilliant, well-informed, and distinguished personage of his day. His poetry and prose were extremely popular, especially among the well-educated public. Part of one of his poems went as follows:

> First, form minute, unseen by spheric glass,
> Move on the mud, or pierce the watery mass;
> Then as successive generations bloom,
> New powers acquire and larger limbs assume:
> Whence countless groups of vegetation spring,
> And breathing realms of fin and feet and wing.

But E. Darwin was more than a fancy writer; he had studied both de Maupertius and de Buffon and accumulated as many facts of the day as he could on matters related to the transmutation of species. He studied comparative anatomy and systematics and was probably the first to use the differences in both flora and fauna between the New and Old Worlds to support his theories. He had unusual qualities of analysis and imagination. By hindsight, the only reason he was not more successful in the scientific community was that his ideas were too far ahead of the limited data that was available at that time. Erasmus Darwin actually originated or developed essentially every mechanism that has ever been conceived for the possible transmutation of species. He thought pluralistically and considered that not one but different mechanisms were probably all acting at the same time either within different organisms or within the same organism under different circumstances. These mechanisms included (1) the random and naturally occurring change proposed by de Maupertius and (2) the direct effect of environment—somewhat like de Buffon's, only not restricted to degeneration. In addition he proposed (3) an indirect effect of environment wherein the organism modified itself in response to some change in the environment, and finally (4) changes brought on by the will of the organism independent of the environment. He felt that the indirect effect of environment was probably the mechanism most commonly used. This was described fully in his *Zoonomia* (1796), where he wrote, "All animals undergo transformations which are in part produced by their own exertions, in response to pleasures and pains, and many of these acquired forms or propensities are transmitted to their posterity."

In spite of their notoriety, it is assumed that if young Darwin had read about these mechanisms before his journey, they would not have meant much to him. They were much too bookish and, worse than that, philosophical. Also it is unlikely that he would have thought that they had any relationship to his observations on the Galapagos Islands.

There remains the influence Lamarck may have had on Darwin's thoughts. Jean de Lamarck (1744–1829) was part of a reactionary group who was opposed to the Newtonian understanding of matter. They considered that all substance including organisms were plastic and constantly changing with time. As a result of his service as a distinguished soldier under Napoleon during the French revolution he was, at the age of 50, elected to a chair in zoology at the Museum. It is remarkable how quickly he was able to become well versed in this field that was essentially new to him. He was soon well-known for his brilliant lectures.

It was as a result of developing these lectures that Lamarck's theories on the transmutation of species occurred to him. This resulted from the natural tendency to arrange the animals in a sequence ranging from simple to complex when describing them. His belief in the plasticity of matter allowed him to extend a simple comparison of these animals into a theory of how they actually changed from one to the other. He first conceived and described to his class how animals as a plastic organic form transmutated from a complex toward a simple type of organization by a process of degradation. He was admittedly under the influence of de Buffon. He refused to recognize the species concept and considered these false snapshots of a continuous process.

By 1802 Lamarck developed his "escalator hypothesis" in his *Recherche sur l'Organisation des Corps Vivants*. This hypothesis stated that there is a continuous flow of plastic organic matter moving up a staircase of existence, dividing into a variety of complex structures that in turn disintegrate back down the inorganic side where new life (organic forms) is constantly being created. Thus in this latter version he reversed the degeneration process of de Buffon and encompassed all the world in one process, a symmetrical and wholistic synthesis he found most pleasing.

But this idea was even too philosophical for Lamarck as it stood, and he felt the need for more detail based on observations. He wrote at this time, "I know full well that very few will be interested in what I am going to propose, and that among those who do read this essay, the greater part will pretend to find in it only systems, only vague opinions, in no way founded in exact knowledge. They will say that but they will not write it."

Finally in 1809, Lamarck published his *Philosophie Zoologique*. In this writing he elaborated on the interaction between the plastic organic form and the raw inorganic environment. Here the apparent diversity of living forms was described as partly due to a shaping process by the physical environment. As the environment is irregular, so the final spectrum of forms will be varied. There is a plastic force in living substance that causes it to continuously flow against the physical environmental molds. And where this environment is most resistant to the flow, there will be more rapid disintegration of the organic matter back into the inorganic state. Moreover, the physical environment is also gradually changing so that its cutting edge shifts and the resultant existing forms of organic matter also shift accordingly (transmutation).

This change in organic form is not just a direct one as indicated by the model above, but there is a response by the organic mass from within adapting structures to fit the new environment. The response starts with a change in some behavior that over a long period of time would become a habit. This in turn would lead to changes in a structure of the organism in order to best carry out the new behavior. This is a slow process with internal balances to resist any sudden changes.

The environmental changes are to be thought of as shifting opportunities to which the organism may respond as a whole, depending on its total form at the time. Those organisms that are able to respond and maintain a close adaptation or ability to exist in their surrounding environment survive; those that cannot, disintegrate. He wrote, "Extensive change in the conditions surrounding the animals will lead to wide modifications in their requirements and perforce to corresponding changes in their activities. Now, if the new requirements become permanent, the creatures then acquire fresh habits which also become fixed like the requirements evolving them . . . (this) in turn (will) entail the use of certain parts of the body in preference to those previously in use and in a similar manner will lead to the disuse of any part no longer necessary . . . (similarly when) requirements render a part entirely useless, the creature will cease to use it ... (and) it will finish up by completely disappearing." In the case of plants the modification would depend on whether certain parts of the plants were used or not for the life processes.

We can envision a modern horse finding itself on an open plain, subject to wolves, adapting the habit of running faster, and with more use of its legs and their muscles, experience changes in these structures. Over an extended period of this experience requiring many generations, the structures become more adapted to the running required by the environment. If the wolves disappear, the horse would become less active and the muscles would shrink to the new requirements over time.

It is unfortunate for Lamarck that he could not express his views more accurately, which led him to be misunderstood, and also that he happened to be a contemporary of Cuvier who was violently opposed to any theory supporting the mutability of species. Cuvier was the supreme scientist of France and he used his power to suppress and humiliate Lamarck. This was also unfortunately easy to do since Lamarck's own supporters misinterpreted his writings. Most effective at this was Geoffrey Saint-Hilaire (1772–1844), another natural historian who had also received a chair because of his support of the French Revolution. St. Hilaire had been influenced in his understanding of the transmutation of species by Erasmus Darwin. As a matter of fact, so had Lamarck, but St. Hilaire in an attempt to defend Lamarck stressed two points mentioned by E. Darwin and denied by Lamarck. One was that the organism modifies itself by its own will toward some ideal purpose. The other was that environment had a direct effect on modifying the organism and that these were acquired and passed on to the next generation. Lamarck had received some of his thoughts on the mechanism of transmutation from E. Darwin's explanation of the possible indirect effect of environment. But history remembers the misguided benevolence of St. Hilaire and has deemed his interpretation as that of Lamarck's—it has since been called neo-Lamarckism. St. Hilaire was not the natural historian that both Lamarck and Cuvier were, and he caused Lamarck and himself to go down in a smashing defeat in a public debate by the able Cuvier. From then on most everyone including Lyell and Darwin laughed at Lamarck. His theory has been remembered with the example in which a giraffe that wanted to eat the leaves on the branch of a tree far above his head stretched his neck and this is why giraffes have long necks today.

Lyell had also explained this neo-Lamarckian interpretation and easily tore it to pieces in the second volume of his *Principles*. This being the interpretation that Darwin had read explains why he probably gained little stimulus in the development of his ideas concerning the mutability of species from Lamarck. And if Lamarck's theory had been discussed with Sedgwick or other natural historians of the day, he would never have been taken seriously. After the debate Lamarck's mutability of species was either avoided as a topic of discussion or relegated to jokes in fashionable parties. It would be wrong to say that Darwin had never heard of the notion of mutating species but its significance was probably ignored and in every respect it was far from the kind of transformation he had in mind when he thought about the Galapagos Islands. He denied any contribution to his thoughts about species mutability from these earlier writers although he acknowledged that he knew of them as evidenced by his statement, "It is curious how largely my grandfather anticipated the views and erroneous grounds of opinion of Lamarck in his *Zoonomia*." Also there is an indication of their effect in his writing, "It was evident that such facts as these (the Galapagos observations), as well as many others, could only be explained on the supposition that species gradually become modified; and the subject haunted me. But it was especially evident that neither the action of the surrounding conditions, nor the will of the organisms (especially in the case of plants) could account for the innumerable cases in which organisms of every kind are beautifully adapted to their habits of life . . . for instance, a woodpecker or a tree frog to climb trees, or a seed for dispersal by hooks or plumes."

THE SEARCH FOR A MECHANISM

Soon after Darwin returned to England in July 1837, he opened his notebook in order to collect all the facts he could from fossils, geographic distribution, systematics, comparative anatomy, embryology, ecology, and genetics to deny or support his hypothesis that species are mutable. But this was only one part of his proposed project. No collection of observations that supported the fact of the mutability of species would mean anything to him unless he could also discover the mechanism by which the change takes place. To Darwin, these two, fact and mechanism, could not be separated; they had no meaning apart. He hoped that the facts once collected and correlated would reveal the mechanism for their occurrence.

In this respect he was more like Lamarck than he realized. Both needed fact and mechanism, and both had to face the problem of adaptation. For once one takes the stand that species can change, it is necessary to know what role or meaning adaptations play in the change. Lamarck settled this in his own mind by allowing the variation of adaptation to be of such an extent or extreme that there was no need for a species concept. But Darwin could not give up the species. To him they were fundamental. He was quite aware of the difference between species. Variation had its limit. Also, for Darwin, species were not so pliable that environment could mold them—certainly not directly. There must be some way that a new slightly different form can replace the older ones and how this new form could arise was of most concern to him.

Further review of Darwin's own fossil finds and those of others shed no light on any mechanism at work. They all supported the fact of transmutation of species, especially in there being more similarity between recent fossils and present living forms in any one general area than between either of these and fossils or living forms in a different general area of the world. But fossils were still extremely scarce and as yet no gradation of different forms could be found that might even suggest a gradual change.

Darwin was drawn again to the geographic distribution of organisms. His analysis of mass movements of groups of organisms, especially between continents such as North and South America or between Asia and North America, agreed with that of Lyell. For some reason a number of the incoming forms could replace those already present. This was apparently consistent with the notion that forms do tend to migrate from centers— as if created there. Barriers to this migration, such as bodies of water for land animals, may restrict certain of these forms but others more adapted to do so could cross over. The Galapagos and other islands were testimony to this possibility. But the type of replacement resulting from such migrations are not transmutations of species where one replaces another similar one. There was no evidence of the new forms being modified.

Wherever he could find indications of a transmutation, the more recent form seemed to have moved into that region from a different region. The Galapagos Islands were an example. He also noted that the two regions always had a barrier between them. He made the following generality, "That migration has played an important part in the first appearance of new forms in any one area and formation; that widely ranging species are those which have varied most frequently, and have oftenest given rise to new species; ... it is probable that the periods, during which each underwent modification, though many and long as measured by years, have been short in comparison with the periods during which each remained in an unchanged condition." These thoughts also opened a new question that had to be answered. Why did the ancestral form of the present living organisms now found on the Galapagos Islands not suffer a transmutation while still in South America? What is the difference between the two places?

A breakthrough as to a possible mechanism finally came to Darwin, not from his analysis of these matters above or those facts accumulated from the voyage but from an entirely new train of thought. His familiarity with rural life, and especially riding horses, recalled to him that indeed man has modified both plants and animals. The differences between racing horses and work horses, Newfoundland and Pekinese dogs, and cabbages and cauliflowers were all managed by selective breeding. This was done by selecting an organism with a particular characteristic from a group of organisms having a range in variation of the characteristic. Two such individuals were then crossbred. This would tend to produce more offsprings similar to their parents than if a random pair of parents were crossed. If this is repeated over a number of generations, always selecting offspring for an extreme of the same characteristic, the resulting offspring will become increasingly different from the original parents for the selected characteristic. This device, given enough time, could produce quite different organisms. It did raise several questions to him. Where were all the intermediate forms of this organism? What in nature could be doing the selection in place of man? And, what is the source of variation in offspring that allows a new set of limits or extremes to occur depending on the parent types being crossed?

Darwin wrote that the answer to the second question came to him while reading Malthus's *Essay on Population* in 1838, an essay, incidentally, that was encouraged to be written by Erasmus Darwin. It was the rapid growth of man as calculated by Malthus which resulted in a strong competition for a limited food supply which reminded him of Linnaeus' and Lyell's emphasis on the economy of nature. Other organisms must also be competing for a common food and only a limited number of the offspring produced in the world could survive. He was quite moved by his own calculations to find that even the slow-breeding elephants, who can only have 6 offspring between their thirtieth and ninetieth years of age, could produce 19,000,000 progeny per pair within 750 years. Obviously a large number of offspring are not surviving or are not having offspring.

Lyell's emphasis on the need of the organism to fit his environment or perish suggested the source of the selection in natural breeding—those individuals with adaptations that fit their environment best were the ones most likely to survive and produce offspring; thus there would be a selection in turn for these adaptations. And as the environment varied, those varieties of organisms with the particular adaptations to fit the new environment best would be selected for it. There was no direct effect of environment. If the environment changed slowly enough, the generations of changing organisms could possibly keep up with the change and a transmutation of species would take place. If they could not, the species would become extinct. Hence there was an answer for a possible mechanism for extinction of groups of organisms as well as for the modification of the species. Lyell was close to this idea but he could not agree to the shift in variation that would have to accumulate over many generations to account for change in species. He also overestimated the destructiveness of the struggle between organisms. Nevertheless, he had been able to guide so much of Darwin's thoughts that Darwin himself wrote, "I feel as if my books came half out of Sir Charles Lyell's brain."

CONTRIBUTORS TO THE
MECHANISM OF NATURAL SELECTION

Darwin called the process above *natural selection* in contrast to the artificial selection determined by man. Again there is some question as to whether he thought of it on his own. If one looks, he can certainly find the idea long before Darwin mentioned it. For example, even Lucretius wrote, "Many were the monsters that the earth tried to make; some without feet, and others without hands or mouth or face or with limbs bound to their frames. It was in vain; nature denied their growth, nor could they find food or join in the way of love ... many kinds of animals must have perished then, unable to forge the chain of procreation ... for those to which nature gave no protective qualities lay at the mercy of others and were soon destroyed."

But a far more serious and available source was that from de Maupertius; "Chance one might say, turned out a vast number of individuals. A small proportion of these were organized in such a manner that their organs could satisfy their needs. A much greater number showed neither adaptations nor order, these last have all perished. Thus the species which we see today are but a small part of all those that blind destiny has produced." This was the source E. Darwin used for one of his explanations of a possible mechanism for species transmutation.

There were also three physicians, William Wells (1757–1817), James Pritchard (1786–1848), and William Lawrence (1783–1867), who each published in 1813 essentially identical essays all in rebutal to the popular interpretation of Lamarck. They emphasized that any acquired condition of the body of man ended with the life of that person. They also suggested that, as far as man was concerned, a mechanism equivalent to natural selection rather than the direct effect of food or climate could result in a modification of his characteristics. Lawrence's book discussed the effect of artificial breeding and implied that it should be applied to the royalty with benefit to all. Because of this and unrests in the government and church, the book was considered too revolutionary. Under pressure he refrained from publishing the book until 1848 when it was received with resounding popularity. But, its theories were not noted or perhaps fully understood at the time. It is known that Darwin read this book, but perhaps man was too limited a part of life for him to take much notice. In a different book, *Lectures on Physiology, Zoology, and Natural History of Man* (1822), Lawrence actually went beyond mere modification and proposed that man had evolved by natural selection. He compared the skulls of different races of man to support his hypothesis. But this idea was neither popular nor probably well heard of at the time.

There was also Patrick Matthew, who even Darwin admitted had written about natural selection before himself, but his arguments were relatively weak and his works little known. There was a French botanist, Charles Naudin (1815–1899), who proposed that natural selection-like mechanisms were at work in the transmutation of species. He based his hypothesis on the results of selective breeding in plants, but Darwin as well as others ignored his work on the grounds that his experiments were not well designed or conclusive.

Finally, there was Edward Blythe (1810–1873), a young natural historian of some promise, who wrote articles in the *Magazine of Natural History* in 1835 and 1837—a time when this journal was energetically read by Darwin. Blythe wrote, "Among animals which procure their food by means of their agility, strength, or delicacy of sense, the one best organized must always obtain the greatest quantity; and must, therefore, become physically the strongest and be thus enabled, by routing its opponents, to transmit its superior qualities to a greater number of offspring. The same law, therefore which was intended by Providence to keep up the typical qualities of a species can be easily converted by man into a means of raising different varieties." This was called a "localizing principle" by Blythe. He attributed natural selection to divine process, a requirement that Darwin never considered necessary. But Blythe, as did Darwin, went beyond their acknowledged mentor, Lyell, to suggest, ". . . as man, by removing species from their appropriate haunts, superinduces changes on their physical constitution and adaptations, to what extent may not the same take place in wild nature, so that, in a few generations distinctive characters may be acquired, such as are recognized as indicative of specific diversity . . . ? May not then a large proportion of what are considered species have descended from a common parentage?" This is a very clear statement on natural selection. But, as brilliant as Blythe was, he did not as yet have the background of facts at his command to verify his own hypothesis, and he later rejected it. But, it is hard to imagine that Darwin would have missed this article and not have gleaned meaning from its contents.

None of the above is meant to discredit Darwin. It is to try to be as complete as possible to establish all the ideas that were being cast about at this critical time. It is one thing to look back at old publications and see the first inklings of some long established theory. It is another thing to understand the full meaning of some new hypothesis, before it is "taught" to you; especially on the basis of someone's remarks that may have little to do with your own belief-web at the time. Darwin was known for being so enthralled in his own data that he could well have read any of the above and missed their import to his own study. Darwin wanted to be an independent thinker—this is impossible, but at least one can set a style of work toward this end. Although a historian could easily make the statement that both evolution and natural selection were in the air of the scientific-minded, Darwin always denied this. He himself wrote much later, "Innumerable facts were in the minds of naturalists ready to take their place as soon as any theory would receive them was (proposed and well documented)."

SOME FINAL QUESTIONS

By June 1842, Darwin had convinced himself that natural selection was the mechanism responsible for the transmutation of species and described this hypothesis in a 35-page sketch, *Version 1,* that he did not

THE GENEOLOGY OF THE IDEAS RELATED TO EVOLUTION

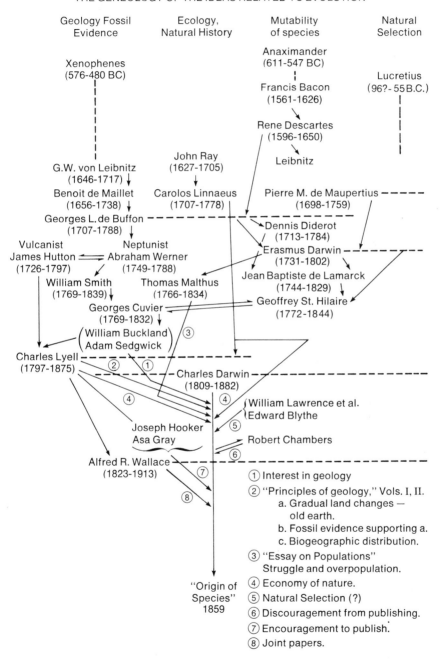

"Origin of Species" 1859

① Interest in geology
② "Principles of geology," Vols. I, II.
   a. Gradual land changes — old earth.
   b. Fossil evidence supporting a.
   c. Biogeographic distribution.
③ "Essay on Populations" Struggle and overpopulation.
④ Economy of nature.
⑤ Natural Selection (?)
⑥ Discouragement from publishing.
⑦ Encouragement to publish.
⑧ Joint papers.

publish because of certain unanswered questions. It was still unclear why migration was necessary for the changes in form, why there were no intermediate forms about, why there were so many forms produced from one ancestor, and the mystery of the source of variation in offspring.

The Galapagos Islands continued to be on his mind. He wondered about the amazing fact that 14 species of finches seemed to have arisen from one ancestral form. Why this divergence on a new island group? It was not until about 1844 that an answer came. He wrote, "I can remember the very spot in the road, whilst in my carriage when to my joy the solution occurred to me; and this was long after I had come to Down. The solution, I do believe, is that the modified offspring of all dominant and increasing forms tend to become adapted to many and highly diversified places in the economy of nature."

No better example gave proof to this fact than these 14 finches. For among them were 6 ground finches that fed on seeds and lived in the arid coastal region. Four of these lived together on most of the islands, 1 having a long beak to feed on the seeds of prickly pears, the other 3 having beaks of three different sizes, each adapted to a particular seed size. The other 2 species, one large, the other small, lived on outer islands and had special beaks to supplement their seed diet with cactus parts. There were also 6 species of tree finches that lived in the moist forest. Among these were 3 that differed in size and had beak sizes adapted to different sizes of insects: 1, a vegetarian with a parrot-like beak that ate buds and fruits, 1 that lived on mangrove swamp insects, and 1 that used its beak or a cactus needle in its beak to probe tree holes for insects. This last one also climbed trees and had other woodpecker-like behaviors. There was a warbler-like finch that lived on insects. It had a thin warbler-like beak and even flicked its wing like a warbler. And, finally, there was the lone Cocos Island species that lived on insects in the tropical forest.

Thus, an answer came. The barriers happened to allow only a few birds like the finch to reach the Galapagos. Yet, sufficient plants and insects could find their way there to supply more food and niches than were necessary. Thus, apparently varieties of this early finch under competition with one another for their common kind of mainland food diverged with the new opportunities. Those varieties with beaks closest to insect-eating types survived by eating insects and produced offspring even more closely adapted for eating insects. This change in beak size and shape continued as long as the competition for similar food continued. And with a variety of foods located at different places among the islands, the beak types diversified until all these foods had corresponding predators. It is interesting that of all the characteristics beak types and behavior displayed the greatest differences among these finches. Plumage, calls, nests, eggs, and even display behavior were quite similar among the different species.

The reason that mainland ancestors of the birds did not evolve or diverge in form in their original habitat must have been due to the lack of opportunities. There were a woodpecker and warbler adequately filling their respective niches on the mainland. Migration is only effective if a barrier prevents some organisms from migrating to the new area, thus allowing unfilled niches to exist.

The question of intermediate forms was also settled as far as Darwin was concerned. They could not remain long in a halfway less fit state. There is a best fit for any niche. The normal source of variation will supply this best adapted organism, which will successfully compete with and eliminate other less fit intermediate forms. And as to the reason these are not found in the fossil record—it is simply because of the paucity of fossils.

Darwin imagined that throughout the history of life there were many cases where a few basic forms reached a spot where many new opportunities were open to them, and great divergence took place. He called such phenomena "adaptive radiations." Figure 4–3 is a drawing that Darwin used to describe the erratic occurrence of adaptive radiations along the pathways of evolution. Most of the diverging branches would become extinct as their forms could no longer follow the changes in their environment for one reason or another.

PUBLISHING THE "ORIGIN"

Darwin summarized a more complete version of his understanding of the transmutation of species and natural selection in an essay written in July of 1844 (230 pages, *Version 2*), but this too he did not publish. There were several reasons for this. He still wanted to be sure by ever testing more facts against his hypotheses. There was also the question of the source of variation in species, which he was no closer to answering as the years went by. He felt this was a severe weakness to his whole study. He was far more critical of his own work than critics might have been, but he did not realize it. He was also fearful of ridicule from others and

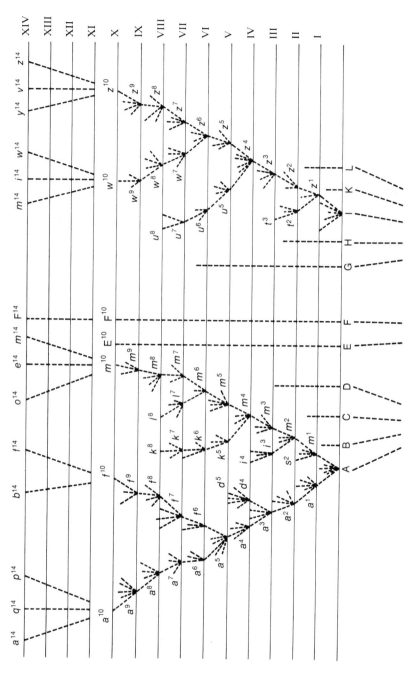

Figure 4-3. Adaptive radiations (Darwin's understanding).

with a good reason. He remembered the defeat of Lamarck. He was also aware of the unpopularity of any evolutionary idea, as witnessed by the fate of Lawrence's books. He mentioned his work only to a few friends, mainly Lyell and Joseph Hooker. Also, he had married well, to his own cousin with a handsome dowry, and he could afford to take the time to feel more content about his ideas. Worried about his health, he set money aside to have his *Version 2* published in case of death and decided to amass much more data into a final tome—*Version 3*.

Then something happened that distracted him considerably. This was the publication of the *Vestiges of Creation*, also in 1844, by Robert Chambers (1802–1871), a bookseller and journalist. It was Chambers who attempted to answer Sedgwick and other critics of Lyell. These critics needed an explanation of how he, Lyell, could explain the apparent progression of organic forms in view of his own uniformitarian ideas. For, although Lyell's arguments were temporarily effective, much of the best geology of this intervening period was being done by catastrophists such as Sedgwick. Very significant work on glaciers and the characterizations of the Cambrium and Silurian epochs allowed these workers to ask questions of Lyell and to be heard.

It was in this climate that Chambers put forth his views on evolution and wrote, "That the simplest and most primitive type, under a law to which that of like-production is subordinate, gave birth to the type next above it, that this again produced the next higher and so on to the very highest, the stages of advance being in all cases very small—namely from one species only to another, so that the phenomena has always been a simple and modest character." He based his evidence on geological and astronomical evidence.

Chambers wrote with elegance and was popularly read, but he was highly criticized by the entire scientific community on a variety of grounds. One was that he had no mechanism proposed for evolution other than "the fundamental form of organic being is a globule, having a new globule forming within itself," which was not acceptable. Also he relied too often on sweeping generalities. For example, evolution should be thought of as gravity and both of these probably arise from one law. He also related to the Diety with glowing sweeping ways to antagonize theologists and scientists alike. In addition many of his facts were biased or incorrect. The outburst against this book also became an outburst against the idea of evolution. Darwin became discouraged and turned from his writing and indulged in an intensive study of barnacles.

Darwin waited. He also fortified himself in the eyes of his peers. He had published articles on coral reefs, barnacles, and other aspects of his voyage. These were enough to award him a Royal Society Medal in 1853. He was on good terms with most contemporaries and had been a member of the council in both the Geological Society and Royal Society for a number of years. By 1854 he had published eight books and several dozen articles. It was September 9, 1853 that he "began sorting notes for my species theory." Lyell mentioned in the spring of 1856, "You will be anticipated. You had better publish." Darwin pushed ahead on his *Version 3*.

It was at about this time that Darwin heard about a young natural historian, Alfred Russell Wallace (1823–1913), who had been collecting in the East Indies and on the Malay Peninsula. He was curious as to Wallace's finds and wrote him on May 1, 1857, mentioning that he had spent 20 years collecting material on how "species and varieties differ" and that he was writing a book on the topic. Wallace in reply sent to Darwin early in June of 1858 an essay titled *On the Tendency of Varieties to Depart Indefinitely from the Original Type*. Wallace wrote, "for that balance so often observed in nature . . . a deficiency in one set of organs always being compensated by increased development of some others . . . powerful wings accompanying weak feet, or great velocity making up for the absence of defensive weapons; for it has been shown that all varieties in which an unbalanced deficiency occurred could not long continue their existence. The action of this principle is exactly like that of the centrifugal governor of the steam engine, which checks and corrects any irregularities almost before they become evident; and in like manner no unbalanced deficiency in the animal kingdom can ever reach any conspicuous magnitude, because it would make itself felt at the very first step, by rendering existence difficult and extinction almost sure soon to follow." Wallace had also conceived of natural selection as a mechanism.

It seems to be more than coincidence that Lyell, Darwin, and Wallace were all field naturalists who thought first geologically and then ecologically. Wallace was also greatly influenced by Lyell and came to his conclusions as a result of the study of animal variations on islands and by questioning why these varied from those on the nearest mainland. Evolution and natural selection were more and more in the air.

This was clearly the same conclusion that Darwin had reached. His first reaction was to withdraw his book and encourage Wallace to publish his essay. He wrote, "I would rather burn my book than that he (Wallace) or any other man should think that I had behaved in a paltry spirit." Wallace had requested Darwin to send on his article to Lyell if he thought it worthy. Darwin did send it on, along with his own intentions to withdraw his book, but Lyell, together with Hooker, convinced him that he and Wallace should give a joint paper. Wallace with much respect for the work of Darwin, acknowledged his priority and readily agreed. The papers were given before the Linnaean Society on August 20, 1858. They consisted of a letter from Darwin to Asa Gray (1810–1888), an extract of Darwin's work, an essay by Wallace, and a letter by Lyell and Hooker explaining the nature of the joint publication.

The immediate effect was mild but excitement started to build up and in September both Lyell and Hooker urged Darwin to cut short his *Version 3*, which was about half completed and publish soon.

Darwin with "hard labour" completed a *Version 4* that was published in November, 1859, as the *Origin of Species*. By the time it was published great discussions had been aroused and the first edition of 1250 copies were sold out on the first day—a second edition of 3000 were also sold soon afterward. It has been said by many that this book caused more stir than any book that has ever been written. Most of the stir was not rejection, but discussion, for it was overwhelmingly accepted by most of its readers. Only much later did some of them begin to worry about what they had so easily accepted. But Darwin was much relieved, and all thoughts of finishing *Version 3*, which would have been a dull unreadable mass, were forgotten.

The *Origin of Species* was successful for many reasons, mainly because it did have such a resounding mass of evidence and good logic to support his argument. Fortunately, it won the support of many of the leading scientists such as Asa Gray, Hooker, and Thomas Huxley (1825–1895), who became its greatest defendant. Huxley was said to have uttered, "How extremely stupid not to have thought of that," and Howett Watson (1804–1881), well thought of botanist at the time, wrote, "Your leading idea will assuredly become recognized as an established truth in science, i.e., natural selection. It has the characteristics of all great natural truths, clarifying what was obscure, simplifying what was intricate, adding greatly to previous knowledge. You are the greatest revolutionist in natural history of this century if not of all centuries."

As probably expected, Lyell was not able to accept the grandchild of his own thoughts until quite late in life. In fact, even Hooker, as well as Lyell, thought that Darwin should acknowledge predecessors such as

Lamarck. But Huxley summed up the general attitude of most scientists when he wrote, "I took my stand (in support of Darwin's theories) upon two grounds; firstly that up to that time, the evidence in favor of transmutation, even if it were often talked about, was wholly insufficient; and secondly, that no suggestion respecting the causes of the transmutation assumed, which had been made, was in any way adequate to explain the phenomena. Looking back at the state of knowledge at that time, I really do not see that any other conclusion was justifiable."

A second reason for the success of *Origin of Species* was that it avoided any direct reference to the Diety. No one could rightfully accuse Darwin of introducing antireligious notions, even if later he was interpreted as doing so. He had anticipated such reactions and even avoided mention of the evolution of man. He held off 12 years for the first reactions to settle down before he published his *The Descent of Man*. Another reason for this delay was the lack of evidence about man. Only a few scattered bits of fossils that may have been ancestors of man had been found by this time and their significance was not known until later. There was a handful of irate theologians who felt "Darwinism" undermined the will of man denying him purpose, but this was not a universal feeling by theologists. It was Thomas Huxley who nobly defended "Darwinism" before these churchmen. Misinterpretation of Darwin's ideas also served to incite biting sarcasm by such gifted writers as Samuel Butler and G. Bernard Shaw.

It has been said that Darwin's evolution could be accepted by many scientists because of the ambiguity with which certain key points were presented. Darwin did not write in this way intentionally but rather to keep all possibilities open rather than to be dogmatic or to oversimplify. He had never settled in his own mind where the source of variation in species was. Therefore, the exact relationship between adaptations of the organism and the effect of environment was far from clear to him. He would refer to "external conditions" as a cause for variations, but he was never definite whether these were the same as those external conditions that did the selecting or not. Darwin was concerned whether in fact even use or disuse of limbs might not possibly play some role in directing variations and perhaps cooperate with natural selection.

Even in his first edition Darwin wrote, "that natural selection has been the main but not the exclusive means of modification." By the sixth and final edition he stated, "This has been effective chiefly through natural selection of numerous successive, slight, favorable variation; aided in an important manner by the inherited effects of the use and disuse of parts; and in an unimportant manner, that is in relation to adaptive structures, whether past or present, by the direct action of external conditions, and by variations which seem to us in our ignorance to arise spontaneously."

Thus Darwin ended up with four factors as possible causes of transmutation including those by Lamarck whom he originally could not accept and, ironically enough, those of his own grandfather. He had made a great circle, but he had put the idea of evolution in a favorable light— permitting its mechanism to be understood when the time was right. He never completed what he had planned to do, but he had made the giant step that was absolutely necessary for anyone to begin to complete it.

Darwin was well aware of this shortcoming and had no answers for what he considered to be the most legitimate criticism of his hypothesis of natural selection as put forth by Fleming Jenkins, Professor of Engineering at Edinburgh. Jenkins questioned how a variation could assume importance in succeeding generations, rather than dilute down as is the case where characteristics blend in the offspring of two dissimilar parents (the accepted manner of heredity at that time). And this would happen in too short a period of time for it to be selected and favored by natural means as required by Darwin's theory. The catastrophists were quick to jump onto this point and felt that only through sudden large jumps could modifications be taking place. They, plus others, argued that it is mathematically improbable that finely adjusted adaptations could result from random variations being acted upon by natural selection; the amount of change is too complex. And how could an organ as specialized and complex as the eye have ever functioned well enough in a rudimentary state to have been selected for seeing? Thus it was that evolution as a fact was generally accepted, but new questions as to the details of its mechanism, natural selection, became the center of focus.

THOUGHTS

Most of the styles of thought used in this search are discussed above. Both the conception and development of the concept evolution and its possible mechanisms required a great deal of gestalt observations and correlative analysis quite similar to that applied to the order of fitness. It could be said to be an extension of the same style of thought spread over a great deal more material spread out farther in space and time. An

interesting point is that for this order to be discovered and for the development of its related concept, the search required a broad synthesis of
geology and biogeography, two seemingly unrelated disciplines. This
reminds us that an explanation for one phenomenon may have to be found
by looking at a completely different phenomenon.

In spite of the fact that the same style of thought is used here as for
explaining the order of fitness, the comparative degree of success is
ambiguous. On the one hand this search has netted one of the most
generalizing concepts of biology. Its power is in its inclusiveness. It is
most interesting that with the added complexity of space and time, the
order discovered added some clarification to the orders and concepts
associated with diversity and fitness. For now when one classifies and
considers relatedness of species, he thinks of the species (if they are living)
as tips of the branches of a phylogenetic tree. In fact, the dendrograms
should be phylogenetic trees with the different heirarchal ranks related
to branching points down the tree toward the base. In this context, species
are closely or distantly related cousins with common ancestors. Along
with this advantage comes the problem of how to define a species in terms
of the added dimension, time. How does one mate an offspring with his
great-great-great grandparent to decide if they are the same species ?
Evolution has also added something new to the concept of niche and
fitness within the natural area. Physical changes in the environment and
new species that have evolved both create new niches and thus opportunities for further evolution. Thus any definition of niche must also
include historical time with an addition of meaning and source of questions
as it did for species. There are these questions but, on the whole, evolution
has been extremely beneficial in ordering our thoughts.

On the other hand, the concept of evolution cannot be considered a strong
scientific explanation for the presence of the diverse forms of life in space
and time. It must remain as a rather low level explanation to any purist.
This is because the data must be used circumstantially and no fine analysis
of biogeographic distribution or of the fossil record can directly support
evolution. Biogeographic data are so complex that even with island studies
only the rough generalities similar to those Darwin found for Galapagos
can be stated. The fossil record is very uneven and so scarce that fine
points regarding the pathways of evolution remain obscure. Even with the
best records, such as those for the horse as shown in Figure 4–4, which
are over long time spans, no systematic order can be found. The number
of adaptive radiations, the regularity of their occurrence, the number of

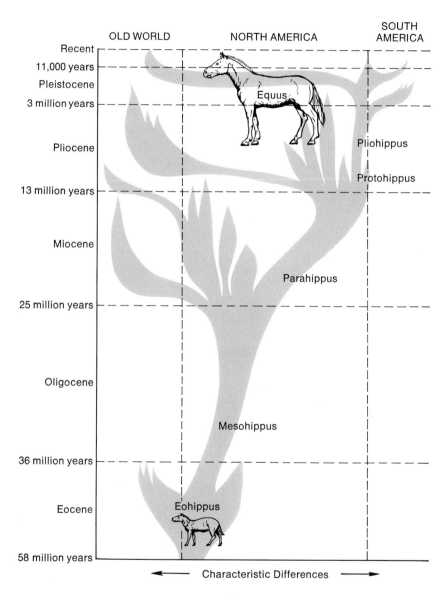

Figure 4-4. Evolution of the horse (general).

branches resulting from each adaptive radiation, the resulting lengths (extinction points) of any pathway, and the rate of change of any given characteristic are sporadic. Even the direction of change of a given characteristic is inconsistent and may reverse at any time. One could never guess from looking at today's horse the tortuous history it has suffered. No Galilean-like equations with any sense of predictability can be found.

If there is no regularity in these historical records, what is their meaning and where are they going? Are there any characteristics that somehow influence the directions of change? One can say that in general the biomass has increased, the numbers and kinds of species have increased, organisms have increased in complexity in terms of number and diversity of cell types, and food-webs have become more complex. There are always those cases, however, that violate each of these general statements, e.g., extinction, degeneracy, and man's cultivated fields. The one characteristic that may decide direction of change is that of efficient incorporation of energy for the earth ecosystem. One could never tell whether this was the case for a single organism such as the horse. One would have to study efficiencies over history. Any guesses as to this must be viewed with skepticism, at least with our present knowledge.

Perhaps the greatest contribution of the fossil record to the support of the concept evolution is the absence of negative observations. If a mammal could be dated with certainty at 500 million years, the concept would suffer considerably. The weakness of the concept (as a scientific explanation) makes it extremely vulnerable. The fact that exceptions of this sort have not held up under careful scrutiny, and the fact that there is a tremendous amount of circumstantial evidence that is consistent within itself means that the explanation is still a good one though it may not satisfy the purists.

Another difficulty with this search is the inability to experiment in order to verify the historical record. This is an obvious limitation we must face when dealing with such space and time ranges. Purists could say that evolution could never be proved. It may mean that it could not be disproved too.

These deficiencies were all well-known to Darwin and this is why he needed a logical mechanism to lend support to his concept evolution. The assumption was that if he could conceive of some process by which species could change, then evolution could be accepted. He needed a mechanism that was both logical and consistent with all known properties of life. A good mechanism is also one that can be tested. The best Darwin could do was satisfy the first criterion. He could imagine no experiments to try and had to rely on the circumstantial evidence of breeders and cultivators. With all due credit for his attempt, and even in spite of the fact that natural selection has been successfully tested as a means to modify species, there remains a major difficulty for our search. Even if natural selection is proved to be a way in which evolution could have occurred, it does not prove that it was the way or the only way that it did occur. No mechanism can stand alone without applying it in a test to the actual phenomenon the mechanism is supposed to produce. This is impossible to do and so the mechanism itself must remain as circumstantial evidence—to the purist.

My point is that apparently different explanations can have different values. Evolution and natural selection may not explain in a certain way the presence of diverse living and fossil forms of life, but it is still a good explanation, at least until something better comes along.

V

order: in form

Though Human ingenuity may make various inventions which, by the help of various machines answering the same end, it will never devise any inventions more beautiful, nor more simple, nor more to purpose, than Nature does; because in her inventions nothing is wanting, and nothing is superfluous, and she needs no counterpoise when she makes limbs proper for motion in the bodies of animals.

LEONARDO DA VINCI

Oft the beholder marvels at the wealth
Of shape and structure shown in succulant surface—
The infinite freedom of the growing leaf.
Yet nature bids a halt; her mighty hands,
Gently directing even higher perfection,
Narrow the vessels, moderate the sap;
And soon the form exhibits subtle changes.

JOHANN WOLFGANG VON GOETHE

An organism by its very name is a form of order—the order of form. And man being an organism, there is no order that has been as attractive to explore. Besides, the organism is a balanced integrated whole that grows and responds to its environment with a great deal of autonomy. It has been considered the epitome of life versus nonlife. Because of these reasons no order of life has enjoyed more developed styles of thought or more diverse explanations for its being. Medical needs started the search and even added the device of experimentation, earlier than with the search into any other order. Here one thinks of the parts, pokes at them a bit, and watches what happens. With this approach, even plants that were rarely considered important eventually were admitted into the organism concept and added to its explanation. As this style of looking at the parts developed, it was later joined by a completely different style that looked at the whole with classical geometric logic and physics. Toward the end of the nineteenth century the concepts derived from these styles were enriched by those that emerged from diverse thinking about the growth of the organism and its evolution. Most of the styles used were soon employed to explain the organism in terms of evolution and in so doing supported evolution.

The organization of the chapter follows this historical pattern. There are many names and dates but hopefully the intimacy of the explanations conceived with the metaphysics of this period will be clear and there is value in following, at least once, the detailed rationale for the substitutions of explanations.

EARLY CONCEPTS OF THE ORGANISM—ARISTOTLE'S MODEL

We are impressed with the complex and logical understanding of an organism that already existed in the early times of ancient Greece. Perhaps it should not be surprising that when such writings did appear they were those of the physician and the anatomist who needed an explanation of the organism for their practice or their goal. Hippocrates (467–357 B.C.) was not original when he summed up the attitude of all cultured Greece toward organisms, especially man. The organism was a harmonious whole, perfect in structure and balance, representing the ultimate in order versus the external chaos of erratic environment. A lengthy series of observations by physicians of the natural healing processes convinced them that the living organism contained organs that not only worked in harmony with one another but synergistically as a whole to preserve the organism from harmful attacks by the environment—disease, accident, and battle wounds. There was in essence a self-correcting reaction when the normal balance of the organism was upset by external abnormalities. The whole organism possessed this self-preserving principle, but occasionally a part of the organism would not be in balance with the others and it had to be removed—surgery. One of the roles of the physician was to ease the self-preservation process by applying ointments and herbs.

The anatomists needed to know how and where the vital processes functioned. One of many was Aristotle, who became noted, although not without dissenters of opinion, for his large number of careful dissections (over 500 animals) and thoughtful interpretations. He developed a method of asking functional questions of the organism that he answered by examining the arrangement and structure of the organs. His logic demanded that the function of any one organ must fit with that of the other organs, all for a "unity of purpose" for the organism. For Aristotle this was best understood by dissecting not only many similar animals to see the degree of variation that is always present but also by dissecting different kinds of animals to note the various forms an organ may take and how their function is correspondingly modified. He also believed that it was necessary to dissect recently killed organisms to be more accurate. He thus discovered the value of comparative anatomy.

Aristotle had learned from Plato that form is an active principle, a cause and a developing agent that produces all objects as they are perceived and not a static shape or contour as usually considered at that time. But to Plato these active forms were the expression of ideas, mental constructs that existed in the minds of men. The result of form as it appears to one's eyes is an imperfect fulfillment of an idea. Man as well as many natural forms were considered to be close to perfection and a blend of several of these ideas—harmony, beauty, usefulness, and goodness. An artist is a craftsman who builds as form does, e.g., not only by playing the lyre with beauty but by constructing the lyre in the same way that natural forms are made—in tune, in usefulness, in beauty, and in goodness. All Greece seemed to agree that the organism was symbolic of all that was virtuous. Aristotle, however, could not feel that the sinew and muscle he fingered was only some metaphysical idea. These were virtuous but they were real. Rather he considered that active form required a working substance, matter, to be molded. Matter was the clay the potter formed and shaped. This we can almost imagine.

It took little dissection for Aristotle to direct most of his attention to the heart. He would only be agreeing with a long history of observers that the life of an organism was related to the warmth of the body and to the beating heart that was engaged somehow in the movement or pressure of the vitally necessary substance, blood. From many observations and simple experiments Aristotle reasoned the following model of the vital

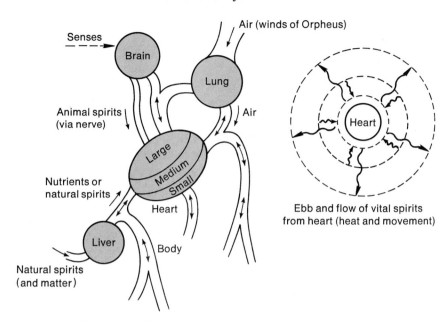

Figure 5-1. Aristotle's model of the vital centers of man.

centers of an organism, summarized in Figure 5–1. The heart is the life-giving organ of the body in the same sense as a limb is an organ of loco-motion. It does so by being the generator of heat motion (vital spirits) required for all life motion. Blood produced in the heart is a flow of this vital spirit. Thus, to Aristotle, active form, growth, and function are one thing produced in the animal by means of vital spirits that are generated and caused to flow by the heart. There is a harmonious pulsation of vital-spirited blood extending throughout the organism in one wave, interspersed with a relaxation ebb as the less spirited spent blood recedes to be revitalized by the heart. Those organs closest to the heart, the lungs, liver, and brain, receive and need more of the vital heat for their more significant activities.

In a like manner, the heart cannot exist alone; the organism is syner-gistic. The liver transports nutrients (natural spirits) derived from food to the heart where they mingle with blood and are carried to all parts of the body for the material needs of the organism—the source of matter. The lung likewise takes some of the "winds of Orpheus" and transports them through vessels to the heart, where they play several roles. This air is necessary for "preserving" the "fire" of the heart. No other element—

water or earth—can do this, as indicated by their ability to suffocate flames. Also air heated by the heart rises in the vessels like a boiling action causing the pulsation of the vessels and hence part of their motion. And finally the brain is the receiver of senses and sends these as "animal spirits" to the heart by way of filamentous cords, the nerves, where they are stored and acted upon. There is a reason for each part and there is order and interdependency among them. All is harmonious, functional, and beautiful.

GALEN'S MODEL OF THE ORGANISM

But Aristotle's model of the circulation of these vital spirits was less thought of then than it was much later. At this time there was always another anatomist to challenge some small point. Few of these left a lasting impression, but there was one who not only challenged Aristotle but superseded him in the eyes of most later workers. This was Galen (131–201), friend and physician to Marcus Aurelius. Galen was a prodigious writer and left us with a record of some of his many dissections and experiments. His ability is attested to by the fact that his model was considered the final one by most for well over a 1000 years, certainly one of the most successful models ever devised for the living organism. Galen depended even more than Aristotle on comparative similarities to establish a correct structure-function relationship. He disagreed with Aristotle on a number of points. He claimed that there were two large equally sized cavities in the heart, ventricles, each with a small subcompartment, the auricle (instead of a total of three as did Aristotle). The ventricles were interconnected by fine pores in the dividing septum. The heart is not muscle because of a difference in "taste, touch, appearance, hardness, softness, density, and colour," and because the heart beats independently, while muscle moves voluntarily. The heart can beat when "removed from the thorax." Also, as shown in Figure 5–2, there are two distinct kinds of vessels, distinguished by their wall thickness and type of blood carried. He called the thick-walled ones *arteries,* the aorta extending from the left ventricle being the largest of these; this latter branches to all parts of the body. He called the thin-walled vessels *veins,* with the largest of these, the vena cava, extending from the right ventricle. These veins also branch to all parts of the body. There is also a smaller vein running from the left ventricle to the lung, and a small artery running from the

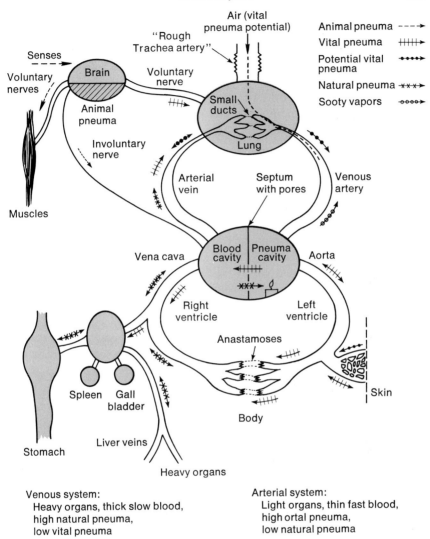

Figure 5-2. Galen's model of the vital centers of man.

right ventricle to the lung. Both of these vessels branch a number of times before ending as small branchlets at the lung. There is also another large "rough artery" that runs from the lung to the larynx and throat. The vena cava has a main branch ending at the liver. The liver has its own veins extending to different parts of the body, the largest of these to the stomach. The blood in the arteries is thin and quick to move while that of the veins is "thick, heavy, and slow to move." Throughout the body the branched endings of the arteries and veins come close together and Galen considered that they were interconnected by fine and indivisible orifices called *anastomoses*. In addition there were fine branch endings of the arteries near the skin.

Galen also experimented with this circulation system. He noted that both arteries and veins would collapse if cut, the arteries losing their pulsation. If a tube is inserted into the artery, blood would continue to spurt for some time indicating flow. He also found that pulsation was limited to the arteries and that this could be blocked by a ligature, the pulse continuing proximal but not distal to the ligature. He hypothesized that anastomoses exist by noting that when a severed artery collapsed, nearby veins did likewise.

Thus Galen found enough radical differences in the anatomy of the circulation and respiration system that he could modify Aristotle's model considerably. Galen felt more strongly than Aristotle that a different structure indicated that the organ served a different function. He also decentralized the many functions that Aristotle had assigned to the heart in search of an even greater interdependence and synergesis between organs. For example, he wrote, "In dispatching to the lung the nourishment which it draws from the liver, the heart appears to repay it in kind and recompense it for the air with which the lung honors it."

Galen thought as an engineer and considered the independent rhythmic motion of the heart as a mechanical property independent of its vital heat-producing role. Its action was like that of a bellows, drawing in or "attracting" fluids or air from outside as it expands, the active phase, and the reverse during contraction, a more passive process. The lung acted similarly but independently of the motion of the heart. They cannot be exactly alike in action since their structure is quite different; the heart beats involuntarily, while the lung moves voluntarily, agreeing with the difference in type of innervation and, of course, the functions of the organs. Similar kinds of arguments distinguish the heart and lung from the muscles—the only thing in common being the fact that they all move. Thus the kinds of motion are different from one organ to the next and each can occur independently of one another. This is quite unlike the continuous wave-like nature of motion in Aristotle's model.

Food is treated in the stomach and then is carried in a vein to the liver. The material and natural spirits (or natural pneuma as Galen preferred to call them) are derived from the food and initially rendered in the liver.

The liver also generates the blood of the body, a function of the heart in Aristotle's model. This liver-rendered nutrient and material may go from the liver directly to some organs by way of veins extending from the liver, but most of it goes to the right ventricle or blood cavity by way of the vena cava where the heart both perfects the natural pneuma and heats (adds vital pneuma to) the blood. The waste products of the liver functions are deposited in the spleen and gallbladder.

Thus it would be expected that the venous blood is thickest and slowest to move because it contains an excess of nutrients and material. Heavy and thick organs such as the liver and stomach need to be supplied by venous blood. The walls of the veins do not need to be thick as the substance within cannot escape easily and yet nutrients must be able to seep through to the heavy organs. The blood ebbs and flows in this venous system depending on the contractions of the heart, thus supplying pneuma both natural and spiritual to all associated organs.

In addition, the heated venous blood flows into the artery extending from the right ventricle to the lung. As venous blood is present in this artery, Galen called it a "venous artery." It must be thick-walled, however, in order not to be compressed in the thorax when the lung contracts, assuring that blood and nutrients reach the vital lung and also preventing the heated blood from escaping easily beforehand. There is a one-way valve at the junction where the venous artery departs from the heart preventing backflow when the heart expands. This is not only to assure the constant blood and nutrient supply to the lung but also to force a little of the blood into the fine-branched endings at the lung and through "small ducts" into the corresponding fine endings of the arterial system. This is one way in which blood may flow from the venous system where it is generated to the arterial system. Blood and natural spirits also enter the arterial system through the small pores in the septum of the heart and across the anastomoses throughout the body. That blood flows from the right to left ventricle must be true as the vena cava entering the heart is larger than the "arterial vein" leaving.

Air is "attracted" to the lung through the rough trachea artery, a vessel also necessary for the voice. The air, or some part of it, is treated by the lung, rendering it with pneuma potential. And as the lung collapses, the pneuma moves down the thin-walled artery, which Galen called the venous artery, to the left ventricle. It is also during the time of this collapse that a process of "exchanging it (pneuma) instantly for the particles of blood" from the venous system takes place. Hence some blood is also present in the venous artery. The walls of this venous artery must be thin in order that they too can collapse when the lung does in order to produce a flow of gas and blood to the left ventricle, attract some blood from the venous side, and also attract from the heart the "effervescent particles of nature burnt and sooty"—the products of its fires. It is the left ventricle, the so-called pneuma cavity, which houses the "principles of heat (vital pneuma)." Although the right ventricle does some heating, the heat is mainly generated in the left side, "the heart (pneuma cavity) being like the hearth and source of the innate heat that vitalizes the animal." The pneuma of the air mixes with the blood that can now be vitalized (heated) and this vital pneuma is taken to all the lighter organs of the body by this thinner fast-flowing vitalized blood in the arterial system. The walls of the arteries must be thick in order to retain the vital pneuma until it reaches these organs. There is also a valve where the venous artery extends from the left side of the heart, but with only two flaps compared with the three on the right side. This weaker one-way valve allows the sooty residues of the vitalizing process of the heart to escape at all times. The arteries need only go to the lighter organs to supply their vital needs since there is an exchange of pneuma with the venous system across the septum of the heart and the anastomoses throughout the body. This assures that pneuma will reach all organs in sufficient quantity. Likewise, the natural pneuma can seep in exchange through these same pores to supply even the small amount of natural pneuma needed by the lighter organs. The arteries also ebb and flow as the heart relaxes and contracts. The fine-branched endings of those arteries that near the skin's surface can also draw in air when these arteries dilate, helping the lungs bring air to the heart.

Finally there is the brain, the storehouse of the stimuli from the senses. Again this is unlike that of Aristotle's model in which the brain but passed on these senses for the heart to store and discriminate. Galen distributed this labor to the brain and considered it the seat or generating center for the animal pneuma that are sent through the nerves to cause the motion of the lung and muscles at the voluntary discretion of the brain.

Galen's model disclaimed the radially pulsing heart model of Aristotle almost completely. Galen's model was not as symmetrical or harmonious with a single-purposed rhythm, but his parts were balanced and interdependent and of almost equal division of labor with a close fit of form and function all for the purpose of the whole organism. This model was too beautiful to want to be questioned. It could explain almost all that needed to be explained about the organism.

THE RENAISSANCE AND GALEN'S MODEL

No serious discrepancy was found in Galen's model until about 1250 when an active school of Moslem physicians, keen in experiment and observation, were not able to find the pores that Galen claimed existed in the septum of the heart. Among those most concerned about this discrepancy was the great physician Ibn Nafis. To Nafis it was reasonable that the blood had to get from the venous or right side of the heart to the arterial or left side, as blood is generated only in the venous system. He also was convinced that in order for the blood to be "fit for the creation of the spirit (vital spirit or pneuma)," it would have to be mixed well with air. This was being consistent with Galen's model. Nafis concluded that the only logical conclusion is that the blood in the right ventricle must flow through the arterial vein to the lung and mix with the blood there before flowing through the venous artery to the left ventricle. He was aided in this modification by Galen's "small duct" passages. These were merely enlarged conceptually to permit a flow rather than a dribble of blood as Galen had suggested.

With the Renaissance a rebirth in interest in the organism and hence in anatomy ushered in Aristotle and Galen in gusto. Galen was championed by Andreas Vesalius (1514–1564) who retold Galen's explanations in 1543 along with a detailed atlas of both the arterial and venal systems based on his own observations.

It was also at this time that Michael Servetus (1509–1553) discovered the same pore discrepancy as had Nafis, whose contribution lay excluded from the Christian world until much later. Also Servetus was a devoted believer in God and considered that the three spirits or pneumas of Galen were one, a divine spirit breathed into man by God as stated in the Genesis.

The pulmonary system that Servetus proposed was perfected in detail by a brilliant student anatomist of Vesalius, Resaldus Columbus (1510–1559) who was said to have dissected over 1000 cadavers. Columbus found blood in the venous artery with no trace of any gases and concluded that not only does the blood flow rather freely through the lungs from the right to the left ventricle but that air mixes with and "thins" the blood in the lung. Columbus did not disagree with the contention that the vital spirits recognized as heat are generated in the left ventricle, but he questioned whether the process was analogous to that of a flame, for he found no waste gases in the venous artery. He concluded that flow in this vessel was only in one direction, into the left ventricle, and that the role of the air was to prevent the heart from overheating. He supported his one-way flow theory by more carefully describing the one-way control of flow of blood by the valves of the heart.

It was Andreas Caesalpinus (1519–1603) who denied most vehemently that air was necessary for the production of vital spirits (heat). He also denied that the heart could be equated with a flame. Accepting Columbus' analysis of pulmonary flow, he concluded that air in the lung served to cool the blood.

Hieronymus Fabricius (1537–1619), a most thorough anatomist and physician to Galileo, elaborated further details of the pulmonary system and discovered "valves" in the veins. He noticed that if one tried to push blood down a vein it was held up at bumps that turned out to be one-way valves. He attributed their purpose to preventing the seepage of blood to the feet, which caused them to swell otherwise, and also to assure that blood with nutrients could reach the upper regions of the body.

HARVEY'S MODEL OF THE CIRCULATION SYSTEM

The Renaissance anatomists had thus rediscovered both Aristotle and Galen, but they had begun to find fault with some of the elder's explanations of organisms. Blood and the circulation system remained a key characteristic of animals and one with which there was still strong disagreement among the anatomists as to just how wrong the ancients were.

A student of Fabricius, William Harvey (1578–1657), was especially disturbed. He wanted to keep the explanations of both Aristotle and Galen and still reconcile these with the more recent claims. He found that the contradictions were too many to do this completely. One of the main sources of confusion was the function of the venous artery. Both Aristotle and Galen had air passing through this vessel to the heart for the purpose of generating spiritous blood in the heart; yet the lung needed spiritous blood for its function and existence. Galen had solved this by claiming the spiritous blood could go back up through the same vessel or that some spirits could reach the lung from the right ventricle by way of the arterial vein. Also Galen claimed that waste vapors can return from the heart to the lung through the venous artery. Harvey, a strong traditionalist in that function is specifically related to structure, could not see how the same vein could carry both blood and gases, especially in opposite directions at the same time. The suggestion that there is a pulmonary flow in one direction would partially solve the problem but leave unanswered how spiritous blood can reach the lung or how the "sooty vapours" return to the lung. This could not happen by way of the discredited pores in the septum. There was also the difficult question of how air (or Galen's pneuma) was involved in the generation of vital spirits (heat) and where this activity took place. Did air preserve the fires or cool them, and was this in the heart or in the lung?

Harvey's recourse was to push through his belief in the principle of single structure-single function regardless of the opinions of others. He drastically concluded that the two ventricles must serve the same purpose. He wrote, "Why, then should we imagine their function to be so different when the action, movement (structure), and beat of both are the same?" Likewise, why distinguish arteries and veins when they are of similar structure, size, and extent of branching? And as he cut into all the different arteries and veins, he concluded, "arteries contain the same blood as veins and nothing but the same blood." There was no gas, air, or "sooty vapours," and all blood had both nutrients and vital spirits. By his belief in this principle he was thus able to simplify the system and concentrated his full attention on the mechanics of the heart and the distribution of the blood.

Another belief that aided Harvey in his study of the mechanics of the heart was that one could transfer what was learned from one species to the others. Realizing that the heartbeat was too rapid in warmblooded animals, he felt no compunction about relying on the observations of toads, frogs, serpents, lizards, and fish, which have slow rates of heartbeat. He noted that in these animals the blood is squeezed from the heart with its lower tip striking the chest to produce the beating effect rather than in the process of "attracting" with dilations. "The heart does not act in diastole but in systole for only when it contracts is it active." These studies also helped Harvey accept the pulmonary circulation of the blood, for in fish the blood flowed from the ventricle by way of a vein through the gills and from the gills by way of an artery to the rest of the body. The gills were for all purposes equivalent to the lungs.

Having satisfied himself with an understanding of the movements of the heart and the manner in which the blood flows through a circuitous pulmonary system, Harvey turned to the flow of blood to the body, that which starts in the aota. He was apparently influenced by his contemporary Galileo into making measurements of the rate of flow of the blood. Repeating Galen's experiment, he found that it took about half an hour for the blood to empty from the body of an animal through a cut in either a large vein or artery. Convinced that blood cannot be generated or destroyed at the rate it flows, as the arteries would burst and the vein would drain "unless it somehow got back to the veins from the arteries and returned to the right ventricle of the heart. I (therefore) began to think there was a sort of motion as in a circle." Pulmonary flow in one direction also paved the way for such a concept. He tested this idea by cutting a vein and finding that the corresponding artery deflated, by heating the blood in the artery locally and feeling a warmth in the corresponding veins, and by the logic of calculations. He estimated that the left ventricle of man could hold from $1\frac{1}{2}$ to 3 ounces of blood when filled. He assumed that from $\frac{1}{8}$ to $\frac{1}{4}$ of this was expelled during each contraction, or from 1 dram to 1 ounce, 3 drams per beat. In half an hour the heart beats from 2000 to 4000 times. Thus from 2000 drams to 5 ounces move through the heart in this time, which is much more blood than exists in the body! Therefore, he asserted as diagrammed in Figure 5-3, "It is

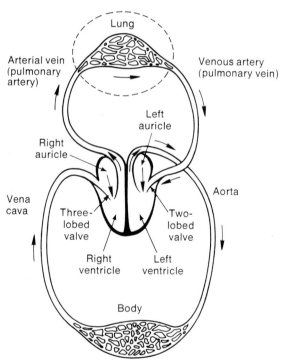

Figure 5-3. Harvey's model of the mechanics of the circulation system.

apparent that the entire quantity of blood passes from the veins to the arteries through the heart, and likewise through the lungs."

He also examined Fabricius' valves of the veins and found that they were always pointed away from the heart preventing blood from flowing back up the veins during heart contraction. He demonstrated with some ingenious experiments the action of these valves. He tied an arm with a tourniquet in order for the veins to stand out. If this is done, nodules appear along the veins indicating the presence of the valves. If the blood in one of these nodules is pressed up the vein to the next nodule, it will remain there after the finger is removed. Moreover, a nodule of blood cannot be forced downward. If a finger is held across a vein and the blood in several nodules above the finger are pressed up the vein several nodes, the empty nodules will remain empty as long as the finger remains across the vein. But as soon as it is removed, blood will move up the vein and fill the nodules from below. "Thus it is clearly evident that blood moves through the veins toward the heart and not in the opposite direction." Harvey's experiments and logic were accepted by all and that blood circulates through the body became fact.

The circulation system is a rather good example to demonstrate how man can use the same principles and derive different conclusions. All anatomists used the principle that structure and function are specifically related, but there were obviously different criteria of how exact or what kind of fit this must be. Harvey was also convinced that "Nature makes nothing in vain." But in many respects he was extremely lucky that his procedure of oversimplifying, especially the comparison between the arteries and veins and between arterial and venous blood, netted results. If it had not been for his convincing experiments, his basic premises would have never been accepted. It did not seem to bother him when many called his attention to the difference in color and other properties between the two bloods. He would answer with an air of indifference that arterial blood was "admittedly more spiritous and heavily endowed with vital force," or "by the movement of the blood, all parts are fed and warmed by the more perfect, more spiritous, hotter, and I might say, more nutritive blood. But in these parts this blood is cooled, thickened, and loses its power so that it returns to its source, the heart, the inner temple of the body, to recover its virtue. Here it regains its natural heat and fluidity, its power and vitality, and filled with spirits, is distributed again." He could say these things without feeling any contradiction in his basic premise of there being but one blood. The quotations above were apparently later remarks after the question of circulation was clarified, for he did not let the problems of vitalization and the role of air or the lung interfere with his hypothetical model during his experiments, and could say nothing about them afterward.

Harvey's model of circulation could not be verified until much later. With the discovery of the microscope, Marcello Malpighi (1628–1694) found bronchi or "small bladders" in the lung that inflated with air and were surrounded by small capillary vessels, but he could not demonstrate that they were associated with veins or arteries or whether air passed through the walls of the bronchi. It was necessary to await the observations of a self-trained amateur and one of the greatest scientific explorers of all time, Anthony van Leeuwenhoek (1632–1723) to discover in 1688 "particles" (red blood cells) within the capillaries of an eel's tail. He was able to time three or four of these particles moving $\frac{1}{15}$ of an inch in $\frac{1}{72}$ of a minute. From this measurement, he calculated that the blood would move 288 inches or 13 times around the eel in an hour. Thus Leeuwenhoek found the necessary connection between artery and vein that Harvey had predicted existed. He also reasoned that the walls of the arteries and veins were too thick to pass "subtile juices" to the tissues and that this must occur at the capillaries, "whereby the blood, when returning in the veins, being deprived of those juices which are taken away, it will appear blackish."

VITAL SPIRITS AND AIR

As the circulation system became mechanical in concept, there was a shift in concern about the action of the vital spirits for an understanding of the animal. Assistant to Newton, Robert Hooke (1635–1703) settled the question of whether the movement of the lung per se was required for the intake of air and life. He developed a double bellows that could be attached to the pharynx of a dog and by alternately blowing air into the lungs, which had fine holes pierced in them, it was possible to keep them distended and full of fresh air. In this way he showed that movement of the lungs per se was not necessary for life and that a dog could be kept alive for at least an hour if and only if fresh air was used. It was also Hooke who built the first air pump for and assisted Robert Boyle (1627–1692) with his many experiments on gases. They found that both a candle and a mouse required fresh air to continue to burn and live, respectively.

Likewise both the candle and mouse extinguished at about the same time under a partial vacuum. Boyle concluded that only part of the air was involved. If the air is expelled from around the mouse and other external (fresh) air allowed in, the animal will live for another hour. But if another mouse is added to the chamber without allowing new fresh air to enter, this mouse will last only about 3 minutes. It was not clear to Boyle whether some part of the air was being used up or damaged by the "steams" that were emitted from either the exhaling animal or the candle. His experiments caused him to favor the latter explanation. One thing that was clear to Boyle was that "atmospheric air . . . is not, as many imagine, a simple elementary body, but a confused aggregate of effluviums . . . that consists of corpuscles of different sizes and solidities restlessly and very variously moved." Boyle's experiments with gases brought the elemental nature of the Greek's "air" under severe question.

Another contemporary of this group was Richard Lower (1632–1691) who considered the challenge made to Harvey—how can blood be one and yet apparently different in the arteries and veins? He noted that venous blood stirred for a long time in the presence of air turned brighter red, the color of arterial blood. Also, blood leaving the lung is brighter than that entering unless fresh air is prevented from entering the lung, in which case the two bloods are identical. Even a recently sacrificed dog, through whose lung fresh air is blown and through whose pulmonary system blood is forced from the right ventricle, will yield blood from the pulmonary vein as brightly colored red as that from the living animal. Lower concluded that air modifies the blood in the lung and no heating or other treatment by the heart is necessary. He proposed that "nitrous spirits" of the air were responsible for the color change. Regardless of the interpretation, it was clear from these experiments that the color of blood, considered to be one of the indications of vitalized blood, can be modified by the air in the lungs. Lower also questioned that the heart was the sole source of heat, for "there is nothing in the heart which is sufficient to produce so much heat." Lower's belief in the simplicity and similarities of blood led him to perform the first blood transfusion experiments. He found and cautioned that different animals including man had inherent differences in blood and that they could be mixed with varying degrees of success.

The final member of this group was John Mayow (1643–1679). He contended that the "aerial element" necessary for life was "nitro-saline" (a molecular form of Lower's nitrous spirits), which is "subtle, agile, and in the highest degree fermentative, (and) is separated from air by the action of lungs and conveyed into the mass of the blood. It is probable that nitro-aerial spirit, mixed with the saline-sulphureous particles ("motive particles") of the blood (or muscle), excites in it the necessary fermentation (heat)." This reaction could occur anywhere in the blood or muscles and was necessary for every kind of movement including the heart itself. The heart if exposed to air would beat because of the combination of the two necessary elements. Nitro-saline was the same ingredient of air that was necessary for a candle to burn. Mayow's explanation was quite original and speculative, but too far ahead of his time to be developed or tested. It was also an open rebellion against the spirits of the ancients that were being questioned, but not as much as Mayow would have liked them to be. He caused some stir, but his ideas were not accepted by many and he rather overstated his impact when he wrote, "Nitre has made as much noise in philosophy as in war."

The main issue concerning the vital role of air was not able to be understood until methods for separating the different parts of the air as suggested by the work of Boyle were developed. The "steams" of Boyle, those products of respiration from an animal or candle that inhibit further respiration and burning, were renamed *phlogiston*. Fresh air (vital air) or that part of air that contained the "spirits of Nitre" was called *dephlogis-ticated air* or *dephlogiston*. Another part of the air became known as "fixed air" as it could be absorbed by a caustic soda solution.

It was the Unitarian minister Joseph Priestley (1733–1804) who re-peated the experiments of Boyle on animals and extended them to plants. He placed a sprig of mint in a confined volume with some water for an extended period, assuming it would expire as had an animal placed in the same situation. But, it continued to grow for "some months," and, what was even more surprising, the air around it did not "extinguish a candle or was it at all inconvenient to a mouse." Further experiments led him to write, "I have been so happy, as by accident to have hit upon a method of restoring air, which has been injured (made phlogistic) by the burning of candles, and to have discovered at least one of the restora-tives which nature employs for this purpose. It is vegetation." In about 1774 Priestley was also the first to heat mercuric oxide to produce another gas that he found would not extinguish flame or life as phlogiston would do.

This process was further investigated by Antoin-Laurent Lavoisier (1743–1794), later to become a martyr to the French Revolution and of whom it was written, "it took but a moment to cut off that head, though a hundred years, perhaps, will be required to produce another like it." By this time gas volume measurements had been perfected well enough to produce fairly accurate results for gas exchange experiments. He found that a part of air, the pure air (dephlogiston), was removed by a burning candle or animals and that a different gas that had the identical absorption properties of fixed air was produced. He found that there was a fixed ratio of fixed to pure air volumes exchanged, and this was essentially constant for either a burning candle, burning carbon, or the respiration of a guinea pig, being about 0.75 to 0.81 by volume. He could now say that what was thought to be phlogiston is nothing but fixed air and that fixed air does not inhibit respiration or the burning of a candle. Rather it was pure air that was being used up and was necessary for these activities. Thus the phlogiston concept had to be discarded. Pure air later became known as oxygen and was the same gas emitted from heated mercuric oxide.

Even more significant were the calorimetric measurements of combustion that Lavoisier was able to carry out. He devised a chamber surrounded by ice enclosed in a jacket chamber so that the heat given off by a reaction in the inner chamber would melt the ice to water. The amount of this water could be measured and was an indication of the total heat change as shown in Figure 5–4. He placed a guinea pig in the chamber for 10 hours, absorbed all the fixed air produced by the animal during this time, and measured the amount of water collected from the melted ice. He found that 13 ounces of water was formed. When he then burned carbon in this chamber until the same amount of fixed air was released by the reaction and absorbed, he found that 10.38 ounces of water was produced. Lavoisier considered these two values close enough to conclude, "respiration is therefore a combustion, very slow to be sure, but perfectly similar to that of carbon. The heat which is liberated in the transformation of pure to fixed air by respiration may be regarded as the principal cause of the conservation of animal heat, and if other causes are involved, they are of lesser importance. It (respiration) occurs in the interior of the lungs. The heat developed by this combustion is transferred to the blood as it passes through the lungs, and thence is transmitted throughout the animal system. Thus the air which we breathe serves two purposes equally necessary for our preservation; it removes from the blood the base of fixed air, an excess of which could be injurious; and the heat which this combination releases in the lungs replaces the constant loss of heat into the atmosphere and surrounding bodies to which we are subject." Thus

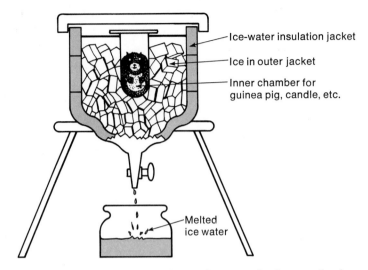

Figure 5-4. Lavoisier's calorimeter for measuring heat production.

Lavoisier's experiments rather conclusively proved that the heat of the body is produced by a general physical energetic process not involving vital spirits. Air is involved in this physical process, and the physical heat produced is somehow necessary for the vital processes of the organism. Lavoisier transferred the vital reaction center from the heart to the lung and considered that the purpose of the circulation system was to transfer this heat to all parts of the body.

In brief summary, the many pages above concentrated for the most part on the history of the replacement of the concept "vital spirit" with physical equivalents. This was a very slow process, for this spirit was not only close to a metaphysics of this long period but it was also a good explanation in that it was general and encompassed heat and motion as well as "life." It was necessary to show that "vital spirits" was too simple a concept by demonstrating the mechanical-like motion of an animal's parts and then that the heat of the organism is derived from a combustion of carbon that could be in a non-living state. Lavoisier did not demonstrate the relationship of this heat to living activities. He assumed it served as a driving force for all bodily functions. It was not until much later when the "fermentation" processes of individual tissues could be studied that it became clear that heat is a by-product of these reactions rather than an initiator. We can see by this history how little one person can do in the replacement process. The substitution is not sudden and in a sense never completed. Some part of the explanation (the concept vital spirit) could still reside in present explanations cloaked in other language. It may be too easy for us to imagine that what an ancient thought about some process is quite different from our own.

Regardless of these arguments, the spirits have been replaced by blood, heart, lung, respiration, metabolism, brain, nerves, hormones, liver, kidney, etc. Animal and natural spirits suffered the same kind of displacement as did the vital spirits. The interesting thing is that with this substitution views on the integration of these parts with synergistic activities for the whole organism were not lost but strengthened. With greater knowledge of organ-organ interaction a powerful new concept, homeostasis, emerged. This concept was conceived by Claud Bernard (1813–1878) who is considered to be the greatest general animal physiologist of all times. He contributed to every field of animal physiology, especially those related to the maintenance of a correct (optimum) balance of nutrients and other factors necessary for metabolism. He conceived that there is an optimum internal environment of the tissues in terms of concentrations of salts, nutrients, hydrogen ions, trace elements, vitamins, and gases, as well as an optimum temperature at which metabolism will occur with a maximum amount of efficiency. And if there is deviation in any of these factors from the optimum, some specific organ is alerted by nerves or hormones that in turn corrects the deviant back toward the optimum, i.e., a feedback self-regulatory activity to maintain the status quo. Often mechanical motion is involved as diagrammed in Figure 5-5. Bernard also

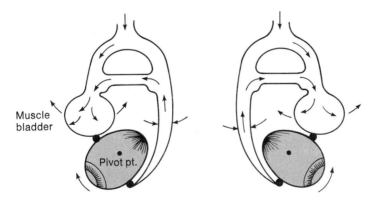

Figure 5-5. Descartes' model for the reciprocating action of the eye muscles—a method of organ response to "nerve" action.

extended this concept of regulation to include response of the organism to its environment, again for the optimum operation of the organism. This would include not only response to germ diseases and injury but also to the need for food, drink, and probably sensual stimulation. The organism may respond by either internal changes or by changing its external environment, this latter including relocating oneself. The greater the degree of homeostasis in an organism, the greater independence it will have from its environment and/or the better it will fit its environment.

THE PLANT AS AN ORGANISM

There is added insight into the nature of an organism when one compares explanations that have been given for a plant versus those for an animal. There was no complex model given for the plant by the ancients. All agreed that plants are alive and thus had a vital spirit although weak it must be. Plants definitely had a natural spirit as they required nutrients, but there was disagreement as to the existence of any equivalent animal spirit. Plato thought that there was as they must desire nutrition, but Aristotle reminded all that plants had no sense reception and therefore no sensitivity—a requirement for animal spirits. It was generally considered that the element earth, in the form of nutrients, arose and was transformed into the plant structure by the fire (heat) of the sun and possibly air was involved.

It was not until the fifteenth century that any experimentation on plants was carried out. These and subsequent studies revealed that the earth in the pot surrounding the growing plant did not decrease in weight and that it was water alone that was converted by the fire of the sun. As plants were still considered to have much of the earth element in them, these experiments challenged the whole elemental system of the Greeks by suggesting that one element (water) was being transformed into another (earth). Likewise there was no consideration of the main body of the plant consisting of separate organs with specific functions until the end of the seventeenth century when Nehemiah Grew (1628–1711) by means of the microscope defined various fibers and felt that they served specific roles in the distribution and conversion of "juices" into the solid parts of the plant. Even so, there was no effort to use the integrated-whole model, as applied to animals, to define the plant.

It was not until thoughts about and analysis of air in its role to supporting the life of an animal was considered that plants took on a more significant role in defining the organism. It was first considered by Mayow that nitro-saline was a basic necessity for all forms of life. He assumed that it was the nitro-saline presence in fertilizers as well as in air that added the vitality to plants. He concluded that "even plants themselves seem to have a kind of respiration and necessity of absorbing air (nitro-saline)." Stephen Hales (1677–1761) came to the same conclusion in 1727 by accurate weight measurements of the earth and water involved in plant growth. He concluded that the extra weight that was found in the plant must have come from the air. The experiments of Priestley and Lavoisier made it clear that plants could somehow convert fixed air (CO_2) into pure air (O_2). In 1778 John Ingen-Housz (1730–1799) discovered that sunlight was necessary and only the green parts of the plant were involved in the conversion above. He also noted that water (juices) moved up the stem most rapidly when the conversion was occurring. Finally N. T. de Saussure (1767–1848) demonstrated in 1804 with careful gas measurements that this process (photosynthesis) resulted in the assimilation of carbon into the plant during the release of oxygen— a process that has the opposite effect of respiration, a process that was also noted to occur in plants in the dark.

These results indicated specialized activity in different parts of the plant and a relationship of this activity to the rest of the plant (juice movement and growth). There were no obvious agents, however, such as the nerves or circulation system in animals, that could account for the integration of the plant and thus allow a universal definition of the organism as a result. In terms of agent structures, this distinction for the most part remains today and the integration depends on "diffusible juices" (hormones) that diffuse (though sometimes actively) to different regions of the plant where they are modified and their actions effect the growth of the different parts of the plant depending on their concentrations. This idea was first stated in a general form by Johann Wolfgang von Goethe (1749–1832) but not really demonstrated by any experiments until the 1870's by Darwin and his son Francis. They found that if a blind prevents light that is coming from one side from striking the tip of a growing seedling, the seedling did not bend. If the blind covered all the seedling except the tip, however, the plant did bend toward the light. The interesting thing is that the bend occurs some distance down the stem in the shaded region as shown in Figure 5–6.

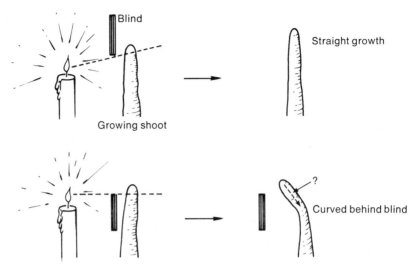

Figure 5-6. Phototropism: experiments of Charles and Francis Darwin.

Darwin proposed that some substance produced in the growing tip could diffuse down the stem and produce the bending of the stem. These experiments suggested that there might be plant hormones (the diffusible juices) produced in one part of the plant that influence a different region of the plant. It was just such a response that led to the discovery of the proposed plant hormones during the twentieth century.

It may be a mistake to think in terms of homeostasis when dealing with a plant. It is not clear what the internal reference medium would be. There is compensatory growth wherein growth of branches and leaves on one side of a stem is influenced by the amount of growth on the opposite side. This does involve hormones; however, it is due to the total concentration of hormone present rather than a homeostatic process as we imagine it in animals. It is probably unwise to expect to find the same mechanisms at work in the two classes of organisms. It should be possible to accept differences in mechanism and still find a single general definition of the organism.

THE GEOMETRY OF THE WHOLE
ORGANISM—DA VINCI AND DÜRER

The approach above has been one of separating the parts and attempting to determine how these parts interact to produce the whole. A completely different method of attempting to understand the organism has been the reverse: to look at the whole in order to understand the organizing principles that contain the parts. This approach is closely related to that of the purists among the ancient Greeks, e.g., Plato's form and its product, the harmony of natural contours. These products are thought to reveal an inner truth as a triangle can be shown to have laws relating its sides. According to this approach these laws are the only proper explanations of the form and shape of "the triangle," and when a mathematical relationship can be found to exist, it is perfection itself with a complete truth, beauty, and goodness.

In the more recent times, no one attempted more to find this perfection for organisms than did Leonardo da Vinci (1452–1519). One of the most gifted men of all times, he was artist, sculptor, architect, engineer, and also a naturalist and an anatomist. He was also not adverse to using the methods of the dissectionists. He, as did Michelangelo, sensed the value of knowing the anatomy of the parts of an organism in order to understand its form. He insisted that the artist must think in terms of the bodily functions of his subject. He read extensively from Galen and the Moslem physicians. By his own dissections he discovered parts of the bronchial arteries, the thyroid gland, and many details of the urogenital tract.

Da Vinci's real search was for the whole, however, and he preferred the purists' goal. About plants he wrote, "The space between the insertion of one leaf (of an elm tree) to the next is half the extreme length of the leaf or somewhat less, for the leaves are at an interval which is about one third of the width of the leaf. And these leaves are so distributed on the plant so that one shall cover the other as little as possible, but shall lie alternately one above the other ... ." He was thus perhaps the first to note what was later called the phyllotaxy or the regularity of leaf arrangement.

Da Vinci likewise measured and asserted strongly that certain proportions of the body of animals must be maintained in their artistic reproduction. For example, he wrote concerning the necessary profile of a person's face, "The length from the parting of the lips to the bottom of the chin, from the eyebrow to the junction of the lip with the chin, and the angles of the jaw and the upper angle where the ear joins the temple, will be a perfect square, and each side by itself is half the head," as shown in Figure 5-7. As far as known, however, he did not square the hypotenuse of the perfect explanation of an organism.

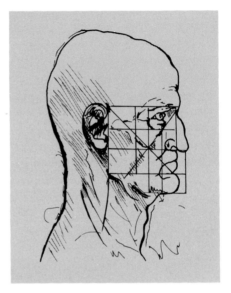

Figure 5-7. Leonardo da Vinci's square of a profile.

Da Vinci did influence other painters to attempt the same goal though, and one among those was Albrecht Dürer (1471–1528). Dürer also studied profiles of faces and elaborated the squares of da Vinci into a Cartesian coordinate system with a grid of lines to keep track of more characteristics along the side of a head. He drew lines through certain cardinal points of the head, such as the end of the eyebrow and the corner of the lips, for a number of different individuals and noted the changes in the spacing between the lines of the grid from one individual to the next. He even found that some faces produced an oblique-angled grid as pictured in Figure 5-8. From such studies Dürer tried to find generalities that might apply to the proportions and patterns of all heads, but he was not successful.

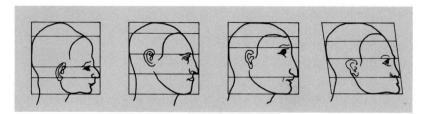

Figure 5-8. Profile studies by Albrecht Dürer.

PHYSICAL LAWS OF THE
WHOLE ORGANISM—GALILEO

The development of physics added an additional mode of analysis for explaining the organism. This was beyond the number-geometry style of the ancients. It was Galileo who first noted, while describing the necessary balance of forces in any supporting structure, that the counterpart of these structures for an organism cannot keep the same relative proportions with an increase in size of the organisms. The reason he gave for this is that in order to support the increase in weight of the larger structure, the dimensions of the supporting member must change disproportionately to that of the body. For example, if we consider an animal to be the shape of a sphere and supported by four cylindrical-shaped legs, then if the animal increases in weight by a factor of eight, the diameter of the body need only double. The pressure on the legs must be balanced by a proportional increase in cross-sectional area, however; hence the diameter of the legs must increase by the square root of eight or about 2.8 times. This is easily seen by comparing the relative dimensions of the legs of a dog with those of an elephant.

There are natural ways found that help compensate for these disproportional changes. For example, the leg or neck of an animal may not only become thicker but also shorter with an increase in size, thus keeping down the weight of the leg or neck. There would also be a tendency for the supportive tissue to be in a shape that would be the stronger per weight, e.g., I-beam shapes in the wing bones of birds, etc.

Galileo concluded from such calculations that there would have to be a maximum size for any structure, living or not, since regardless of the attempt to modify the supportive elements a limit would be reached beyond which the added weight could not be supported. Even stationary structures with a solid base such as a tree would have a limit. Galileo estimated that the diameter of a column must increase proportional to the $\frac{3}{2}$ power of its height to account for the increase in weight and bending moment and that a tree could not grow taller than about 300 feet. It is true that tapering of the trunk as found in the oak tree helps extend the height some, but not a great deal. This maximum height for a tree has been approximated in nature.

Galileo extended this principle, which has been called the *principle of similitude*, to the motions of an animal. He wrote, "A dog could probably carry two or three such dogs upon its back; but I believe that a horse could not even carry one of his own size." Jumping was another case in point. A flea can jump quite high in relation to its own height, a dog less, and a bull even less. Jumping depends on sudden impulses rather than weight alone, and the work done in raising the weight is equal to its mass times the height raised (mh), while the power to raise the animal is proportional to the mass of the animal (the muscle mass) (m). Therefore animals with similar body structures would jump to about the same height (h) and this is roughly the case.

The principle of similitude is not as the purists would like as there is no inherent mathematical relationship contained within it, yet it applies to the whole and involves abstract laws. Perhaps the difficulty is that it supplies an explanation in terms of physical reality instead of being a self-contained abstraction. Physical laws seem to have one foot in each camp, abstraction and reality. But granted this weakness or strength depending on how you look at it, Galileo could find organisms to which this principle applied, which indicated that organisms are subject to the same physical laws as the nonliving.

MORE ON PHYLLOTAXY—GOETHE

In spite of the necessary reliance on physical law, man continued to search for mathematical regularity in natural form, and he found some. In 1754 Bonnet brought attention to the existence of a perfect geometric spiral in the cone of a pine tree. Goethe set his creative mind to seek a basic body plan for all plants, only to discover in its place that there was no one plan for all plants but patterns of leaves and their derivatives arranged in regular order along a stem. The leaf unit was the simplest expression, and flower parts and fruit were modified patterns of this unit. What was of most importance was that these units were invariably arranged around the stem as a spiral or helix; hence one more advancement in the concept of phyllotaxy.

The study of phyllotaxy progressed extensively during the nineteenth century, with many workers rewarded by the kinds of regularity they found. Phyllotaxy never strayed from being in the category of a spiral, and the spiral fascinated many. True, it does not possess a true symmetry as defined in crystallography where mirror images of the repeating units must exist. Yet it has a kind of symmetry with the repeating units being complex and possessing an irregularity of shape.

Empirically, it was first found that the leaf arrangement around the stem came in certain patterns. The simplest case is that in which the leaves alternate on sides of the stem, being 180° apart with every other leaf being over the top of one another. Each spiral cycle would have two leaves. The phyllotaxy of this arrangement was called $^{1}/_{2}$ (one turn having two leaves). The next simplest arrangement was one in which the leaves were spaced 120° apart, there thus being three leaves in a cycle with a phyllotaxy of $^{1}/_{3}$. The next most complicated arrangement that could be found was one in which five leaves occupied two cycles; hence the phyllotaxy is $^{2}/_{5}$. Extending this to all cases, the phyllotaxies found were $^{1}/_{2}$, $^{1}/_{3}$, $^{2}/_{5}$, $^{3}/_{8}$, $^{5}/_{13}$, $^{8}/_{21}$, $^{31}/_{34}$, $^{21}/_{55}$, etc. The series of the numerators or denominators is known as the Fibonacci series and is formed by the numerator or denominator being equal to the sum of the two preceding equivalent parts of the fraction. The limit of this series is equal to 0.38197 with an angle of 137° 30′ 28″ between succeeding leaves. This is known as the "ideal angle" because at this spacing no leaf will ever be directly over any other leaf. The argument is that this would be an ideal arrangement both because the leaf veins that run down the stem would not interfere with one another and also because the leaves would exert a minimum amount of shade upon one another. There is also the appeal to the mathematical purists since the fraction 0.38197 is known as the "golden mean." By definition this is the fraction between the parts if a line is divided so that the ratio of the smaller part is to the larger part as the larger part is to the whole original length of line (0.38197:0.61803 = 0.61803:1.0). This is excitingly close to squaring the hypotenuse of plants.

D'ARCY THOMPSON—STUDENT OF FORM

Studies of form leading to a purely mathematical description such as phyllotaxy or those that reveal adherence to a physical law such as Galileo's principle of similitude have been exceedingly rare. Perhaps this means that few forms follow such abstract descriptions but it could also mean that as yet man has not developed the sensitivity to perceive or the abstraction to describe the order that exists in living forms. Among the few who have tried to carry on this tradition has been one individual who stands out far above all others. This is D'Arcy Thompson (1860–1948). Perhaps his uniqueness lies in the fact that he is almost unprecedented in being equally well trained in the classics, mathematics, and natural history and accomplished enough to have received academic chairs in all three. And what is more, D'Arcy Thompson could somehow combine all of these in their purest form. He has been said to be the last and one of the greatest natural philosophers. He put down most of his ideas in one book, first published in 1917 and revised in 1942, *On Growth and Form*, which has been said to be the finest work of literature in all the writings of science that has ever been published in the English language.

That D'Arcy Thompson sought the Grecian goal is evident in his writing, "that in the study of material things, number, order, and position are the threefold clue to exact knowledge; that these three, in the mathematicians hands, furnish the first outlines for a sketch of the Universe," and "moreover, the perfection of mathematical beauty is such, that whatsoever is most beautiful and regular is also found to be most useful and excellent." But even for Thompson, the purest mathematical form that he could find in nature was the spiral. There was not, however, only the phyllotaxy of leaves, florets of the sunflower, and elements of the pine cone. There were also coils of hair; the shape of the elephant's trunk; the chameleon's tail; and the many spirals of shells, horns, teeth, beaks, and claws—all of which can be described mathematically. Of especial interest to him were those cases of accretionary growth caused by the regular deposition of shell, enamel, chiton, and other inanimate products formed by the organism. Many of these produce equiangular or logarithmic spirals, which are spirals in which each new increment along the curve is proportional to the distance from the center of the spiral or the distance already traversed along the spiral. The shell of the mollusc *Nautilus* is perhaps one of the most spectacular examples of this equiangular spiral form as can be seen in Figure 5–9. The spiral may occur in a plane as

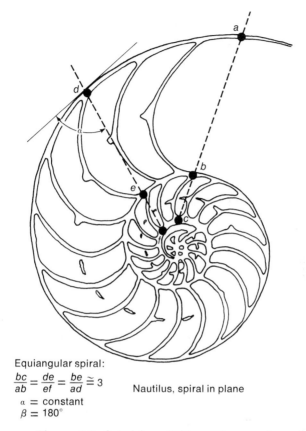

Equiangular spiral:

$$\frac{bc}{ab} = \frac{de}{ef} = \frac{be}{ad} \cong 3$$

$\alpha = $ constant
$\beta = 180°$

Nautilus, spiral in plane

Figure 5-9. Spiral form (D'Arcy Thompson).

it does for *Nautilus* or at a variety of angles depending on the species as also shown. Regardless of the angle, if the spiral is an equiangular one, there is a constant ratio of lengths of curve subtended by any two adjacent and equal angles of rotation. This ratio is that of the golden mean, 0.61803. Thus this curious constant is found again, even if it does always seem to be involved with spirals.

PHYSICAL LAWS AFFECTING SIZE

D'Arcy Thompson found that for some observation he needed to combine this mathematical approach with the laws of physics. He wrote, "Shell and bone, leaf and flower, are so many portions of matter, and it is in obedience to the laws of physics that their particles have been moved, moulded, and conformed. My sole purpose is to correlate with mathematical statement and physical law certain of the simpler outward phenomena of organic growth and structure and form." And more specifically he felt that "in an organism, great, or small, it is not merely the nature of the motions of the living substance which we must interpret in terms of force, but also the conformation of the organism itself, whose performance of equilibrium is explained by the interaction or balance of forces, as described in statics." The "form of an object is a diagram of forces."

Thompson derived many of his methods of explanation from Galileo and Newton. It should be no surprise that he was especially influenced by the principle of similitude and extended it to include physical factors other than weight. He noted that the speed of an animal depends on the strength of the muscles that in turn is a function of the muscle's cross-sectional area or a squared power of a linear dimension, while the mass of the body depends on the third power. Thus with increase in size, the muscle cross-sectional diameter must increase proportionately larger than, say, the length or girth of the animal. Hence there would be an upper limit to the body weight that would be able to be moved, and the smaller animal is faster with respect to its length than the larger animal.

Also, walking involves the leg swinging like a pendulum with the amplitude of each step varying as the length of the leg and usually the height. But the time of the swing of the leg varies as the square root of its length. Therefore, the speed of walking, the amplitude per time of swing, will vary as the square root of the length of the legs; speed equals amplitude: $(\alpha L)/$time $(\alpha L)^{\frac{1}{2}} = \alpha''(L)^{\frac{1}{2}}$. The smaller the animal, the faster it walks in terms of its size. A fly walks roughly 3 inches in a $\frac{1}{2}$ second. If the fly be considered about $\frac{1}{4}$ inch tall, then according to the assumption above, man should be able to walk 3 inches per second $\times$ (70 inches/$\frac{1}{4}$ inch)$^{\frac{1}{2}}$, which turns out to be about $5\frac{1}{2}$ miles per hour. This is a little fast for man but close enough to support the relation.

Diffusion has even more applications to organisms than does weight. Both heat loss and gas exchange are diffusion phenomena and dependent on surface area, a square dimension. Heat production as a result of metabolism on the other hand is proportional to the mass of the organism, as almost all the tissues are involved, and varies as the third power. As there is a necessity for mammals to maintain a similar body temperature, the rate of metabolism would have to increase per unit tissue with a decrease in the size of the animal because the heat loss becomes proportionally larger along with the surface area.

D'Arcy Thompson also reminded us that there was a direct effect of size on the functions of certain organs. For example, the generation of sound or the sensitivity to vibrations depend on a vibrating structure and, with a decrease in size of the organism, the vibrating structure must also decrease. This would shift or limit the vibrations of this structure to those of a higher frequency. Also with smaller organisms, the smaller eyes would be less able to resolve two objects and suffer a loss of acuity because of the decrease in amount of light.

Thompson concluded from such considerations that there was an optimum-sized organism for a particular way of life, its niche. Each kind of habitat medium, water, land, or air, would allow a maximum-sized organism depending on its type of supporting structure. There would also be a minimum-sized organism for a particular composite of body functions and rate of metabolism. Thus he imagined that this necessary correlation of size with pattern of body functions would result in a group of discrete organisms that in turn would fit together into a complex pattern to form the ecological communities.

This idea was, by necessity, left vague because of the inability, as D'Arcy Thompson saw it, of being able to define as yet all the ways physical laws actually influenced all the structures and functions of an organism. He did feel that enough was known to imagine what life would be like on a planet where the force of gravity was double to that on earth. He felt that it would be impossible for the bipedal form to exist. Rather there would be only short and many legged creatures and even serpent-like methods would be common as a means of locomotion. Trees would be shorter with shorter branches. Less if any organisms could fly. On the other hand if one considered a planet that had less gravity than that of earth, the organisms could be much larger as they would require less energy, and, therefore, less of the body space would be required for respiration and circulation. These are but a few of the changes one could imagine with a change in only one physical factor, gravity.

ALLOMETRY

The relative disproportionate change of dimensions with change in size of the organism that led to the principle of similitude was studied and extended by Julian Huxley (1889–   ). Huxley was concerned with how to display the relative changes of such dimensions for analytical purposes. He found that if a log-log plot of any two dimensions is made, the resultant curve is usually a straight line. Hence the two dimensions Y and X can be related by the equation $Y = aX^b$, where a is a conversion constant and b, a relative power term. If logarithms are taken of both sides of the equation, $\log Y = \log a + b \log X$ results, the equation of a straight line with $\log a$ an intercept and b the slope of the line. If b is equal to 1.0, then the two dimensions vary proportionally to one another, and the organisms of different size are similar as far as these two parameters are concerned. This treatment of two parameters with variation of size of the organisms is known as *allometry*. When b is greater than 1.0, the correlation is said to follow a positive allometry; if less than 1.0, a negative allometry. It is remarkable how widely allometric relationships can be found. The comparative organisms can be from a variety of sources. They may even be from different species if the organisms have somewhat similar basic body plans.

Several examples of allometric relationships are shown in Figure 5–10. Note that pituitary weight, brain weight, and heart weight show negative allometry with respect to body weight, 0.70, 0.57, and 0.82, respectively. Such empirically derived results do strongly suggest that these organs are being influenced by physical factors analogously to the way legs are affected by gravity. The difficulty, however, is to determine what physical factors are operating. In the case of the pituitary, for example, with increase in weight one might imagine that less hormone is necessary or that the circulation system is more effective or that the pituitary is more efficient in production or excretion of the hormone, but it is virtually impossible to tell which of these are involved by this analysis. Perhaps with further comparisons of power factors b, classes of factors can be discriminated from one another. There is also uncertainty as to the meaning of the intercept $\log a$. The difficulty of interpreting the factors involved has discouraged the use of allometry. This tool almost demands that there be an attempt to work out its problem, for it is almost the only insight known that demonstrates the close correlation of form and function in a quantitative manner. It is striking proof of the interrelationship of parts within the organism and the existence of a self-regulating system working for the unity of the whole. It has all sorts of possibilities and should be extended to other levels of organization such as those found on a sub-cellular level. Hopefully allometry will not be forgotten.

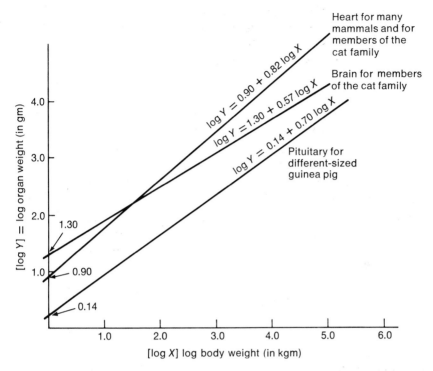

Figure 5-10. Allometric studies in mammals.

THE DETERMINATION OF SHAPE

Although variation of dimensions with size was important to D'Arcy Thompson, it was only part of a grander plan of the organism he needed to understand. Such variations demonstrate the significance of physical forces on the size but not on the specific shape of the organism or its parts. If there were cases of mathematical purity, such as the spiral, there must be other shapes that physical forces demand.

The search for physical forces that influence shape led him to surface tension. In droplets of water it is surface tension that becomes of gigantic relative importance and produces the spherical shape. To be sure there are disadvantages in being that small, for with less mass the organism will have less momentum and be more subject to the movements of its physical medium—witness Brownian motion. But there are also advantages in being spherical, for this shape would contain the maximum volume for the minimum surface area; hence there would be a minimal loss of diffusible substances per volume. Even more significant to D'Arcy Thompson was the fact that the spherical shape demonstrated by surface tension is a simple but good example of the "principle of minimal work (or least energy)." This principle recognizes that an equilibrium state of forces will exist and only exist where there is the most efficient force with a minimum cost of energy expended to a shape, in this case, the spherical shape of the organism. In a sense this is stating in the sophisticated and measurable way of physics what Aristotle and Galen and others meant when they stated that no part of an organism can exist unless it has a purpose to the organism and that it is the most efficient agent possible. In a sphere the forces of surface tension are all tangential and at a minimum since the surface area is at a minimum. Also these forces balance each other during the maintenance of the spherical shape. This principle usually involves surfaces and volumes with the surface area tending toward a minimum for a fixed volume.

Thompson pointed out many examples among living organisms that obeyed this principle. For example, there are many cases of spherical symmetry among the protista, e.g., bacteria, algae, and protozoa. When heterogeneous contents exist within microorganisms, they will affect its shape. The resulting shape will still be determined by the minimal work principle. A common irregularity is one that would tend to form an elongated structure, which would mean that the organism had a radial symmetry. Thompson was able to describe the elongated shape by the solutions of a general equation known as Plateau's surfaces of revolution. He found that examples, such as *Vorticella* and *Stentor*, fit these descriptions. Also ink drops, which must obey this principle, falling through water disperse themselves into shapes similar to the medusa of jellyfish as shown in Figure 5–11.

D'Arcy Thompson also used this principle of minimal work with tightly packed cells of tissue by comparing their shapes with those of soap bubbles. The soap bubbles tend to form 14-sided (hedra) figures in such a way that 3 of these faces meet in an edge with equal angles of 120° between the faces, while 4 edges meet at a point with coequal angles of 109° 28'. Studies of the pith cells in plants revealed that there is a mixture of different hedra cells, but most of them range between 12 and 16 with an average close to 14. It is interesting to note that variation also exists in the hedra of soup bubbles but not to such a great extent. He attributed this difference to the heterogeneous nature of plant cell contents.

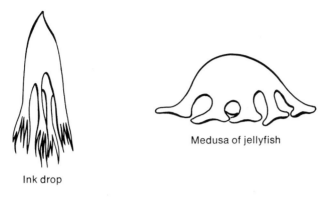

Medusa of jellyfish

Ink drop

Figure 5-11. Ink drops and jellyfish.

Thompson was likewise able to apply this principle to solve for the diameter and the angle of branching of capillaries. In his words, "The general principle, then, is that the form and arrangement of the blood vessels is such that the circulation proceeds with a minimum of effort, and with a minimum of wall surface, the latter condition leading to a minimum of friction and being therefore included in the first." One could imagine many other places where the principle applies to structures involving the flow of liquids or gases. He summarized his feelings about the application of this principle to these problems as he discussed the circulation system: "To prove that it is the very best of all possible modes of transport may be beyond our powers and beyond our needs; but to assume that it is perfectly economical is a sound working hypothesis. And by this working hypothesis we seek to understand the form and dimensions of this structure or that, in terms of the work which it has to do."

In addition to surfaces and volumes, the principle of minimum work can be applied to the forces of stress and strain. As an engineer balances tensions and compressions with structural elements of strength in line with these stresses, so D'Arcy Thompson was able to show that the trabeculae, the lattice work of calcium deposition within bone, align in the most efficient orientation to withstand the stresses exerted on the bone. He compared this internal structure of the bone with the cross structures in a hoisting crane, as shown in Figure 5–12. He also noted that if a bone is placed under abnormal stresses during its formation, the deposition of these trabeculae alters accordingly. Similar analogous changes have been noticed in the deposition of wood in a limb or stem placed under abnormal

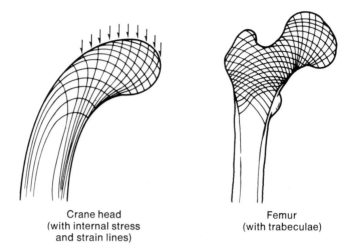

Crane head   Femur
(with internal stress   (with trabeculae)
and strain lines)

Figure 5-12. Cranes and bones.

stress. Here is one more proof of the necessary balance of internal with external forces in shaping the final form of either the living or nonliving structures.

CRYSTALLIZATION AND THE SHAPE OF LIVING ORGANISMS

In his search for the heterogeneous factors and their resulting internal forces, Thompson critically examined the contribution made by internal deposits such as crystals of salts and the spicules of sponges. But after much study he came to the conclusion that although there is specific shape found with these elements, they varied considerably and did not conform, for the most part, to the laws of crystallization. Writing about the depositions found in the radiolarians, he said, "all the more remarkable that we should meet with the whole five regular polyhedra, when we remember that, among the vast variety of crystalline forms known among minerals, the regular dodecahedron and icosahedron, simple as they are from the mathematical point of view, never occur. Not only do these never occur in crystallography, but it has been shown that they never occur owing to the fact that their indices (numbers expressing the relationship of the crystal faces to the three primary axes) involve an irrational quantity; where as it is a fundamental law of crystallography, involved in the whole theory of space-partitioning, that the indices of any and every face of a crystal are small whole numbers." If crystallization, one of the more obvious ways in which complex shapes can be built from piling small regular pieces together, cannot account for the overall shape of even the microscopic organisms, then it must not be a very significant factor in shaping any organism. He also concluded from this study that although heterogeneous depositions of submicroscopic structures play some role in shaping organisms, the main influence must be by general physical laws such as those discussed above and others yet to be understood in their application to forms and shape.

TRANSFORMATION OF COORDINATES— DIAGRAM OF FORCES

With this accumulation of studies D'Arcy Thompson was convinced that physical forces were not only involved in the shaping of the parts of an organism but also in balance with one another to form the whole organism. He wrote, "We may study them (muscle, bone, etc.) apart, but it is a concession to our weakness and to the narrow outlook of our minds. We see, dimly perhaps but yet with all assurance of conviction, that between muscle and bone there can be no change (in form) in the one but it is correlated with changes in the other; that through and through they are linked in indissoluble association; that they are only separate entities in this limited and subordinate sense, that they are parts of a whole which, when it loses its composite integrity, ceases to exist." Thompson saw that there must be a balance of form as Bernard saw that there must be a balance of function defining the unity of the organism.

D'Arcy Thompson's broad background gave him another tool to sketch in his necessary diagram of forces of the whole organism. He examined the grid line patterns used by da Vinci and Dürer and the Cartesian coordinate system developed by Descartes. He recalled the meaning of this latter tool to Descartes when he wrote, "He had in mind a very simple purpose; it was perhaps no more than to find a way of translating the form of a curve (as well as the position of a point) into numbers and into words." This is what he was looking for. As with the principle of similitude, he used this tool to compare the total form of one organism to another. A taxonomist compares a few specific characteristics. Thompson's diagrams could compare them all; all involved in shape, that is. He conceived of the idea of transforming the shape of the one organism into that of another by changing the grid patterns until cardinal points and the profiles of the two organisms conform and noting the pattern of changes that were required in the coordinates to accomplish this transformation. Figure 5-13 shows a simple transformation in which

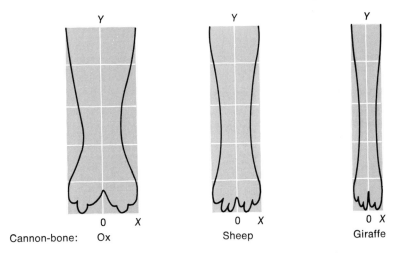

Figure 5-13. A simple Cartesian coordinate transformation.

the common bone of an ox is transformed into the corresponding bone of a sheep and a giraffe. First the relative size of the three bones is standardized by adjusting their heights to the same value. Then a similar Cartesian coordinate grid is drawn over each bone. Now the width of the ox bone display is shortened, grid spacing and bone together, until the outline of the ox bone coincides with that of either of the other two bones. If this is done, it is found that a perfect fit is produced with the sheep's bone if the ox system is narrowed to about two-thirds of its original size, or with that of the giraffe if it is narrowed to about one-third of its original size.

There are of course other more common types of transformations possible. Some involve a change in the coordinate system according to a specific function such as that of a logarithm. The more common is one with distortions to the lines and even the shear or twisting of the lines with respect to one another as shown in Figure 5-14. If the form being studied is radially symmetrical, then radial coordinates are useful. A special case

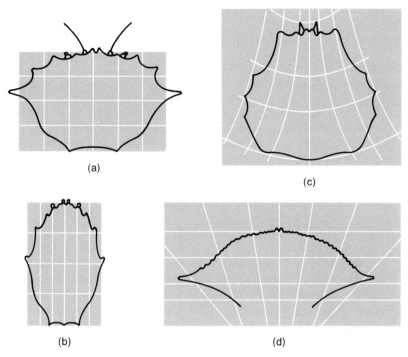

(a)

(b)

(c)

(d)

Figure 5–14. Common coordinate transformations.

of interest is the transformation of a conical shell into a spiral-shaped form very much like *Nautilus* by merely changing a straight line radial into a logarithmic spiral coordinate system.

D'Arcy Thompson also used transformation to predict intermediate forms. For example, as shown in Figure 5–15, the head of the fossil horse of the Eocene can be transformed into present-day *Equus*, with the complex change in the grid pattern subdivided into even steps in such a way that intermediate forms of the heads of horses should be indicated. This interesting idea was supported almost exactly for the case of *Mesohippus* and *Protohippus*, as indicated. *Parahippus* is an example, however, that does not fit. The problem very likely is that the changes that occurred during evolution were not orthogonal, i.e., varying in a constant linear fashion. There would be expected to be a series of changes in the direction and rate of the coordinate system during this evolutionary transformation. There is also always the possibility that some of the fossil intermediates were branching off the main line of evolution between the Eocene ancestor and *Equus*. Perhaps it is more remarkable that any

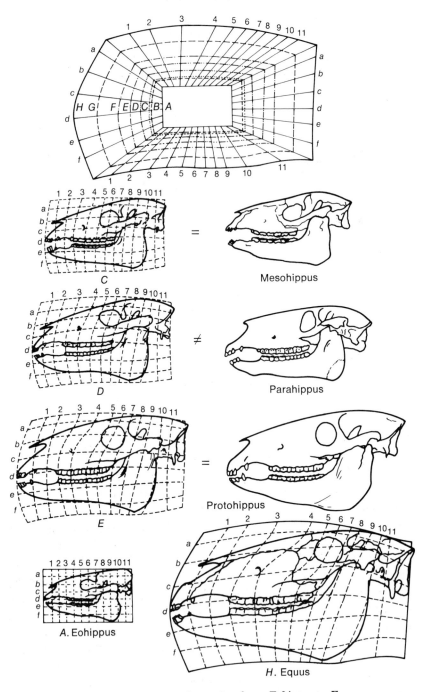

Figure 5–15. The transformation from *Eohippus* to *Equus*.

intermediate fossil fits this hypothetical product. Probably the main reason for the two that do fit is that of all the characteristics the head change is the one that was most linear.

D'Arcy Thompson became quite convinced from his studies of transformations that the whole organism is involved in the change. That is, there are never any sharp breaks in the grid patterns due to the change of only one characteristic of the organism. Rather, all adjacent lines in the grid are affected in a gradual manner extending across the grid. Moreover, the more closely related the organisms were to each other, the more integral the transformation was. If the two organisms were un-related, the transformation was either almost impossible or quite illogical. For this reason, Thompson emphasized that the ease of transformation between species should be used as a measure of their relatedness. This idea has not been followed up because it is difficult to describe the changes in the coordinate systems during a transformation, but interestingly enough this is almost exactly what some taxonomists are now doing by scanning the outlines of the organisms and feeding these data into computers and comparing them with similar scans of other organisms. The fact that transformations do involve the whole organism is one of the impressive proofs of the complex coordinated balance of forces that exist throughout the organism.

D'Arcy Thompson was just beginning to use the transformation of coordinates tool. He imagined that it could be used in other ways, ways that we would do well to think about. It could be extended to include a third dimension in order for volume transformations to be considered. Also possibly dynamic changes of form may be followed with time as one of the coordinates.

COMPARATIVE ANATOMY AND PHYSIOLOGY—CONTRIBUTIONS TO EVOLUTION THEORY

At first it might seem strange that the search for unity that produced a Bernard and a Thompson did not also find the idea of evolution. This is not so difficult to understand, however, when we realize that from the time of Aristotle and Galen, when organisms were compared, it was to fortify an understanding of the organism. The search was always for commonness and not for the differences between species. The verte-brate body plan was too similar on the one hand, and those organs that were different were too different with no apparent sequence in their structures from one organism to the next to suggest any ordered change being responsible for their differences. Harvey encouraged this comparative method of study for the same end.

Those who studied form were even more drawn inward to think of the details of shape and correlations within. Goethe came the closest to discovering the idea of evolution because of his method of arranging organisms or the parts of organisms into sequences based on similarities of form. He also considered that changes in form could exist easily. Goethe certainly conceived of the transformation of one form into another, but there was no thought that one actually arose from the other with the slow pace of geological time. His purpose of putting organisms in a sequence was to find an ideal common body plan. He was reinforced in this method of comparison by finding that if some organisms had certain parts, then others also had them. In spite of the disagreement by leading anatomists, Goethe concluded that there must be an intermaxillary bone (bone supporting the incisors) in the upper jaw of man since it is present in all other mammals. And search he did, and he not only found it in humans but also in birds, amphibians, fish, and reptiles (turtle). He could say this without any thought of evolution. It was not until Lamarck suggested the idea to him that he considered it. He quickly accepted it and was a strong supporter of Lamarck, but as Goethe's ideas were generally too philosophical or new, he too was for the most part ignored.

It is perhaps more surprising that the great Cuvier, who was not only the most comprehensive comparative anatomist of his day but had also added fossils to his studies, did not conceive of evolution. But Cuvier was also concentrating on the organism. He considered that a small part of a fossil organism could be used to reconstruct the whole. No greater faith in the unity of the organism can be demonstrated. Apparently a search for this unity was too strong and appealing and full of evidence and, if anything, spoke to the consistency of the species and opposed any thoughts of change in that organism. Why would it change when it already has the most efficient combination of parts possible?

Even after the naturalist skillfully supplied evidence that a change of the species had occurred and reminded the anatomist that his gaze was too narrow and inward, it was most difficult for this latter group to accept evolution. For one thing they did not want to be told that they were ignoring the forest for the trees, especially when the tree was magnificent as it stood.

D'Arcy Thompson, for example, could never feel content with the notion of natural selection as a mechanism for evolution. If form is determined by physical forces and physical law is inherent in the arrangement of all matter, it is difficult to see the need for natural selection in the shaping of organisms. Also heredity is important, but if it acts in discrete ways, e.g., tall versus short plants and brown versus blue eyes, this does not agree with the integrated changes that occur to the whole organism as revealed by studies of the transformation of Cartesian coordinates. "Characters which we have differentiated by heredity studies insist on integrating themselves again; and aspects of the organism are seen to be conjoined which only our mental analysis had put asunder. The co-ordinate diagram throws into relief the integral solidarity of the organism, and enables us to see how simple a certain kind of correlation is which had been apt to seem a subtle and a complex thing." This criticism was an important one to make, for it did cause the evolutionist to think a little harder.

Another and far more serious concern, however, that Thompson has highlighted was that involving the source of form. For it has become assumed that the organism is built by the hereditary apparatus which makes the pieces which in turn fit together naturally much as a crystal grows. But his studies revealed that crystallization could not be playing a significant role in form production. Heredity is important he admitted but he insisted on reminding us that we could not ignore the physical laws of matter that affect the form of the parts and the whole of an organism.

The long delay in converting the anatomist to accept evolution seemed to cause a greater swing with more devotion once it did occur. Anatomists have figuratively married evolutionists by directing their entire purpose toward supporting the existence of evolution. Before this switch, similar structures within different organisms, such as Goethe's intermaxillary bone, were called *homologues* and were used to indicate a similarity of function for the different organisms. Now, however, a homologue is a structure in one organism that can be compared with a structure in a different species both believed to have been derived from a common structure found in an evolutionary ancestor. Now it can be said that the goal of comparative anatomy, plant or animal, is to find homologues.

It can be said that, in general, comparative anatomy and physiology lend support for evolution and help add a logic for deciding what some of the pathways of evolution might have been. They do not add proof of its existence or give any further insight into the mechanisms involved in evolutionary change. One of the interesting ideas that comparative physiology has suggested, however, is that the choice of variant organism during the selection process is the one that makes the most efficient use of the energy from its surrounding environment.

EARLY CONCEPTS OF THE DEVELOPMENT OF
FORM—PREFORMATION VERSUS EPIGENESIS

In any search for the unity of the organism and the relationship of form and function, it is necessary to think of when and how these forms and functions came into being. The Greeks, for the most part, did not separate existence from the process of coming into existence. Aristotle's heart was not only pulsing vitality for the continued function of the organism, but it was supplying vitality for the growth of the organism as if these two phenomena were one. It was popular to study this growth by observing egg development. When it came to an explanation of this remarkable process, most agreed with that of Democritus (*ca.* 460 B.C.) and Empedocles when they wrote, "The chick is formed before the hen broods upon it." This seemed the most logical conclusion, that the future animal is preformed. But there was not absolute agreement on the point. Aristotle happened to gain much of his philosophy regarding form and matter from his empirical studies of the egg. His descriptions of the developmental processes were so complete that it has been said that "almost two thousand years were to roll by before it was to be equaled or surpassed." He noticed that the heart of the chick was the first organ to appear and move and thus live. It was because he first saw blood appear in the heart of the chick that he assigned the blood's origin to the heart and naturally conceived of the heart as the vital producing and radiating center of the body. Such careful observations convinced Aristotle that the preformationists were wrong. He found no such thing as a preformed chick. Rather he assumed that form was generated—or more correctly—generating matter into the organism. The active principle of form came from the male parent and shaped female matter. This he assumed proceeded by progressive steps; once the heart was formed, it in turn would shape organ *B*, then organ *B* would fashion organ *C*, etc., as an epigenetic process.

Thus started two completely different explanations of the developmental process of an organism that has in one form or another dominated all explanations of this process ever since. Da Vinci felt that the development of an embryo was one of the most beautiful of all processes of life and far ahead of his time described in fine details the uterus, placenta, and the processes of birth of a human. He considered that the limbs were already present and only needed the "soul of the mother" to pass by way of the umbilical cord to complete the form and awaken the soul of the child. Fabricius wrote the first systematic treatment of chick development and compared its development with that of fish and mammals, describing the fetal membranes with great accuracy. He did much to encourage preformationism by drawing the 3- and 4-day old chick far too advanced for its age.

Harvey devoted more time to the study of development than he did to the circulation system. In fact he used the former to understand the latter. He summarized a vast amount of his work in *On Generation in Animals* (1651). He supported many of Aristotle's observations on the chick egg. He emphasized most strongly that the egg and womb are analogous and are the necessary medium within which all animals need to develop, "ex ovo omnia." He wrote, "Like a potter, first she divides her materials, and she allots to the Trunk, the Head, and the Limbs, every one their share or cantlin: as Painters do, who first draw the Lineaments, and then lay on the Coulors." Harvey was ambiguous or perhaps broad in his thoughts, however, when he also wrote, "the manner of the generation of Animals is diverse," and some were "made intire at once." Thus his writings could be used to support either preformationists or epigeneticists.

One would think that with the advent of the microscope the decision as to mechanism of development would be settled, but apparently it only helped convince oneself of one's own convictions and further polarized investigators into the two opposing camps. Malpighi unwittingly supported the preformationists by drawing rather well-formed chicks as a result of his early observations of the egg. It was about this time that Bonnet discovered that aphids developed parthenogenetically, i.e., without the need of the male organism, and he wrote of Malpighi's observations, "(They are) one of the greatest triumphs of rational over sensual conviction." Almost everyone agreed including Descartes, Grew, Leibnitz, and Buffon. It was also about this time that metamorphosis in insects was described, the type where a small form gives rise to an almost identical larger form in steps—a rather convincing argument in favor of preformation.

Preformationists were for the most part ovists who claimed the preformed organism is within the egg. Confusion resulted, however, when Leeuwenhoek discovered sperm under the microscope in 1677 and maintained that it was necessary for this sperm to enter the egg before it could become fertilized. And what was even more, he claimed to have seen an animiculae, or microscopic organism, in the head of the sperm. It had been known that male semen was necessary for the development of the egg but it had been thought that semen played a protective role or stimulated the egg to grow. Some preformationists felt that the male must play a bigger role in the fertilization process and had hypothesized that a preformed human, a homunculi, was present in the semen and entered the egg and was the future organism. Naturally Leeuwenhoek's discovery was hailed as a successful fulfillment of that prediction and a school of animiculists formed in opposition to the ovists. Buffon was in support of this claim and suggested that the preformed organism was actually a "living organic molecule" (the sperm was the molecule) which could multiply in semen many times to account for the large number of sperm which were found. Between 1694–1699 several other microscopists reaffirmed Leeuwenhoek's observations and drew even more extensive drawings of the homunculi, usually with a large head.

It was Lazare Spallanzani (1729–1799) who first tried to settle the question by experiments. He had managed to artificially induce fertilization in silkworms, amphibians, and dogs by supplying semen. He first tried to clarify whether it was semen or vapors which may arise from the semen which was responsible for the fertilization. The vapor hypothesis was being put forth by physicians who claimed that no semen was reaching the ovaries of some of their pregnant patients. He found no support for this hypothesis and he was unable to induce fertilization by the evaporated steams from seminal fluids of either toads or frogs. He next tried diluting the semen with a "very large quantity of water" and found that this diluted semen could still induce fertilization of the amphibian eggs. This fact plus the inability to find any homunculi in the sperm heads convinced him that the preformed organism could not be associated with the semen or the sperm and that it must be found in the egg. He showed that semen was necessary for the egg to develop by reducing the ability to fertilize an egg with seminal fluid that had been filtered through several "blotting papers," and yet finding that the substance filtered out of solution was able to aid the development. Spallanzani's conclusion did much to help the ovists' cause. This is of interest because about 50 years after these experiments were carried out they were essentially repeated yielding the same results but were interpreted almost exactly oppositely and this conclusion was also accepted by many.

DYNAMIC MORPHOLOGY EMERGES—
WOLFF AND GOETHE

Perhaps because of the internal strife, preformationism dominated the explanation of the developmental process even more than ever. Only one voice seemed to object to the explanation and this was not until 1759, a good hundred years after the microscope was in use. It was in this year that Caspar F. Wolff (1738–1794) published his *Theoria Generationis*, which was primarily aimed at denouncing preformationism. Wolff thought in terms of the flux and flow of matter with time. These thoughts became of significance to him when he observed the way that specialized structures, the leaves and flowers of plants, were derived from the homogeneous material of the stem. He concluded that there must be a form-generating process at work in living matter. He supported this contention by rediscovering the observations of Aristotle and Harvey that the blood vessels of the chick were not present in the early stages of the chick embryo. With these observations in hand he set out to bombast preformationism. Most observers were unaffected by his writings, however, mainly because it was inconceivable how something as vague as a form-generating principle could work. Also, Wolff could offer no logical explanation in place of preformationism. Under the influence of Newton, most required mechanical models for their explanations and no machine can be changing its shape and still function. The only other objections to preformationism were the mild comments of a few who considered that the first rabbit must have been gigantic to contain all the preformed rabbits that have resulted ever since.

It was not until later, after more philosophers had written about change and minds began to relax about mechanical explanations, that the creative energies of a Goethe could be realized and develop a concept of dynamic morphology. He too was stimulated by observations of plants. He noticed that buttercup leaves grown under water take on quite different shapes than those grown in air. As he watched the palmetto leaf pass through its different forms at different stages of development, he concluded that it is the leaf and not the whole plant that is the basic unit of all plant forms. He wrote, "All is leaf. This simplicity makes possible the greatest diversity." His imagination allowed him to transform the leaf into a variety of different forms "from node to node." The flower was a repetition of these growing leaf units with the nodes close together, thus forming calyx, sepals, petals, stamens, the nectar parts, pistil, and even the fruit and seed capsule, each a burst of transformed leaves. Goethe considered that there was a "cycle of growth" within which the end products of the repetitive growing units varied in size and expression in a definite pattern. At first there would be an expansion phase starting from the earliest leaves, the cotyledons, followed by the other leaves increasing in size as they are formed. This phase continues until a flower is produced that begins by a contraction phase as the calyx is formed. This is followed by an expansion for the corolla (the sepals and petals), a contraction for the stamens and pistil and a final expansion for the fruit. Goethe had arrived at his conception of growth independently of Wolff, but when he heard of the latter's work in about 1807, he found himself in complete agreement except for minor points such as Wolff's contention that a flower is a degenerate leaf. As with Wolff, preformationism was out of the question in Goethe's thought. Any observations of the continual growth of plants with no apparent adult end point would discourage one from such a concept.

Goethe also did experiments to try to understand which factors influenced these transformation processes. He exposed the plants to different environmental conditions. He cut the plant into pieces and noted the changes in the leaves as the cuttings regenerated their lost appendages. He found that he could prevent flowering if he overfed the plant. He was in search of a mechanism for the changes but could only conclude that diffusible plant juices were responsible and that these required or acted on metabolic activities somehow.

Goethe's descriptions and ideas were considered too philosophical by most. His writings did have an influence on a capable observer who was respected by all, however, mainly Karl von Baer (1792–1876). Von Baer studied the details of the development of some organisms with large eggs such as those of the amphibians. His descriptions were of the finest in detail and they caused many to reevaluate the preformationists' explanation of development. He summarized his finding and general conclusions in 1828, the latter becoming known as the "laws of von Baer." Two of these laws that had been influenced by Wolff and Goethe and were a direct attack against preformationism were (1) in development from the egg, the general characters appear before the special characters, and (2) from the more general characters the less general and finally the special characters are developed.

MOSAIC VERSUS REGULATIVE DEVELOPMENT— ROUX, DREISCH, AND SPEMANN

It was not until 1888 that the technical skill became available and, more importantly, the embryologists were beginning to ask questions about the developmental process that could be answered by experiment. The birth of experimental embryology can be said to have arisen as a result of the experiments and thoughts of Wilhelm Roux (1850–1924). Roux's approach was to divide the developmental process into separate but still complex functional processes and then attempt to analyze these in terms of known physical principles. Roux considered that the age-old problem of preformation versus epigenesis was one of establishing whether the parts of the embryo are completely self-differentiating or need to interact with one another in order for differentiation to occur. This clarity of defining the problem enabled Roux to devise experiments to answer the issue. He wrote, "(The problem is) to determine whether, and if so how far, the fertilized egg is able to develop independently as a whole and its individual parts. Or whether, on the contrary, normal development can take place only through direct formative influences of the environment on the fertilized egg or through the differentiating interaction of the parts of the egg separated from one another by cleavage."

At first Roux attempted to alter the environmental conditions around amphibian eggs. He found as others had that abnormal raising of the temperature produced monsters, and yet heat and oxygen were necessary for the process. He then placed the eggs in a rotating system in which he could cylically alter the influence of gravity, heat, light, and even magnetic fields. He found that normal development resulted, suggesting that the eggs "do not need any formative influence by such external agencies." He also skillfully separated the neural tube from the digestive tube in the early embryo and found that they both could continue to develop normally, thus independently. Hence, there was independent development of the parts in the embryo, but he realized he had not solved the main problem. There remained questions as to whether or not these parts existed in the egg as independent units, and how dependent development was on the internal environment of the egg.

Roux devised a method of puncturing and killing with a hot needle one of the blastomeres that result from the first cleavage. Although the results were inconsistent, in a fair number of his experiments he obtained half embryos at blastula, gastrula, and even later stages of development.

In the advanced cases, he found smaller archenterons, half-sized blasto-coeles and notochords, and incomplete neural tubes. He concluded, "In general we can infer from these results that each of the two first blasto-meres is able to develop independently of the other and therefore does develop independently under normal conditions. Each of these blasto-meres contains therefore not only the formative substance for a corre-sponding part of the embryo but also the differentiating and formative forces. We can say: cleavage divides qualitatively that part of embryonic material that is responsible for the direct development of the individual by the arrangement of the various separated materials which takes place at that time, and it determines simultaneously the position of the later differentiated organs of the embryo." Thus it was that Roux's results supported a refined extention of preformationism that he called *mosaic development*. Roux considered that the main question that he was never able to solve was the nature of the "formative forces" existing in each part of the mosaic.

Roux's results were soon challenged by the work of Hans Dreisch (1867–1941) who first published in 1892. Dreisch used sea urchin eggs, which are also quite large and easily available. He separated the early products of cleavage by shaking the cleaved eggs or placing them in calcium-free water. From his results with sea urchins he wrote, "It cleaves as if for half-formation (or as he showed later, even with one of the blastomeres from the eight-cell stage), but forms a whole individual of half-size." He thus concluded that Roux's theory "at least in its general form," must be discarded. This ability of the part to reform and produce the whole was called *regulative development* and replaced to some extent the earlier description of the epigenetic process. But the new description proved even more difficult to explain. Even Dreisch, who was a con-firmed mechanist and had no difficulty imagining all sorts of mechanical models for living organisms, could not conceive of an explanation for regu-lative development. How could a part produce the whole as if a totipotency existed uniformly throughout the egg? He also found that the fate of any cell within these early embryos is a function of its position within the embryo, as if there was some mysterious field of influence that deter-mined the course of development independent of the living matter per se. No machine has the ability to be divided into two parts and have the parts reconstruct the missing pieces nor can a machine have its parts rearranged and somehow alter these parts to function as the ones that were in their original places. These results led Dreisch to only one conclusion and one that was against his own better judgment. It was that there must be some very unique property of living matter, a vital principle, which can never be explained by physical laws. His results had such an impact upon himself that he gave up his experimental studies and became an active professor of philosophy effectively promulgating vitalism, the existence of metaphysical forces in living matter.

The replacement of preformation by mosaic development and epigenesis by regulative development produced an increasing interest in experimental embryology in the ensuing years. Many workers began to trace individual blastomere products of cleavage in order to determine their fate. It was found that for many invertebrates, such as the earthworms, flatworms, and mollusks, there were very similar homologous patterns of cleavage with the early blastomeres resulting in predictable end product organs or tissues, suggesting a mosaic-type development. On the other hand Hans Spemann (1869–1941) showed in 1901–1903 that if the first two blastomeres of the amphibian newt egg are separated by a loop of baby's hair, two whole embryos may result. This refutation of Roux's early work was supported by finding that the dead remains of the pierced blastomere still attached to the other blastomere influenced its development process and produced the apparent half-embryo. Spemann also noticed, however, that if the hair loop was not in line with the first cleavage plane, only one of the two blastomeres survived. The successful half was found to be the one that contained most of the grey crescent. This was true even if the product was much smaller in size than the other part that did not contain as much of the grey crescent. He then hypothesized that the grey crescent was involved in the regulative process, or organizing process as he preferred to call it.

By 1918 Spemann found that if he transplanted small pieces of ectoderm from the early gastrula of newt to other regions of another similar gastrula, the ectoderm graft is accepted and it becomes incorporated into the embryo as if it were there from the beginning. Here was another case of the cell's fate being determined by its new position. If he transplanted the grey crescent from an early gastrula to another gastrula or even blastula at an abnormal region, however, the tissue surrounding the transplant and the transplant itself differentiated into neural tube, notochord, and somites, in addition to those tissues involved with the host's normal grey crescent. His conclusion was that this organizer tissue, the grey crescent, probably differentiated before the other tissue. After it invaginates and comes into association with other tissues, it induces these tissues to differentiate into neural tissue.

This meant that in a sense both mosaic and regulative types of development seem to be present at different times within separate tissues of the same embryo. The organizer is able to self-regulate after the first cleavage, as indicated by its ability to become two embryos when the grey crescent is divided equally in half. On the other hand, this tissue becomes self-differentiating or "determined" by the gastrula stage. The remaining tissue is however still regulative with a potency to become other than its normal fate until the organizer induces it to self-differentiate (become determined). The sea urchin seems to be an organism with a delayed time of determinancy, while many of the mollusks and similar invertebrates may be determined before the first cleavage. Also, from plant studies, equivalent processes of determinancy (mosaic) and indeterminancy (regulative) development are found, although both of these are occurring at the same time in different parts of the plant throughout the life of the plant. A summary of the history of these opposing mechanisms is diagrammed in Figure 5–16.

To discover that these two types of development occur in the same organism does not explain their differences, i.e., details of the processes. What is regulation? What happens during the embryonic induction process to cause determinancy? More will be said about these later within a different context (order), but the point is that the mosaic-regulatory controversy remains.

We do not have to know all about the details of these mechanisms, however, to think about how the form of the whole organism develops. The controversy reminds us that the parts do interact in some complex manner to produce the whole during development. Also, these mechanisms may lead to explanations of other developmental processes. There is the process of regeneration, for example, where a part of an organ, or an appendage, or, in some organisms, a whole major portion of the organism is restored by the remaining parts of the organism when these lost parts are separated. This restoration process can be thought of as a breakdown of the determinancy with some kind of regulatory process generating the lost parts. This occurs to some degree in all multicellular organisms and even in some unicellular ones. No phenomenon attests to the continued integration of parts of the organism to maintain the whole as does regeneration.

There have been other styles of thinking applied to the development of an organism that follow the D'Arcy Thompson approach of looking at characteristics of the whole developing organism in order to say something about the mechanisms involved. Curiously enough D'Arcy

(a) Preformationism

(b) Epigenesis

i_a. Preexistence of organism in egg.
{Empedocles and Democritus}

i_b. Final shape created by form gradually.
{Aristotle}

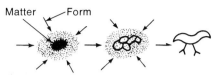

ii_a. Ovists versus Aiminculalists.
{Malpighi, Leeuwenhoek}

ii_b. Dynamic morphologists.
{Wolfe, Goethe}

iii_a. Mosaic development.
{Roux}

iii_b. Regulative development
{Dreisch}

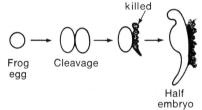

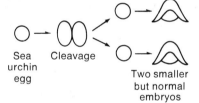

Figure 5-16. Models of the mechanisms of development.

Thompson did not apply his technique of the transformation of co-ordinates to growth studies. This and allometric correlation studies have been applied to growth by J. S. Huxley and many others since. Allometry is especially useful as there is a considerable change in size with age. It is quite easy to find log-log relationships of sets of dimensions with growth, for example, the growth of different gourds as shown in Figure 5-17. There have also been elaborate theoretical field theories devised to explain patterns of differentiation occurring in different parts of the organism. These usually depend on the presence of hypothetical diffusible agents, which form gradients across different regions of the organism with different tissues resulting, depending on the relative concentrations of the agents at a particular region. John T. Bonner who has surveyed these

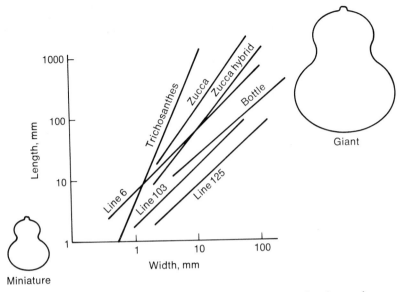

Figure 5-17. Allometric correlations in the growth of gourds.

whole-organism studies and theories and has carried out a number of his own on all sorts of organisms has concluded that "with increasing sizes of organisms, those parts (and their associated processes) which vary as the square of the linear dimensions will show corresponding increases in their division of labor (kinds of differentiated cells)." In other words those organs involving functions of diffusion, linear motion, or integration (nerves) must differentiate with an increase in number of kinds of cell types proportional to the square root of their linear dimension. This conclusion, whether right or wrong, does not reveal the mechanisms at work but it may suggest that there is a simple process that determines differentiation and that this is regulated by the whole organism. Bonner goes on to say, if we knew how this regulation was achieved, "then we might not be so bothered by ... we might even explain ... the wholeness of the organism."

DEVELOPMENT AND EVOLUTION

If we look at the historical relationship between the concepts development and evolution, it is essentially identical with that of comparative anatomy and evolution. Several early anatomists, such as Fabricius, compared embryos in the same way they did adult forms and the similarities in structures to confirm their definition of the organism. This approach also seemed to oppose any consideration of a change from one organism to another that might have suggested that evolution had occurred.

Harvey did introduce a unique idea that may have led to the concept of evolution. He wrote, "Nature, by steps which are the same in the formation of any animal whatsoever, goes through the forms of all animals, as I might say egg, worm, embryo, and gradually acquires perfection with each step." This might have suggested a transformation from one animal to another as any one animal develops but Harvey did not indicate such a thought. "The forms of all animals" apparently referred to general shapes rather than an exact appearance of the adults of the different animals. With Goethe, development was one other form of "metamorphosis" and did not suggest evolution anymore than did his comparison of species.

Even after Lamarck had introduced the idea of evolution, others wrote about development in a way that could suggest a link between the two, but no association was made. For example, J. F. Meckel (1761–1833) and E. R. A. Serres (1787–1868) rediscovered Harvey's idea and added more specificity to it when they wrote in 1812, "Man only becomes man after traversing transitional organizatory states which assimilate him first to fish, then to reptiles, then to birds and mammals." This definitely implied that, at least for the vertebrates, the more complex organisms pass through the other vertebrates that are less complex as stages of development. This concept became known as the "law of parallelism" and is shown diagrammatically in Figure 5–18. Von Baer also noticed the similarity of embryos of the vertebrates but disagreed with the law of parallelism and included another (see page 172) "law of von Baer" on the subject when he wrote in 1828, "During its development as animal departs more and more from the form of other animals, and the young stages in the development of an organism are not like the adult stages of other animals lower down on the scale, but are like the young stages of those animals." By scale, von Baer was referring to a scale of complexity and not evolution. His statement is also diagrammed as shown.

Apparently the ideas above were unknown to Darwin nor did comparative embryology play any role in the development of his ideas on evolution until after his return from the long voyage. In his search for support for evolution, he was influenced by von Baer's laws in establishing an affinity between species. He was also aware of and used the "law of parallelism"

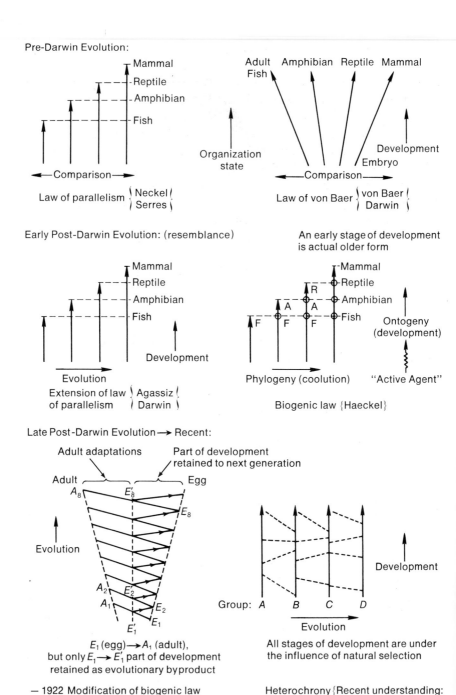

Figure 5-18. Diagrammatic relationship of development and evolution.

in support of evolution as he wrote, "Extinct and ancient animals resemble to a certain extent the embryos of the more recent animals belonging to the same classes." Thus Darwin introduced geological time and completed the analogy between development and evolution. This notion was appealing to Darwin as it supported his understanding of adaptive radiations wherein new structures that arose during the radiation came from one basic or common type.

The widespread acceptance of evolution including the statement above by Darwin encouraged other workers to compare embryos as supportive evidence of evolution. Among these the one most noted was Ernst Haechel (1834–1919) who further extended the relationship between development and evolution with most positive and explicit statements in 1866 and again in 1875. He was convinced that adult stages of ancestors are repeated by descendents and exist as earlier stages of development. The similarities others noted and expressed in the law of parallelism were not merely similarities but were absolute fact, and he demanded this in his remembered phrase "ontogeny recapitulates phylogeny." He offered as evidence the presence of "gill slits" in the embryos of the chick and mammal indicating the adult stage of the fish. He also contended that all metazoans, multicellular animals, pass through a gastrula stage whose ancestor he called the "gastraea" and can be found represented by the coelenterates. Haeckel's theory, the "biogenic law," stated that the earlier stages of development *must* be passed through in order for further phylogenetic changes to be able to take place. Any evolutionary novelty must occur as an addition onto the adult form of the ancestor as an additional developmental stage. But what is more important, the additional stage of development, hence evolution, is caused by and influenced in its direction of change by the preceding stages of development. Thus Haeckel supplied the force that caused evolution by emphasizing that the development process is this force and the significant factor in the expression of living forms. Evolution is merely a historical record of the complex energies and expressions of the developmental process. If we could understand the forces of development, we could understand the diverse ways of evolution. He was not denying the mechanism of natural selection but considered it only one side of the process of shaping living matter. Natural selection was helping to shape from the outside while development was shaping and pushing from within the living matter. It is understandable how this "law" was accepted by many. It not only fell in the wake of evolution, which it supported, but it had the same inherent beauty and simplicity of synthesis that the concept of evolution had by encompassing not only all the facts of evolution but all the facts of development in one statement. It had a tremendous influence on sending many workers back into the laboratory to study and compare the details of development in many organisms and among other things paved the way for the experimental embryologist.

There were some difficulties with this law that Haeckel recognized. For example, the membranes associated with the fetus of mammals are not found in the egg-laying vertebrates and vice versa. Haeckel allowed that there could be "inserted adaptations" that were necessary for an organism to fit its particular habitat and that these are not retained in the basic developmental record to be used for the next developmental stage—the next phylogenetic product. These are unstable adaptive conveniences not required for a further developed structure to occur. This leeway permitted the biogenic law to be more easily accepted and many example observations were put forth in its support. These were not only found for animals but also in plants. It was noted for example that the fern frond passes through stages of development similar to the ancestral-like plants, the *Psilotum*, and even the development of the xylem in *Lycopodium* or vessel elements in flowering plants seem to reflect their phylogenetic sequence. Also the steler patterns of many seed plants develop from a protostele to a siphonostele and finally to the dictyostele pattern. Even G. Bernard Shaw was enthralled by the idea and wrote, "During its life history an animal climbs up its family tree."

There were however many objections to the biogenic law. Thomas Huxley was quick to assert that the fossils of most invertebrates did not resemble the embryonic stages of development of any living invertebrates that he was aware of, e.g., the trilobitates and brachiopods, for example. Insect embryologists noted, for instance, that different adult species of *Peripatus* cannot be distinguished; yet their embryos are quite unique and used for classification purposes. As more was learned about the beginning stages of development, it became clear that cleavage patterns may vary as they do between chick and amphibian. The "gill slits" that Haeckel had used as an example were not gill slits in structure of function in the birds and mammals.

Various attempts were made to modify the biogenic law to fit all the exceptions. One of the most successful, diagrammed in Figure 5–18, was offered as late as 1922. It gradually became clear, however, that not the biogenic law, the law of parallelism, nor even those laws of von Baer that related to evolution could apply to all observations. Rather, it appears on further consideration that any stage of development is subject to the same kinds of modifications during evolution as are the adult stages.

This property, known as *heterochrony*, was most clearly demonstrated by a number of examples in which the descendent adult structure resembles more closely an early stage of development of an ancestor rather than the reverse, a special case of heterochrony known as *neotony*. For example, Ctenophores are neotonous-like *Turbellaria*, and the angle of man's head is closer to that of the embryo of the ape than to that of the adult ape's head.

It can be said that Haeckel's biogenic law is not correct but Haeckel served a great service to the cause of evolution by stimulating the students of evolution to examine more than the adult form and above all to make it clear that the forces of evolution are acting on a process and not on some static structure. Apparently it is the process of development that is evolving.

Although heterochrony must be kept in mind, it is still possible to compare embryos of different species, and in cases where there are a number of stages of development in common, there probably is a close phylogenetic relationship. A comparison of the many stages of development offers many more points of reference or homologues to aid in hypothesizing the pathways of evolution. This method has been most helpful in supplying a logic for the hypothetical phylogenetic ancestors of the vertebrates (the echinoderm-protochordate hypothesis), the hypothetical phylogenetic relationship of the invertebrates, and for the hypothetical pathway of many of the Protista (e.g., the flagellates, ciliates, sarcodines, fungi, and especially the algae—this last group being exampled in Figure 5–19). This same logic has also led to the conclusion that all

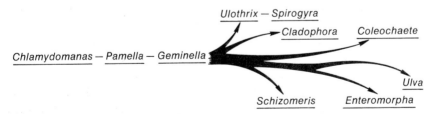

Figure 5-19. Hypothetical scheme of green algae evolution.

organisms evolved from a single cell. These schemes are so debatable that it is far from known whether there were many different single cell types, most of which became extinct but some branched later—the grass model, or whether all forms came from one single cell type—the bush model. In any respect, comparative studies of anatomy, physiology, and embryology, although they add no proof of evolution as a historical explanation of diverse life forms, on the other hand do not deny it and add additional circumstantial evidence to its existence.

THOUGHTS

Here we have an obvious order that demands an explanation. What is clear is that this order is a package that is right in front of us or is us. We can see it, feel it, measure it, watch it change, dissect it, and do things to it. Definitely those who are most successful at this search need to see directly and work directly with hands. This seeing-feeling approach allows on the one hand a rich source of basic assumptions, a source of complex details, and a variety of rapid succession experiments; but it may also tend to confine the resulting models or explanations. One should also reverse the implication of this statement by adding that within this confinement there is a healthy vigor to the development of concepts and models resulting in deep-meaning, highly predictable, and useful explanations, i.e., good ones. Basic assumptions such as symmetry, harmony and rhythm of parts, structure-function unity, synergism of parts, and purpose-maximum efficiency of functions came easily with organisms and were early channeled by observations related to heat, ordered or directed motion, blood, beating heart, need for air and breathing, and of course nutrients. Mechanical-like descriptions, though many in kind, are the types of models that seem to emerge and remain with the inputs above. Even the spirits of the ancients have a mechanical elemental base with the spirits being gaseous, liquid, or invisible ethers and additional properties that presumably only an ancient can still imagine. Mechanical explanations of an organism reached some kind of perfection during the eighteenth and early nineteenth centuries with amazing details of the workings of the parts—including the organism's organs and their physical environment (e.g., heat and gases). Mathematics and physics fit well into these models of either the parts or the whole. It has been the experimental embryologist who has recently reminded all that simple mechanical models may either be incorrect or incomplete. This is not to imply that the methods and ways of thinking of the former searchers are faulty. Far from it, they could well be applied more vigorously now. But the observations discovered by the experimental embryologists will have to be accounted for in addition to these older observations, and this appears to call for some modification of the previous mechanical models.

This intense and rewarding search has revealed a number of interesting points about life-forms and science. One is struck by the complexity of the part-whole question. The organism is the ideal form to examine this organizational relationship. The interrelation of parts in a "community" remains too fuzzy and that of the parts in a cell is either too complex or too full of inherent indeterminacies. With organisms, one quickly moves away from the simple notion that an organism is the sum of its organs towards synergisms. But with D'Arcy Thompson and the experimental embryologists there is concern that the concept synergism may be an oversimplifying trap to our logic. Thompson reminds us that there is a continuity of physical laws running through all nature including life matter; thus the interactions of parts may depend primarily on diagrams of forces of the whole. On the other hand the experimental embryologist reminds us that synergisms themselves must develop and are probably not fixed patterns in the sense of a prescribed fulfillment but subject to modification and reversals. Both of these suggest that something beyond synergism or even homeostasis is operating. The search for this something could be the next goal.

Interesting properties of science are well exampled in this section. These include (1) the role of basic assumptions versus observations in shaping explanation, e.g., the various explanations for the heart or venous artery based upon symmetry, synergism versus structure, organization, and content; (2) the various interpretations possible from the results of an experiment, e.g., Spallanzani's experiment; (3) the misuse or subjugation a new procedure or tool can suffer within a present belief-web of explanations, e.g., animiculae being seen with the microscope, and by the best microscopists; (4) and, finally, the way the narrowness or determinacy of one search can ignore or inhibit a different search, e.g., the search for the explanations of the order of form not conforming with those of the order of space and time.

This search has netted much but this fact should not be construed as the ending of the need. There are still the questions about form and though some feel that the search into other orders will do the answering, this would seem to be an unwise and presumptuous stance.

vi

order: in continuity

"We do not record flowers," said the geographer.
"Why is that? The flower is the most beautiful thing on my planet!"
"We do not record them," said the geographer, "because they are ephemeral."
"What does that mean—'ephemeral'?"

The Little Prince, ANTOINE DE SAINT EXUPERY

It takes little observation to realize that the oak produces an acorn from which we can be sure another oak will grow. Likewise some of the characteristics of the parent appear in the offsprings. But what is it that continues from generation and why is there variation? What is it that goes on? This order in continuity had many weak explanations until the microscope helped discover first that the cell and then some part of it passed on from generation to generation. Eventually the synthesis of hereditary studies of organisms with the continuity of chromosomes occurred and this permitted other amazing deductions to yield the gene theory and properties of the gene. Soon reproduction fell in line and natural selection was theorized and verified as a possible mechanism for evolution. This dip into the microscopic and unseeable abstracts is full of exciting deductions.

EARLY EXPLANATIONS OF HEREDITY

The ancient Greeks, although concerned with transformations, wrote comparatively little about the step from parent to offspring. Democritus considered that particles from all over the body representing the structure and condition of the various organs collected in the semen during the synthesis of a preformed fetus. This theory, known as *pangenesis*, was denied by Aristotle who argued that the children of one-armed men contained all their limbs. His substitute was that a form-generating principle unique for each species was passed on from generation to generation. Of particular interest to the Greeks was the manner in which sex is determined. The general consensus was that the determining factors are related to the semen and sex was influenced by either the amount or the side of the organism from which it came.

Harvey called attention to the egg or womb as the optimum environment for such transformations to take place, but he added nothing new to explain the transformation. Grew's careful observations of plants initiated the idea that pollen is equivalent to the semen of animals and the ovule is equivalent to the egg, thus setting a foundation for the artificial crossing of plants. As a result, several active centers developed during the late seventeenth and eighteenth centuries that produced beneficial hybrids of plants and new breeds of animals. If such practices added no new insight into the mechanisms of heredity, they did add further knowledge of the general structure of the reproductive organs.

Apparently concepts and observations of developmental processes exerted a strong influence on the more elusive explanations of heredity. It was preformationism that helped Leeuwenhoek find animiculae in the head of sperm in 1677 and imagine their ability to carry the miniaturization of the male form, "a thousand times smaller than the eye of a big louse," from one generation to the next. Thus if one was willing to stretch his imagination, he could claim this preformed organism was so small that enough could exist in the present organisms to account for all the future generations. The microscope was allowing the eye and the mind to find a whole new world below the resolving ability of the unaided eye. Why could there not be an almost infinite supply of these most tiny sperms? After all Leeuwenhoek estimated that a single codfish produced at least 150×10^9 such sperm. The only other explanation given for heredity at this time was a modification of pangenesis put forth by Maupertius in

1744. He suggested that not only did elementary particles associated with the various traits of the organism descend and collect in the gonads but that these are carried by the semen and mix with an equivalent set from the female, the blend influencing the traits of the offspring. The battle between the animiculists and ovists prevented any compromise concept such as blending, however, and the general acceptance of Spallanzani's experiments kept the preformationists' explanation of heredity as the more logical one.

DISCOVERING THE CELL

Even after further observations cast doubts on the existence of a preformed organism, no explanations for heredity were offered for some time. More substantial explanations apparently had to await the development of observations of the substructure of the organism. This could be thought of as beginning with Hooke's observations with a microscope of the structure of plant material. He noted the presence of cavities in the solid matrix of cork presumedly for the purpose of allowing juices to seep through the plant. Grew reinforced this contention and considered that these cavities were vesicles occurring as a variety of hollows depending on their location within the solid matrix. Malpighi agreed and added that the "utricles" were connected end to end thus supporting Hooke's original assumption. Wolff was probably the first to find an equivalent irregularity in the solidness of animal organs as he wrote in 1759, "The particles which constitute all animal organs in their earliest inception are little globules, which may always be distinguished under a microscope of moderate magnification." He assumed that these globules arise from small vacuoles somewhat like air pockets and might expand and that this occurred as the organs of the embryo develop.

Such observations were scattered and considered of little significance until the internal texture of the organs began to be used as an indication of organ function. This new significance led to improved methods of slicing and staining the organs for further identification and characterization. The sudden burst of observations was first summarized by Xavier Bichat (1771–1802) in his *Anatomie Generale* (1801), which in turn did much to stimulate further investigations. Bichat not only emphasized the need of these observations in order to understand organ function but was the first to point out that there are specific regions of the organ with specific structures and thus specific functions to the organ and that the differences between these regional appearances were a better guide to organ function than general appearance of the whole organ. By this emphasis Bichat initiated the study of tissues, histology. He noted that the size and shape of the internal cavities as well as specific stain reactions were the best markers to use.

One of the many stimulated by Bichat was Brisseau de Mirbel (1776–1854) who suggested that the cavities were not only markers but an essential part of the tissue and required for their function. He called these cavities *cells* and wrote in 1808, "Plants are made up of cells, all parts of which are in continuity and form one and the same membranous tissue." The cells were cavities separated from one another by thin walls that extended throughout the tissue. Lamarck, a keen microscopist, balanced this concept for animal tissue as he wrote in his *Philosophie Zoologique* (1809), "Cellular tissue is the general matrix of all organization and that without this tissue no living body would be able to exist, nor could it have formed."

THE CELL THEORY

Thus it was that "cells" became studied for their own sake. Several investigators noted that there were fine cracks or spaces within the walls between cells. A curiosity in these led R. J. H. Dutrochet (1776–1847) to devise a new technique of study that he described in 1824, "I place a fragment of plant material, what ever it may be, in a glass tube closed at one end containing concentrated nitric acid; I then immersed the tube in boiling water. In five or six minutes, sometimes less, the tissue is separated from one another, and it can be seen that the wall which separates two adjacent cells is double and not at all single." From such experiments, Dutrochet was the first to conclude that all plant tissues are actually made up of cells "united only by a simple adhesive force" and, therefore, all cells must have an individuality and independence of their own as well as an integrated activity with each other when in the formation of a tissue. He also suggested that the cells were the vital units of the organism and proposed that "growth (of the organism) results both from the increase in volume of cells, and from the addition of new little cells." Dutrochet was not certain where those new little cells came from except that they were "originally (tiny) globular" structures. It can be said that Dutrochet had proposed a cell theory to explain the organism. The cellular basis of plants was further elaborated by J. P. F. Turpin (1775–1840) who noticed that the cells of filamentous algea divide and in 1826 he wrote that each plant cell had its own vital center for "vegetation and propagation" purposes and that each cell was the product of other cells.

Meyen (1804–1840) confirmed these observations in 1830 and proposed that since algae or fungi can be found as either single or joined cells, they are "united together in greater or smaller masses . . . (and) even in this case each cell forms an independent isolated whole; it nourishes itself, builds itself up, and elaborates raw nutrient materials, which it takes up into very different substances and structures." Meyen considered cells "elementary organs" that arose by expansion within other cells. He was also the first to observe circular movements of subcellular substance and attributed it to the "gravitation of plants"—an inherent property of the cellular fluid.

It was Robert Brown (1773–1858), a most skillful microscopist, who extended Meyen's observations of cellular fluid movements, some of which were named after him. He was also not the first to observe but did contribute the clearest description of the nucleus of the cell as a result of his observations on orchids in 1831. Brown considered the nucleus as a vital center for the life of the cell because it was found in every cell he observed. Gabriel Valentin (1810–1883) confirmed that the nucleus was present in animal cells from his observations of epithelial cells and also discovered a granular body inside the nucleus, which he called the *nucleolus*.

In 1835 Hugo von Mohl expressed disagreement with Meyen's description of the origin of cells. From his own observations of division in filamentous algae, he concluded that a circular constriction appears in the middle of the cell that extends until the cell is divided into two equal halves. He found a similar process in complex plant tissues and wrote, "cell division is everywhere easily and plainly seen . . . in terminal buds and root tips."

Thus it was that a firm basis for the cellular composition of organisms was well established by 1835. This so-called cell theory has often been attributed to the work and thoughts of Malthias Schleiden (1804–1881) and Theodor Schwann (1810–1882) as the result of a special meeting in 1838 in which their observations were compared and simultaneous publications were planned. It is probably more accurate to say that, at least in the case of Schleiden who was a lawyer turned botanist, by their aggressive debate they brought the cell theory into the minds of everyone during this time. It is also unfortunate that Schleiden considered his greatest contribution a theory of cell formation that was similar to that of Meyen in which the nucleolus appearing "like a thick gasket or thick-walled hollow globule" enlarges to form the "cytoblast (nucleus)" which in turn enlarges to form another cell.

Schleiden's forceful nature apparently influenced milder mannered Schwann to observe the same process in animal cells. Schwann felt that cells could also arise by a "kind of crystallization" process in the "cytoblastem" from substance that exists between cells. Schwann was a physiologist who became intrigued by the membranes of cells. His creative imagination conceived that these membranes must be involved in the breaking down or metabolizing of substances on the outside of the cell. The cells were also selectively permeable to the breakdown products of this process, thus allowing certain necessary substances to become available for the vital activities of the cell. He also wrote convincingly of the degree of independence of cells. He was also the first to note that the egg and sperm were cells that could break away from the rest of the organism and exist independently. Thus he brought heredity and development together by concluding that the cell was the agent of continuity and the source of a new organism. Unfortunately, this interesting idea was lost because of Schleiden's argumentative manner, which built up an antagonism to any of their ideas except the general notion of the cell theory that had already been established.

In spite of this loss the cell theory was available and did become used in time by the embryologist and student of heredity. Prevost (1751–1839) and Dumas (1800–1884) had repeated Spallanzani's experiments in 1824–1825 and proved the necessity of the sperm for the fertilization process. Von Baer discovered the mammalian ovum in 1827. In 1835 Pierre Peltier (1788–1842) suggested that sperm arose from differentiated body cells but he could find no evidence to prove this. Felix Dujardin (1801–1860) supported this hypothesis in 1837, however, by finding sperm in the seminiferous tubules of the testes. This was followed by the observations of Rudolf A. von Kolliker (1817–1905) who traced the histogenesis of both the egg and sperm and confirmed that they are single cell body products that arise by division from cells within the gonads.

The gap of cellular continuity was closed when George Newport (1803–1854) took time motion pictures of the fertilization process in amphibians in 1853 that demonstrated that the sperm actually fuses with the egg. Carl von Nageli (1817–1891) did a parallel study with plant tissue in 1854 and also could claim with more certainty that all plant cells arise from other plant cells by the process of cell division and not by the methods proposed by either Schleiden or Schwann.

This result plus others of a similar nature led to the sound denouncement of any method of cell formation except by cell division, a proclamation most clearly expressed by Rudolf Virchow (1821–1902) in 1858.

PANMERISM

Thus it came about that cells were accepted as the substance of continuity between generations and that any explanation of heredity must take this fact into account. Where it was clear before that an organism dies and another replaces it by a clear-cut break, the cellular theory of continuity and development brought with it a new notion, mainly that there is an eternal unit of life that accumulates substance, grows, and reproduces its identical self by cell division endlessly through time. Variation in traits from generation to generation or even between species became secondary to this overwhelming and unifying concept known as *panmerism*. Variations in the forms of any organism could be accounted for by the organization or shape of the pile of these common basic units. It would seem that it was merely a matter of working out the details of the organization for any given organism.

Cellular panmerism, in the purest sense, there being one single cell type universally found in all organisms, soon ran into difficulty. For one thing it was found that many protazoans and simple algae and fungi had obvious unique characteristics and yet they were "animals whose organization is reducible to one cell." Likewise it was obvious that the egg differed from one organism to the next. It was clear that cells are not of one uniform type.

But the idea of panmerism was too powerful a synthetic one to let go because of variations in the apparent structures of cells. Van Beneden (1846–1910) answered the discrepancies between eggs as due to differences in amount of yolk, for "though it form part of the egg, it cannot be regarded as an integral part of the egg-cell ... but in every egg there exists one egg cell, one germ which is the original cell of the embryo." Max Schultze (1825–1874) suggested in 1861 that differences in appendages—flagella, cell walls, odd inclusions, etc., were unimportant for they could not be part of the vital unit of the cell, but rather it was the nucleus surrounded by the cell sap or as he called it, *protoplasm*, that was the panmeric constant of life. Others disagreed with Schultze's "protoplasmic theory," for they found too many differences in the protoplasm from cell to cell and were skeptical of the need for uniformity of the nucleus. Ernst Brucke (1819–1892) proposed, also in 1861, that there must exist in the protoplasm many "persistent organs" or "smallest living parts" that could satisfy the panmeric requirement of the ability to assimilate, grow, and reproduce themselves eternally and be found universally in all organisms. This latter idea found diversity of expression. Nageli envisioned that the units were "micelles," a complex molecule that Herbert Spencer (1820–1903) considered to be "physiological units," and even Darwin used them for his "gemmules."

Darwin, frustrated from an inability to find any hereditary mechanism to explain natural selection, turned in 1868 to the only concept of heredity left with the discrediting of preformationism, namely pangenesis. Pangenesis also fit with and may have had some influence on his acceptance of the acquirement of characteristics as a mechanism for evolution. According to this "provisional hypothesis of pangenesis" to explain evolution, small particles, "gemmules," specific for each of the characteristics of the organism could collect as sets within the sperm and egg and could combine upon fertilization and determine the characteristics of the offspring. These particles could reproduce themselves and were panmeric. They were also the sensitive receptors, while dispersed throughout the organism, to acquire the characteristics that their immediate environment imposed upon them. This hypothesis had no direct evidence to support it and was only reluctantly accepted by many as the only mechanism known for heredity. It was, however, championed by Haeckel as late as 1876.

The possibility of the nucleus alone being the panmeric unit did not receive support until long after the resentment against Schleiden, who had emphasized the nucleus in his cell-forming theory, had died down. Several microscopists had noted star-shaped figures near the nucleus during cleavage, but it was not until 1873 when Anton Schneider described a systematic change of the nucleus into a mass of filaments during the cleavage of the eggs of the flatworm, that further observations of the nucleus were generated. He noted that these filaments separate from one another and move toward two star-shaped figures that he appropriately named asters. These all disappear and are replaced by two vesicular nuclei in their place just prior to the division of the cell. This division always occurs between the two new nuclei. Edward Strasburger (1844–1912) confirmed this observation and added further details in 1875 as a result of his observations on the cells of the growing flower buds of spiderwort, *Tradescantia*. These results were sufficient to cause Strasburger, Kolliker, and others to claim that the nucleus was the "ideoplasm" or panmeric unit of life all were searching for and "omnis nucleo e nucleo."

The new look at the nucleus produced many quick results. In 1879 M. J. Schleider (1804–1881) described the initial appearance of the nuclear filaments as appearing as rods inside the nuclei just before both the nucleus and nucleolus disappear, while he was observing division in the cartilage cells of a frog. He also discovered a delicate filamentous structure, named for its shape, the spindle, which connects the two asters. The "nuclear rods" become arranged on this spindle and then divide, their parts moving to the opposite poles of the spindle. By 1882, this process had been characterized into stages by Walter Flemming (1843–1905) as a result of his observations of the epithelial cells of salamander. He also named this process *mitosis*, and the rods were called *chromosomes*.

It was also about this time, 1883, than van Beneden demonstrated that the sperm and egg have only one set of the "half-chromosomes" and that during fertilization the number is restored to a double set. Theodor Boveri (1862–1915) confirmed this and wrote "Not only does one of the nuclei consist of the same number of chromosomes as the other, but it can be shown in many cases, where observations has been possible, that the chromosomes from the sperm have the same size, shape, and identical properties as those from the ovule nucleus. We may say that each sperm chromosome meets its mate in the female ovular nucleus." Boveri also wrote, "We may feel justified in contending that the chromosomes possess individuality."

To complete the thread of continuity of the chromosomes from generation to generation, some earlier work had suggested that there was a reduction in number of chromosomes during the formation of the germ cells. This process was somewhat complicated and unclear, but van Beneden was able to trace the chromosome changes during spermatogenesis and oogenesis in the worm *Ascaris* in 1883. He could say with some definiteness that even without all the details the chromosome number was halved during these processes now known as meiosis and thus the chromosome number remains constant from generation to generation.

Thus it was that Boveri came to believe that the chromosomes were even more basic panmeric units than the nucleus. After all it was a set of chromosomes that could now be said to assimilate, grow, and reproduce themselves apparently eternally as they are clearly passed on from generation to generation by way of the gametes. He wrote, "Omnis chromosoma e chromosoma." Chromosomes differ considerably from one organism to the next, however, both in number and appearance. Thus it was not long before von Beneden and Boveri looked elsewhere in the nucleus for the vital universal unit and found it in the centrosome in about 1887. Was it not the centrosome which was the generating center which produced the spindle and controlled the movement of the chromosomes, and perhaps was involved in chromosome formation? Certainly they satisfied the panmeric criteria, and it could be written, "Omnis centrosoma e centrosoma." As might be expected, small differences were found in centrosomes from one organism to another—especially between plants and animals. This led Boveri to seek ever smaller subunits of uniformity. This he did again in 1895 for a small structure always found within the centrosome and believed to give rise to the spindle, the centriole. He could write with some satisfaction "omnis centriola e centriola."

WEISMANN'S GERM PLAN

Although this endless search, summarized in Figure 6–1, occupied the minds of some investigators, there were those who found the diversity of the chromosomes more valuable than any universal units might be. Strasburger, for example, was led to write, "The constancy of the number of the chromosomes in the nuclei of the sex (germ) cell is doubtless of great importance for it ensures the equal influence of the two parents in the sexual act and the act of fertilization is, in all the higher organisms, the center of gravity of the maintenance and development of the species." And there was the implication in these words that the unique characteristics of the chromosomes were related somehow to the differences in the species.

Another creative investigator who was inspired by chromosomes was August Weismann (1834–1914). He was able to develop a theory of heredity and development badly needed and far ahead of his time. He summarized this theory in his *Das Keimplasma* in 1892. Darwin's ambiguity as to the mechanism of evolution had caused a split among its supporters.

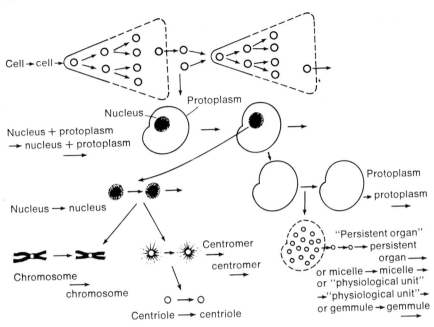

Figure 6-1. Panmerism.

Darwin's model of heredity, pangenesis, embraced the acquirement of characteristics. But Weismann was very much opposed to this explanation, and wrote, "An organism cannot acquire anything unless it already possesses the predisposition to acquire it." He demonstrated this by cutting the tails off of mice for many generations only to find that the offspring had tails of normal length.

Weismann was also opposed to any pangenic model of heredity. In answer to his own question, "How is it that a single cell (the egg) can contain within its self all the hereditary tendencies of the whole organism?" he wrote, "It is impossible for the germ cells to be, as it were, an extract of the whole body. Then there remains, as it were to me, only two other possible, physiologically conceivable, theories as to the origin of the germ cells, manifesting such powers as we know they possess. Either the substance of the germ cell is capable of undergoing a series of changes which, after the building up of a new individual leads back again to identical germ-cells; or the germ cells are not derived at all, as far as their essential and characteristic substance is concerned from the body of the individual, but they are derived directly from the parent germ-cell."

(a) Pangenesis:

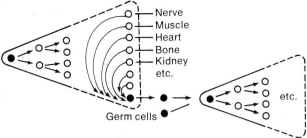

(b) Germ Cell is Differentiated Body Cell:

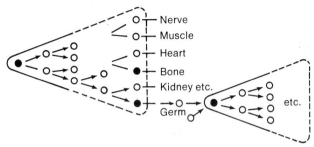

(c) Weismann's "Germ Plasm" Method:

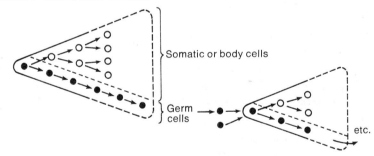

Figure 6-2. Models of the origin of germ cells.

From these alternatives, diagramed in Figure 6–2, Weismann favored the latter one. For one thing it was the much simpler process. Also, "The fact that the embryo grows more strongly in one direction than another, that it's cell-layers are of a different nature and are ultimately differentiated into various organs and tissues—forces us to accept the conclusion that the nuclear substance has also been changed in nature." This contention was later supported by the findings of Strasburger and Boveri that in several invertebrate eggs there was a loss of some chromosomal material from body cells during early cleavage, while the potential germ cells could be distinguished by their chromosomes remaining intact.

The fact that germ cells contained $1N$ set of chromosomes that fused as a $2N$ set during ferilization also emphasized to him the significance of chromosomes in heredity and that within the germ cells "the whole number of (hereditary) elements must be contained collectively in the half quantity of chromosomes." For this reason he also predicted before meiosis had been observed that there must be a reduction division step ($2N \rightarrow 1N$) in the lineage of the germ cells from the fertilized egg to the formation of the sperm and egg in order to maintain the constancy of number of sets of chromosomes from generation to generation.

This theory was known as the "continuity of germ-plasm" theory. Weismann's creative imagination built with the ideas of others as he wrote, "I shall call the inheritable substance (the chromosome complex) of a cell its idioplasm," and the ideoplasm of the germ cells, the "germ plasm." The ideoplasm has a "definite chemical" and "highly complex structure, conferring upon it the power of developing into a complex organism" that proceeds from the chromosome outward, first to the nucleus and then to the protoplasm. This complex is made up of material biological units called "ids." The chromosomes may each be an id if they, the chromosomes, are small and many in number. Or, as suggested by the granular nature of the chromosomes, there may be several ids, an "idant," making up one chromosome. As there is more than one germ cell produced by an organism, the inheritable factors that are contained in these ids must be able to assimilate, grow, and reproduce themselves; thus panmerism must still be satisfied to some extent. For this reason Weismann conceived of each id being made up of a set of "biophores," each one having the same panmeric abilities. Each of the biophores represents the details and organizing ability to produce a specific trait of the organism. Thus each id was a microcosm containing the totality of hereditary factors necessary to produce an organism. The germ plasm flows without change through the individual, while the body cells are differentiating as they divide.

Weismann conceived that differentiation was both the loss and the modification of the ids during somatic cell division. This results in the loss or latency of all biophora except one that can express itself for each determined adult cell. This model is diagrammed in Figure 6–3.

The beauty of this model was that it allowed Weismann an explana-

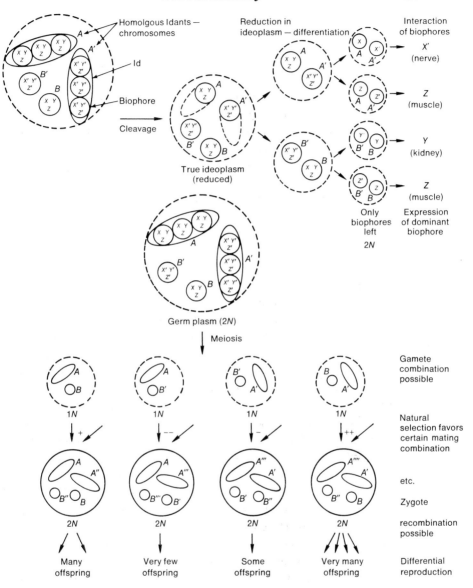

Figure 6-3. Weismann's model of heredity and development.

tion for a variation of hereditary characteristics upon which natural selection, the mechanism of evolution he championed, could act and produce evolution. This was made possible especially by a probable exchange of ids between the parental chromosomes during meiosis, which would result in a variety of kinds of haploid germ cells depending on the particular combination of ids formed. Also variation would result from the particular composition of ids formed during the fusion of the diploid germ cells at fertilization. Thus a very large number of kinds of individual organisms can be produced. During the differentiation process, the homologous ids and biophores of the remaining active chromosomal material compete with one another. There may be a blending between these two homologous ids or more commonly only one of the ids expresses itself with its biophore determining a specific form of the trait for which the biophore is responsible.

Natural selection acts on the overall organism and thus favors those individuals containing the ideoplasm that could produce the most fit organism. This organism would then have a better chance to produce more offspring with similar germ and ideoplasm. There still remains a source of variation at each generation to assure a fit if the environment changes.

Weismann also conceived of the value of death for eliminating the older combinations of less fit ideoplasm as well as the need for new combinations at a rapid enough rate to keep up with the changes in the environment.

This was the first logical model to explain how natural selection could work. There is no wonder that Weismann was considered the foremost contributor of his time to that fraction of evolutionists known as the neo-Darwinists who considered that natural selection was the only method by which evolution could occur. His theory went very far in dissuading the so-called neo-Lamarckists counter hypothesis that hereditary characteristics could be acquired by some kind of pangenesis. It is unfortunate that these particular names were associated with the particular mechanisms of evolution that they were, for this did much to raise Darwin even higher on a pedestal, which no man is entitled to at the expense of Lamarck.

Although Weismann's theory did much to convince others that natural selection was the mechanism of evolution, there were those who were concerned about certain details. For example, there were many organisms that had been studied that did not lose any chromosomal material during cleavage as far as known. Many of the details of mitosis had resulted from such observations. Nevertheless, his theory filled a very empty void and would have to be disproved before it could be discarded, and this did not seem to be an easy thing to do.

STATISTICAL APPROACH TO
HEREDITY—GALTON

Meanwhile a different approach to heredity was beginning to be considered by a few investigators. Breeding for special plants and animals had empirically suggested that the traits of an offspring were usually a blend of those of the parents. This was accepted with little question by most everyone. During the later part of the nineteenth century, statistical methods of handling data were becoming fashionable, mainly under the influence of Quetelet. Francis Galton (1822–1911), a cousin of Darwin, was influenced to apply statistics to hereditary traits. He subsequently put forth his quantitative laws of heredity during 1886–1889.

As a result of studies on 150 families he noticed a tendency for differences in traits to balance each other out during succeeding generations. Thus there would be less effect on the offspring from more distantly removed ancestors. He thus proposed that the parents' traits blended together and contributed one-half of the characteristics; the grandparents together, one-quarter; the great grandparents together, one-sixteenth; etc., of the offspring.

The result of this explanation was to encourage many investigators to begin the laborious task of recording variations of traits from generation to generation. Others had noted before, but with statistics it became obvious, that some of the traits of an organism, especially among plants, did not blend together. Rather they were found in an on-off manner, e.g., red or white petal color, green or yellow seed coat cover. Weismann's theory had suggested a possible mechanism of competition between two forms of a given trait that might explain the on-off effect and hence was of some interest to elaborate statistically.

It was as a direct result of these kinds of studies that predictable patterns of the relative numbers of the two conditions in offspring, on or off, were discovered simultaneously in 1900 by three different investigators in three different countries: Karl Correns (1864–1933) in Germany, Hugo De Vries (1848–1935) in Holland, and Erick Tschermak in Austria.

Before publishing their results, each of these botanists rediscovered a paper revealing that the same results had been found by an Augustinian monk, Gregor Mendel (1822–1884), 34 years earlier. Mendel had published these results in 1866 and had even sent a letter describing his experiments and results to Nageli, but the latter gave Mendel no encouragement and their significance was not recognized at that time. Because Mendel's work was in some ways even more complete than that of the later discoverers, and because his logic has proved to be one of the clearest examples of the kinds of deductions one can make from relatively simple observations, Mendel's experiments are worthy of elaboration.

SEGREGATED CHARACTERISTICS—MENDEL

It happened that Mendel enjoyed working the vegetable gardens of his isolated Tyrolean monastery. The long days of watering and tending led him to notice that, especially among the peas, there were striking extremes in characteristic traits such as flower color and arrangement, seed colors and shape, color and texture of pods, and even stem lengths. He was curious whether there were different kinds of peas with the separate traits growing as a mixture or whether one plant could express itself in such different extremes from one generation to the next. He therefore kept the seeds separate from one another and planted them in one section of the garden. He found that the daughter plants were usually a mixture of traits rather than all agreeing with the parent plant. Because of the disposition of the plants with the different traits, this result could not have been due to differences in the soil or any other environmental factor he could think of and was apparently hereditarily determined.

But how could one plant produce such diverse extremes among its seeds? His curiosity led him to self-pollinate the peas and prevent cross-pollination with any others. He found that for some of these plants the product seeds yielded mixtures of a given trait, while others bred true containing only the trait of the parent plant. He found that those that bred true did so for "over four to six generations," which was as long as he carried out these experiments. In this way he had produced pure strains of a number of "differentiating characters."

The next big question was what would happen if two of the pure strains for opposite expressions of the differentiated characters were crossed? He crossed both pollen from a plant breeding true for yellow seeds with a plant for green seeds and vice versa. He found that regardless of which direction of the pollination, the hybrid offspring plants all had yellow-colored seeds! The obvious thing to do next was to cross this hybrid plant with its own kind to see if it would breed true. The results of this cross was a mixture of plants with the majority having yellow-colored seeds. This pattern of results was found for other characters and led him to write, "Those characters which are transmitted entire or with hardly any change on fertilization, and are consequently in themselves the main characters of the hybrid are termed dominant, while those which become latent are called recessive, this latter expression has been chosen because the characters thus designated withdraw or disappear to all intents and purposes in the hybrids, even though they make their appearance again unchanged in the progeny of such hybrids."

Mendel was curious whether the amount of dominance could be measured and whether dominance differed from one character to another. He therefore carried out a number of experiments and recorded the results. He found, for example, that during one year out of 8023 seeds from the hybrid-hybrid cross, 6022 were yellow versus 2001 green. For a different character such as shape, he found that for the same period out of 7324 seeds, 5474 were round versus 1850 wrinkled. He noted that for all characters tried the dominant to recessive ratios were the same and always close to 3:1, e.g., 3.01:1 for seed color and 2.96:1 for seed shape.

The story goes that because Mendel grew so many peas to perform these experiments, his brothers at the monastery became tired of eating peas and he was obliged to eat most of them. As a result he became very round and took to walking 20 miles through the mountains and smoking 20 cigars each day. This gave him plenty of time to try to explain the curious constant relationship between dominant and recessive characters.

One thing was obvious and that was that the recessive factor that accounted for the recessive character must be present in the first hybrid cross. Otherwise, it could not have appeared in any of the products of the self-crossing hybrid-hybrid. If this is the case there must be at least two factors, one for each character, the dominant and recessive, present in the hybrid plant. It is natural to expect the number of factors present in any one plant to remain constant from generation to generation. If it can be assumed that there are two and only two factors in each plant, since there are only two forms of differentiating characters, there must also be two factors for yellow seed color in the true breeding yellow seed-colored parent and two factors for green in the true breeding green seed-colored parent.

In order to retain the number of factors at two in the hybrid, each parent must contribute only one factor. Since the factors are the same in either parent plant, it would not matter which of the two was contributed and the hybrid would contain one yellow and one green factor. When the hybrid is self-pollinated, however, some daughter plants must have received the recessive factor from both parents and no dominant factors or else this latter would have expressed itself. Most of the daughter plants do apparently receive a dominant factor from one of the hybrid parents. Therefore, the only conclusion is that either one of the two factors of a plant can be passed on to the next generation.

It is simplest to assume that the two factors have an equal probability of representing the parent in the daughter plants. It is also simplest to assume that either of the factors from either parent has an equal probability of combining with either of the factors from the other parent for each daughter plant produced.

Therefore, if Y represents the dominant factor for yellow seed color and y the recessive green seed color factor, then YY would be the factors in the one breeding parent for seed color and yy in the other parent. All the hybrid plants would thus contain a Yy combination of factors. Self-pollination of the hybrids, $Yy \times Yy$, would result in all the possible

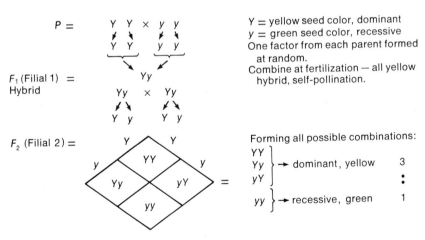

Figure 6-4. Mendel's simple hybrid experiment.

combinations of factors, as best represented by a crossing square shown in Figure 6–4, or YY, Yy, yY, and yy, each with an equal probability. The expression of these factors would depend on the presence or not of a dominant factor Y. Therefore, three-fourths of the products of this cross would be dominant; the remaining one-fourth, recessive. In other words the assumptions above have produced the 3:1 ratio of dominant to recessive as Mendel had found experimentally.

Mendel also considered that if these assumptions are correct, it should be possible to design an experiment in which the theoretical results can be predicted and then test it experimentally. He considered that if the hybrid form Yy is back-crossed with the recessive parent yy, then according to his assumptions the products of the cross would be Yy, Yy, yy, and yy, in equal probability, with a dominant to recessive ratio of 1:1. He then did the experiment and found an 85:81 yellow to green ratio of seed color confirming within the accuracy of such a statistical experiment that his assumptions were indeed correct.

Mendel's assumptions had been that there is such a phenomenon as dominancy, that two and only two factors for any one character are present in at least plants, and that these factors may combine with the factors of another plant with equal or random probability during any cross regardless of which is the male and which is the female plant.

In the course of these experiments Mendel had kept track of more than one character as a matter of efficiency with experiments on such a slow growing organism. He noticed that usually two different characters

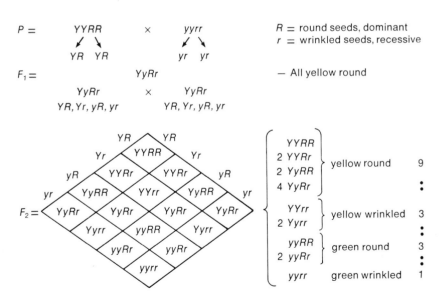

Figure 6-5. Mendel's dihybrid cross showing independent segregation.

acted completely independent of each other. For example, if he kept track of seed shape as well as seed color, he could find all four combinations: round yellow, round green, wrinkled yellow, and wrinkled green seeds in the progeny. And if he started with, say, a *YYRR* plant, one that bred true for yellow round seeds, and crossed it with a plant that bred true for green wrinkled seeds, *yyrr*, the hybrids all contained yellow round seeds presumedly with the factors *YyRr*. If these hybrid forms are cross-pollinated, the result is a 9:3:3:1 ratio of number of progeny with round yellow, round green, wrinkled yellow, and wrinkled green seeds, respectively.

As shown in the cross square in Figure 6–5, this can only result if the different factors from either character and either hybrid parent can segregate independently with an equal probability of combining in all combinations in the production of the progeny.

THE SYNTHESIS OF HEREDITARY FACTORS AND CYTOLOGICAL OBSERVATIONS—SUTTON

By the turn of the century a number of persons had observed that the individual chromosomes retained their apparent structure during cleavage and meiosis in the great majority of organisms, thus disagreeing with Weismann's model of a loss of chromosomal units during differentiation. One of these investigators was Walter Sutton (1876–1916) who took the opposite interpretation, namely, that since there are "constant size-differences observed in the chromosomes (then one is led) to the suspicion ... that the individual chromosomes of the reduced series (haploid set) play different roles in development." In other words, instead of con-sidering that each id had all the characteristics of the organism, as Weismann had, perhaps each separate chromosome was responsible for a character.

Sutton supported this hypothesis by carefully tracing specific chromosomes from before meiosis to the formation of the gametes (the sperm and egg) and their fusion at fertilization. He noted that during meiosis either one of the chromosomes of any homologous pair could end up in any of the gametes. This was independent of whether the parent was a male or female, i.e., whether the gamete was an egg or a sperm. Thus if it is con-sidered that the homologous pair of chromosomes from the male parent is represented by $A_{1\,\male}$ and $A_{2\,\male}$, then either $A_{1\,\male}$ or $A_{2\,\male}$ chromosome will be found with equal probability in the sperm of this parent. Likewise, the equivalent homologous pair of chromosomes $A_{1\,\female}$ or $A_{2\,\female}$ will have an equal probability of being found in eggs of a female parent. ($\male$, a shield and spear, is the symbol for Mars and the male—representing those typical male characteristics of strength, endurance, and aggression; while $\female$, a mirror, is the symbol of Venus and the female—representing those typical female characteristics of vanity, seduction, and beauty.)

Sutton also noted that as a result of this independence of distribution of chromosomes, any combination of parental types of chromosomes can come together during the fertilization process, i.e., $A_{1\,\male}A_{1\,\female}$, $A_{1\,\male}A_{2\,\female}$, $A_{2\,\male}A_{1\,\female}$, and $A_{2\,\male}A_{2\,\female}$, with equal probability. Moreover, two different homologous pairs of chromosomes in either parent appear to act independently of each other and could presumably segregate independently in any combination during the formation of the gametes.

Sutton was struck by the similarity of the behavior of Mendel's factors and the homologous chromosomes as diagrammed in Figure 6–6. Conceptually this was a difficult correlation to make. The factors of Mendel were related to the overall characteristics of the organism, while chromosomes were but cytological particles found in each cell! Yet, nevertheless it is still a remarkable fact, as he expressed in 1902, that "the phenomena of germ-cell division and of heredity are seen to have the same essential features, viz. purity of units (chromosomes or factors) and the independent transmission of the same; while as a corollary, it follows in each case that each of the two antagonistic units (one of the chromosomes of the homologous pair or the dominant or recessive factor) is contained by exactly half the gametes produced ...."

Once the conceptual link could be made, it was hailed as one of the most remarkable syntheses of hypothetical deduction from indirect empirical observations with direct observations that has ever been possible. Both approaches to an understanding of continuity merged and both were greatly enhanced by the support of the other. Now Mendelian factors had substance, and cytologically seen particles could be related to hereditary characteristics. Weismann's model was wrong in detail but had made this synthesis possible. Also the panmeric criterion, which had served its purpose by highlighting the chromosomes, was no longer considered to be a necessary property and the hunt for the basic universal vital units of the continuity of life became sidetracked.

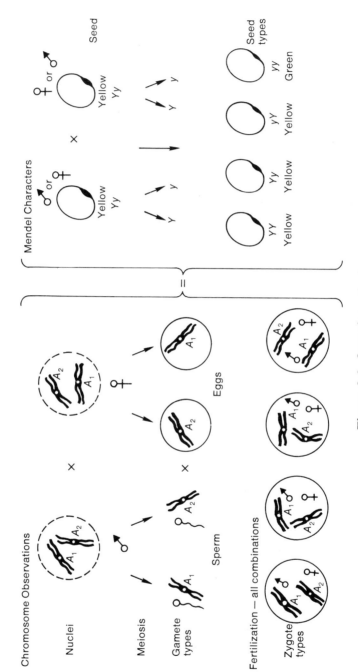

Figure 6-6. Sutton's synthesis.

THE CHROMOSOMAL BASIS OF HEREDITY

There was of course a skepticism that Sutton's synthesis was merely a coincidence and it would be necessary to have more direct support for the role of the chromosome in heredity. Such support was soon in coming from an unexpected source, mainly that of the method of determining the sex of an individual. Sex determination was one of the most baffling questions that had concerned man. There was no dearth of ideas on how it might happen—by 1841 there were over 500 different mechanisms proposed but none of these were generally acceptable or testable.

During the latter half of the nineteenth century one idea proposed by Schultze was considered as a very strong candidate for the correct mechanism. This was that there were two kinds of eggs produced in equal numbers: one that produced a male and the other a female. Correns had noticed that in some plants there were two different kinds of spores, each related to a different sex.

Near the turn of the century, however, observations on the chromosomes of insects revealed that there were apparent differences between the chromosomes of the male and the female. One of the greatest students of the cell, Edmund Wilson (1856–1939) summarized and interpreted the results of these observations in 1905. He grouped the chromosome patterns of insects into two types, *A* and *B*. In the *A* type the male cells all have one less chromosome than the female cells, resulting in an unmatched chromosome in place of a homologous pair in the male. A study of the sperm revealed that half of them have a complete set and half are missing one chromosome, while egg cells always have a complete set. In the *B* type of insect, a similar result is found except that there is a smaller or partial chromosome existing in the male in the equivalent place of the lack of a chromosome in the *A* type.

This association of an odd pair of chromosomes with the sex of the organism strongly suggested that these chromosomes determine the sex in insects in some way and that these so-called sex chromosomes did so by being heterogeneous in the male, as diagrammed in Figure 6–7. As a result of this correlation, Wilson wrote, "The cytological evidence has revealed a visible mechanical basis for the production of males and females in equal numbers and irrespective of external conditions. Phenomena of this kind seem likely to throw further light on the mechanism of Mendelian heredity as well as of sex production, for they demonstrate a disjunction (separation) of different elements in the formation of gametes;

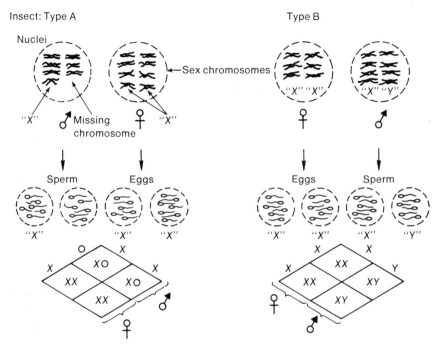

Figure 6-7. Wilson's model of sex determination in insects.

and this is a fact, not a theory." Since then the homo-hetero gametic mechanism of sex determination has been found to be very common. Wilson also made the interesting suggestion that because the Y sex chromosome does not disappear during interphase as do the other chromosomes it cannot be active at this time and thus sex was probably determined by the number of X chromosomes in insects producing different quantities of some metabolic product and has nothing to do with the Y chromosome.

In 1906 the fruit fly *Drosophila melanogaster* (which has the rather undesirable meaning of garbage lover) was introduced as a practical animal to study heredity. It was useful not only because it was easy to culture and had a short generation period, about 2 weeks, but even more so because it contains a giant chromosome in the cells of the salivary gland of its larva. These giant chromosomes result from the duplication of chromatids without cell division. The chromatid products remain adjacent to one another and the homologous chromosome pair remain wrapped around one another producing a very large structure within which characteristic details can easily be seen. There are only four chromosomes in the set in a fruit fly, and one of these is a sex chromosome—a fruit fly having an XX female and XY male mechanism of sex determination.

It was Thomas H. Morgan (1866–1948) and his remarkable graduate students who used this fruit fly to reveal the role of the chromosome in heredity. By 1910 Morgan had found several abnormalities in the flies that were apparently of a hereditary basis. He found, for example, flies with small and deformed wings, which he called "vestigial wing" mutants, as they differed from the normal "wild-type" flies. He found that this "vg" wing was recessive to normal wing in any hybrid cross. He could also demonstrate that it followed a Mendelian pattern of segregation yielding Mendelian ratios 3:1, wild type:vg, for the hybrid-hybrid cross (called the F_2 for second filial generation from the pure strain parent generation, P).

Morgan also discovered a white-eyed mutant, the wild type being red-eyed, which did not follow the Mendelian pattern. Rather the results of any cross depended on the sex of the white-eyed fly. If, for example, a normal red-eyed female is crossed with a white-eyed male, the F_1 (the first filial generation after the pure strain parent generation) is all red-eyed and the F_2 is 3:1 red:white-eyed as expected for a Mendelian characteristic. If the reciprocal cross is made so that a wild-type male is crossed with a white-eyed female, however, then half of the F_1 are females with red eyes and the other half with white eyes! In the F_2 both the male and female offspring are half with red eyes and half with white eyes; there thus resulting a 1:1 red:white ratio. Morgan concluded from this result that the only explanation that seemed plausible was that the X (sex) chromosome was responsible for eye color. Figure 6–8 is a model of how Morgan was able to explain what he called a "sex-linked" trait's hereditary mechanism. This model reinforced the belief that the Y chromosome has little if any influence in the determination of any hereditary trait. Also the fact that eye color could be associated with a chromosome was strong support for the chromosome theory of inheritance.

In about 1913 Calvin B. Bridges (1889–1938), one of Morgan's graduate students, noticed that occasionally a cross in the white-eyed fly studies did not yield the results expected by Morgan's model. This was found to be very rare, only 1 out of about 3000 crosses of white-eyed females with normal males. In these exceptions a white-eyed daughter appeared in place of a red-eyed one. Such an exception would normally be ignored,

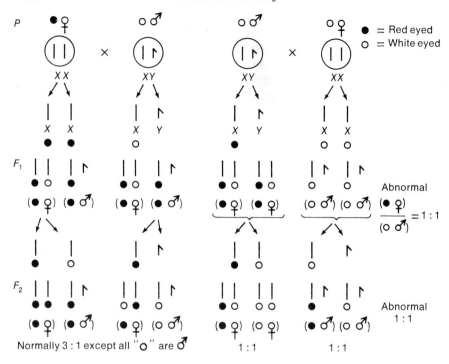

Figure 6-8. Morgan's model of sex-linked characteristics.

but as it challenged the chromosome theory of inheritance and Morgan's model, he pursued the matter at great length.

Bridges also noticed that there were also red-eyed son exceptions produced in the F_1 at about the same probability as the occurrence of a white-eyed daughter—another violation of the model. From this result he hypothesized the creative suggestion that the exceptions could be explained if the two XX chromosomes of the white-eyed female did not separate during formation of the gametes. This so-called nondisjunction, the movement of both XX chromosomes to one of the cells during the division process, results in the two XX chromosomes being found in some egg cells and none in others. This in turn would result in an odd number of sex chromosomes appearing in the fertilized egg—XXX, XXY, X, or Y when fusing with a normal male gamete, as shown in Figure 6–9.

Bridges was able to verify this hypothesis by finding that the white-eyed female exceptions did indeed have an XXY sex chromosome complement, and the red-eyed male exceptions had only an X sex chromosome.

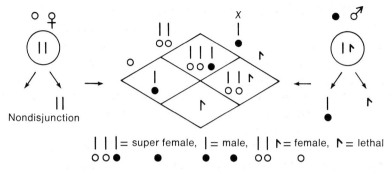

Nondisjunction

| | | = super female, | = male, | | ▶ = female, ▶ = lethal

Figure 6-9. Bridge's model to explain exceptions to Morgan's model.

No flies were found with only a *Y* chromosome, but this was only expected according to the hypothesis that it is the *X* chromosome that influences certain "sex-linked" traits and not the *Y*, and presumedly a fruit fly needs an *X* chromosome to survive. This was truly a case of the exception proving the rule, and the chromosome theory of inheritance was accepted by all. Also as a result of these findings Morgan could add in 1915, "In general it may be said that even an abnormal set of chromosomes, once established in a cell, tends to persist through all succeeding generations. This evidence indicates that the chromosomes are not mere products of the rest of the cell but are self-perpetuating structures."

THE GENE THEORY OF HEREDITY—CROSSOVER

Although dihybrid crosses, those in which two different characters were followed by crossing parents that were each pure strains for the opposite expressions of the two characters, had most often yielded a 9:3:3:1 ratio of characters in the F_2 generation, Mendel did find a case in which the two characters segregated together acting as one and yielding a 3:1 ratio. This was found for seed coat color and flower color—grey-brown seed coats and purple flowers segregating together and dominant over white seed coats and white flowers. Since this was an exception rather than the rule, Mendel did not dwell upon it and offered no explanation for its occurrence. But studies with fruit flies soon revealed that segregation of two or more characters was a common occurrence and called for an explanation.

Again the chromosome theory was helpful, for there are many chromosomes in peas, but there are only four in the basic set in the fruit fly. As there are obviously more than four characters of fly, each chromosome must serve for more than one character and character coupling or linkage of those characters associated with the same chromosome is only to be expected. This was first supported by Morgan when he found there were mutant characteristics other than white eyes associated with the *X* chromosome, i.e., sex-linked.

Mendel also noticed, however, that although these characters may segregate together, they may also appear as if they were associated with separate chromosomes and segregate independently. Others observed the same phenomenon but no one could supply an explanation for this "incomplete coupling."

It was Morgan who hit upon an answer as he was observing the behavior of the chromosomes during meiosis. He noticed that at metaphase I the various chromatids intertwine around one another and at anaphase I there is often some difficulty in the separation of the two chromosomes of the homologous pair. It occurred to him that the different characters associated with the same chromosome are located at different regions along the chromosome; perhaps each character is at one of the granular bumps of the chromosome. Then if the chromatids should "split" as a result of the difficulty of separation of the intertwined chromatids, the resulting pieces could segregate independently of one another. This would depend purely on the random chance of intertwining and splitting.

This hypothesis was supported by the finding that the "amount of coupling" between two characters associated with the same chromosome was statistically constant and differed from one set of characters to the next. This led him to write in 1911, "If the materials that represent these factors are contained in the chromosomes, and if these factors that couple be near together in a linear series, then when the parental pairs (of chromosomes) conjugate (during meiosis), like regions will stand opposed. There is good (cytological) evidence to support the view that during the strepsinema stage (prophase I) homologous chromosomes twist around each other, but when the chromosomes separate, the split is in a single plane . . . . In consequence, the original materials will for short distances, be more likely to fall on the same side of the split, while remoter regions will be as likely to fall on the same side as the last, as on the opposite side. In consequence we find coupling in certain characters and little or no evidence at all of coupling in other characters; the difference depending on the linear distance apart of the chromosomal materials that represent the factors." A diagram of Morgan's model of splitting, or *crossover* as it is now called, is shown in Figure 6-10.

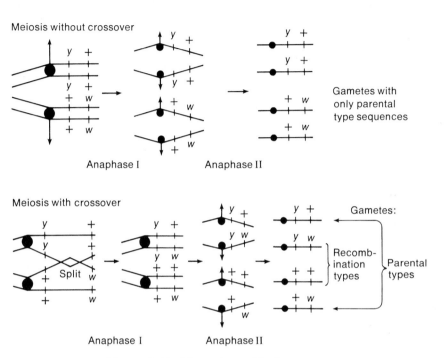

Figure 6-10. Morgan's model of crossover.

Thus Morgan not only put forth a logical explanation for incomplete coupling but, as a result of his model, suggested that the factors related to hereditary characteristics are located in specific regions aligned in a linear fashion along specific chromosomes. As fragments of chromosomes were found in only rare exceptions, far too few to account for the number of "splittings," Morgan also had to assume that the fractured pieces would rejoin so that there would be intact chromosomes that had exchanged equal amounts of material. This was a difficult notion to accept and neither Morgan nor anybody else saw the splitting take place, let alone the rejoining of the fragments. It required the homologous chromosomes to line up exactly point by point along the chromosomes during meiosis I. As far as could be told, this condition was met. Another difficulty in accepting this explanation is the indirect nature of the results or data with which this model must be tested. It must be remembered that this splitting or crossover phenomenon occurs at meiosis during the formation of gametes. The gametes show no differences in traits of the sort being examined with these dihybrid crosses. This means that the organisms resulting from controlled matings over a large number of experiments must be analyzed. In spite of these practical difficulties, however, Morgan's group pushed ahead and collected the necessary data to support or deny the existence of specific loci associated with hereditary factors.

The crossover hypothesis received its greatest support from the consistency of the probability of crossover between any two factors (characters), called the *frequency of crossover*. This suggested the hypothetical loci were at least fixed. For example, Morgan found the crossover frequency between yellow (y) and singed bristles (Sn) to be consistently 21%. He also found that the crossover frequency between y and lorenze eye, a malformed eye (lz), was consistently 27.7%. This permitted him to show that the three loci are arranged in a linear array by finding that the crossover frequency between Sn and lz was 6.7%. He could thus say that the loci are located in a y–Sn–lz sequence. He could not, however, tell the direction of the sequence along the chromosome by this kind of data alone. Regardless, with this tool, it was possible to actually "map" the loci of the different factors along the chromosomes. This has resulted in the mapping of roughly 2000 different loci in fruit fly. As a result of his many studies and those of his students, Morgan proposed what he "ventured to call the theory of the gene" in his classic book, *The Theory of the Gene* (1928). "The theory states that the characters of the individual are referable to paired elements (genes) in the germinal material (chromosomes) that are held together in a definite number of linkage groups; it states that the members of each pair of genes separate when the germ cells mature ... and in consequence each germ-cell comes to contain one set only; it states that the members belonging to different linkage groups assert independently ... , it states that an orderly interchange—crossing over—also takes place, at times, between the elements in corresponding linkage groups; and it states that the frequency of cross-over furnishes evidence of the linear order of the elements in each linkage group and of the relative position of the elements with respect to each other."

How remarkable are the deductive abilities of man that he can make such interesting thoughts by looking long and hard at a little banana-loving fly. The fact that genes could be mapped left little doubt that crossover must occur even when it was not actually possible to prove it cytologically until 1931. This was made possible in both fruit fly and corn by discovering mutant forms of the chromosomes with the cytological characteristic differences at both ends of one of the chromosomes in a particular homologous pair. On the one end of one of the chromosomes of corn was a nob that could take up a specific stain. The opposite end was extra long. With these markers it was possible to find the products of crossover during meiosis by observing that the gametes contained chromosomes with only one of the markers, as diagramed in Figure 6–11.

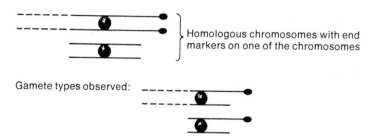

Figure 6-11. Cytological proof of crossover in corn.

CHARACTERISTICS OF THE GENE

Morgan had tried to calculate the size of the gene on the basis of the smallest known frequency of crossover and/or deletions. Basic in the belief of crossover is the assumption that the genes are as beads on a string and breaks occur between the beads; otherwise the genes would not be intact to carry out their function. Morgan considered that the gene was a very stable entity and "when all this is given due weight it is nevertheless difficult to resist the fascinating assumption that the gene is constant because it represents an organic chemical entity." Heredity may indeed be the result of combinations of "hereditary atoms" that can somehow control developmental processes by their unique properties of combination analogous to the formation of molecules and by their great stability. His smallest estimation was based on data from the X chromosome. This chromosome is about 200 microns (μ) in length and has an estimated volume of 0.8 cubic microns at metaphase I, thus having an average cross-sectional diameter of 20 millimicrons. The smallest chromosome bandwidth was estimated to be about 125 millimicrons. This size, 20 by 125 millimicrons may seem quite small by cytological measurements, but it is considerably larger than any molecule and left room for many questions as to the complexity and stability of the band or gene. The large number of bands and the fact that they could often be associated with a deficient gene locus were still the best candidates for any cytological indication of the gene. It was considered that it only remained to clarify the structure of the bands.

Although the concept above concerning the nature of the gene was well accepted, there were recognized properties related to stability that had to be explained. While studying primrose, de Vries had discovered spontaneous changes in phenotype that could not be accounted for by a simple gene theory. Examples of this were when the petal color of a pure strain plant would have one offspring with a change in color. Offspring from this single exception would retain the new color pattern. Apparently the gene changed, or mutated, as de Vries defined it. Apparently the gene had various states of operation, each being comparatively stable, yet interchangeable. These events were so rare that they could not be studied until Herman Muller, another of Morgan's students, discovered in 1927 that X rays could induce gene mutations at several thousandfold their spontaneous rate of occurrence. He wrote, "If as seems likely on general considerations, the effect (X-ray induced mutations) is common to most organisms, it should be possible to produce 'to order' enough mutations to furnish respectable genetic maps, in their selective species, and, by the use of the mapped genes, to analyze the aberrant chromosome phenomena simultaneously obtained." Thus Muller supplied many more mutants for the further analysis of the gene.

Muller also devised a method of determining gene size. He assumed that there was a sensitive target area susceptible to the X ray in which a single hit of an X ray would cause a mutation. Knowing the average rate of mutation induced by a beam of X rays of known density, he calculated the sensitive area, or target area as he called it, of the gene to be between 4 and 40 millimicrons in diameter. Muller had in this way established a new criterion of the gene. It was a mutational unit as well as a crossover unit.

Other properties of genes and gene action were soon discovered. One of these was finding that more than one phenotype may be directed by the same gene. Another finding was that more than one gene affects the same phenotypic characteristic. This latter effect explained the continuous variation in measurable characteristics, such as height and shades of color, wherein the resultant phenotypic expression was the net sum of expressions of all the genes involved.

REPRODUCTION AND EVOLUTION

The search for an explanation of the continuity of life-forms in the sense of a mechanism of reproduction had no influence on the development of the idea of evolution. The constancy from generation to generation would dissuade any such ideas. Once evolution seemed plausible, however, a search for a possible mechanism led Darwin to think in terms of heredity. But so little was known about heredity and the reproductive process that he could not explain the source of variation of the organism.

There has been some conjecture as to why Darwin did not know about the work of his contemporary Mendel. It has been argued that Mendel published his finding in an obscure journal. This is true, but he also wrote letters of his experiments and results to Nageli who could have passed this information on to Darwin or used it himself once *The Origin* was published. It is probably true that the knowledge of Mendel's work never reached Darwin, but there is considerable doubt whether it would have been of much use to him if it had. Mendel's results indicated that there were extreme expressions of differentiating characters, not the gradual shifting of a character from generation to generation. These extreme expressions were hardly the continuous spectrum of variations that would be necessary to match the random changes in environment.

On the other hand, there were no traditions of explanations of reproduction that had to adjust to the concept of evolution. Once evolution was accepted, it became a major source of inspiration to find explanations of evolution in terms of reproduction. It is not an exaggeration to say that the need to support the concept evolution helped produce and did shape the concepts of reproduction. Weismann's model is a typical by-product of this pressure of evolution, mainly through a desire to explain natural selection. His beautiful synthesis of cytology and development for an explanation of the processes of evolution did much to direct study at the reproductive process per se rather than just the overall organism and its adaptive fit with its environment. Sutton completed this synthesis by bringing about a stronger correlation between the characteristics of an organism and the gymnastics of the chromosome. Weismann had supplied both meiosis and the fusion of the gametes at fertilization as sources of variation. The parental sets of chromosomes could be reassorted and the characteristics of offspring could be many depending on the number of recombinations possible. Sutton increased the number of recombinations by suggesting that each chromosome could be responsible for a different characteristic. When it was found that there were about 24 chromosomes in the basic set of man, this meant that there could be 2^{24} or nearly 17 million possible kinds of gametes for each sex formed at meiosis and about 300 billion combinations of chromosomes possible as a result of fertilization. This would certainly be enough variations of characteristics to fit almost any ecological niche. And with the discovery of many genes on one chromosome and many alternate alleles for any gene, and the possibility of crossover, the source of variation is increased beyond imagination. The evolutionist was soon to say that every device and process of reproduction had apparently evolved toward the enhancement of the natural selection process.

Even after most were convinced that natural selection was the mechanism of evolution, however, there was disagreement as to its nature. The main concern was centered around the rate of evolution. It had never been clarified how natural selection could account for the rapid changes that would have had to occur in the history of life to have produced the complexity and diversity of the present living organisms.

An extreme point of view containing shades of progressionism was that championed by de Vries and his group of followers who became known as the "mutationists." They considered that the changes would have to be too rapid for gradual selection to be responsible for or create the necessary beneficial innovations. Rather they proposed that large advantageous mutations had occurred and that evolution took place "invariably ... by sudden leaps without intermediates." There were considered to be many mutations from which some would by chance be advantageous to the organism. As to why the detrimental mutations were not observed, he replied in 1905, "In my opinion it is the struggle for life which is the cause of this apparent rarity; which is nothing else than the premature death of all the individuals that so vary from the common type of their species as to be incapable of development under prevailing circumstances (thus the many) useless mutations will soon die out." De Vries was led to the conclusion that the importance of the environment to evolution has been overestimated and that by such large advantageous mutations organisms could be produced to fit or seek out a niche in which they could exist. He thus relegated natural selection to the secondary role of removing deleterious genes.

It was necessary to develop a more formalistic treatment of natural selection and perform experiments to test its validity before this matter could begin to be cleared up. A theoretical treatment was made possible when a mathematician G. H. Hardy and a physician W. Weinberg independently in 1908 rebutted the claim that if detrimental hereditary factors were dominant, they would, according to Mendelian heredity, soon spread throughout the whole population. They did so by considering the gene as a factor available to the entire species population through crossbreeding. For example, in a population of N individuals, it may be made up of X individuals with a homozygote genotype AA, Y individuals with a heterozygote genotype Aa, and Z individuals with homozygote genotype of aa. Then N would equal $X + Y + Z$, and $(X + Y)/N$ of the individuals would have a dominant phenotype, and Z/N individuals would have a recessive phenotype. On the other hand as X individuals contribute two gametes containing an A allele to each A gamete contributed by the Y individuals, and there are a total of $2N$ allele loci for the gene being considered in this population, the fraction of the total gametes produced by the population containing A would be $(2X + Y)/2N$, and $(Y + 2Z)/2N$ for a. They defined the first fraction as the allele frequency for A and symbolized p, the latter fraction the allele frequency for a and symbolized q, so that $p + q = 1.0$. If there were more alleles of the gene present in the population, their frequencies could be defined in a similar fashion and included for the total of 1.0.

If the normal case of equal gametes produced by the different sexes be considered, then $pA + qa$ sperms will be available and able to cross with $pA + qa$ eggs. Thus there will be $(p + q) \times (p + q)$ or $(p + q)^2$ zygotes formed. The result of these crosses will be a p^2 fraction of the zygotes with the genotype AA, a $2pq$ fraction with Aa, and a q^2 fraction with a genotype of aa; there thus being a fraction of $p^2 + 2pq$ individuals with the dominant phenotype and a fraction of q^2 individuals with the recessive phenotype.

Using this relationship, they found it possible to consider any genotype distribution in a parent population and calculate the zygote fractions in the next generation. For example, if 0.1 of the population has a genotype of AA, 0.2 Aa, and the remaining 0.7 has a genotype aa, then there would be 0.3 of the parent population expressing the dominant characteristic and 0.7, the recessive characteristic. p would equal $(2 \times 0.1$

+ 0.2)/2 or 0.2 and q would equal $(0.2 + 2 \times 0.7)/2$ or 0.8 for the allele frequencies of A and a, respectively. The zygotes or individuals of the next generation would have the genotypes $p^2 + 2pq + q^2$ or $(0.2)^2 + 2 \times 0.2 \times 0.8 + (0.8)^2$ or $0.04 + 0.32 + 0.64$ for fractions of the individuals with AA, Aa, aa, respectively. This would mean that in the descendents 0.36 of the population would have the dominant phenotype and 0.64 would have the recessive one. This is an obvious shift in distribution of both genotypes and phenotypes. If the allele frequencies are calculated, however, i.e., $(2 \times 0.04 + 0.32)/2 = 0.2$ and $(0.32 + 2 \times 0.64)/2 = 0.8$, they are the same as before and therefore the next generation would have the same genotype and phenotype distributions as this one and so would all successive generations.

This conclusion has been called the *Hardy-Weinberg law* or, as explained by Hardy, "(If a population of organisms whose) numbers are fairly large, so that the mating may be regarded as random, that the sexes are evenly distributed among the varieties, and that all are equally fertile, (then) there is not the slightest foundation for the idea that a dominant character should show a tendency to spread over a whole population, or that a recessive should tend to die out." Regardless of how the genotypes are distributed in the initial population, there will, under the conditions as stated above, be a set relationship or stable equilibrium, which will be established and maintained from the next generation on.

When Hardy and Weinberg proposed this quadratic relationship, they did not mean to imply that evolution, which would be a change in the allele frequency over a number of generations, could not take place. Rather they were establishing a relationship that must exist if the alleles are in equilibrium. They recognized that various factors could cause a shift in the allele frequency. With this quadratic relationship as a base, it eventually became possible to derive theoretical treatments of the factors that could alter the allele frequencies. The credit for this goes to two most gifted individuals, Sewall Wright and Ronald Fisher, who both first published some of their conclusions in 1930 and have followed this with numerous publications since.

Their conclusions showed that allele frequencies would change if there were nonrandom combinations of gametes fertilized because of (1) small populations or large inequalities of male and female gametes, (2) gene mutations, (3) migration of individuals, and (4) selection of mating and fecundity of individuals with phenotypes that best fit their environment—this last being the only one producing a nonrandom direction of change in the allele frequencies.

Although these theoretical treatments were necessarily valuable, it was difficult to find examples in nature to verify them. This was finally made possible about 1947 through the work of Theodore Dobzhansky, another of Morgan's students. Dobzhansky had noticed that there were about 15 different strains of *Drosophila pseudoobscura* based on specific inversion differences usually located in the third chromosome. He studied three of these strains in some detail: the Standard, *ST*; Arrowhead, *AR*; and Chiricahua, *CH*. He treated chromosome mutations as alleles. Dobzhansky noticed that the frequency of these different inversions

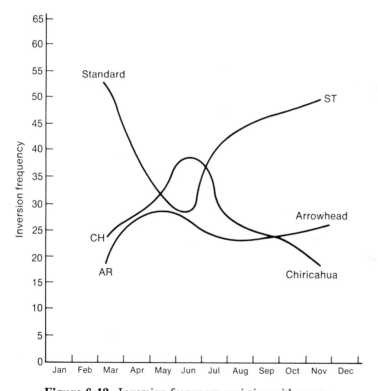

Figure 6-12. Inversion frequency variation with season.

varied with the season, as shown in Figure 6–12. For example, the frequency of *ST* fell from 50% in March to 28% in June and returned to 48% by September. This suggested that natural selection may be operating on these inversions, "but the great rapidity of the observed changes constitutes an apparent serious argument against accounting for them on the ground of natural selection." This was a common criticism of natural selection that was far from proved at this time; however, he wrote, "It should be remembered that very little is known about the intensity of selective forces which operate in natural populations."

Dobzhansky set up so-called population cages with several hundred flies with known combinations of inversion frequencies in each. Under these conditions, the flies multiplied rapidly but there was a severe competition for food. Also sample eggs were taken out periodically and cultivated in bottles with a surplus of food. In Figure 6–13 there is an

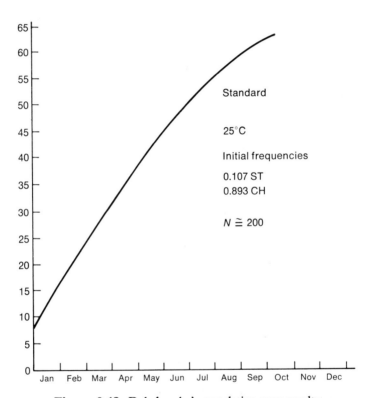

Figure 6-13. Dobzhansky's population cage results.

example of the results of *ST* changes in a representative cage maintained at 25°C. The starting frequencies were 0.107 for *ST* and 0.893 for *CH* in March. By December *ST* had reached 0.71. Wright analyzed this curve and concluded that there must have been a selection for the heterozygote form ST/CH over either of the homozygote forms, ST/ST or CH/CH. He also estimated that a theoretical curve could closely fit the data if the selective values for the different genotypes ST/ST, ST/CH, CH/CH were 0.7, 1.0, 0.4, respectively. If there had been no selection pressure, the selection values would all be 1.0, or the Hardy-Weinberg quadratic relationship. Dobzhansky found that the eggs taken from the cages and grown under nonselective conditions followed this Hardy-Weinberg relationship. It is considered especially significant that such responsiveness was displayed in this experiment. There is no apparent difference between the strains of this fly. Yet there is obviously some advantage in the hetero-zygote. This indicates how subtle the adaptations can be and how every trait of an organism, no matter how small it might appear to be, is probably being maintained by selective pressure.

There have been many other more recent experiments, especially with rapidly multiplying microorganisms, that support the formalizations proposed by Wright and Fisher. Of especial interest is the finding that regardless of the allele or the medium, at least with microorganisms, it takes approximately 200 generations for one allele to replace another if the two alleles start with frequencies close to 0.50.

This and many other experiments since are rather conclusive proof that natural selection can occur and this in turn supports it as the mechanism by which organisms evolve. They have shown that allele frequency changes are extremely fast, thus putting less credence to the belief that large mutations are necessary. It is Wright's contention that the organism is so responsive to selection pressures that it must always be closely adapted to its environment within the limits of its gene complement and that any large mutations must be deleterious, for they would certainly alter the organism drastically away from this best fit. This is a logical conclusion, but the certainty of it remains to be demonstrated somehow.

THE LIFE CYCLE CONCEPT

There are still questions about which part of the life cycle of an organism is most sensitive to natural selection pressures. When Darwin had thought about natural selection, he emphasized the total fit of the adult organism to the environment and considered that survival and reproductive potential of the adult form were the important issues. With the reactions against Haeckel, the developmentalists emphasized that every stage of development is under an equal pressure of natural selection and that these all influence the survival and fit of the organisms. With Weismann, however, the emphasis on the need for meiosis and fertilization as a source of variation became the sensitive part of the life cycle as far as pressures of natural selection were concerned. Most evolutionists accepted this point of view, for it seemed only natural that the part of the life cycle which would be under the greatest pressure of natural selection would be the part which would enhance the natural selection process itself. It was easy to imagine every other characteristic of the organism subservient to the sexual reproduction process. Now survival was considered only important to assure the formation of gametes and fertilization and in some cases to care for the next generation in order for them to be able to produce their gametes. The most successful organisms could be said to be the sexiest. Evolutionists have more than accepted what had been meant to be a criticism of natural selection when Samuel Butler wrote, "A hen is apparently nothing more than a means contrived by the egg so that another egg may be hatched."

There has been dissatisfaction between the developmentalists and the evolutionists concerning the meaning of life cycle. The former has been accused of missing its significance and ignoring the value of death, while the latter is accused of using circular and therefore fallacious reasoning. J. T. Bonner, a student of the life cycle, has suggested a way to resolve these differences of opinion and permit a fresh new look at the life cycle in his *Size and Cycle* (1965). Bonner conceives of the life cycle as a multitude of submicroscopic processes, presumably molecular, that flow as parallel trains through time resulting in the expansion and contraction of matter in a cyclic fashion with each cycle being a life cycle. He extends the heterochrony of the developmentalist to every single step in this flow of processes and "therefore, the life cycle may be thought of as a complex of chains of steps being continually encouraged and discouraged by a complex of environmental opportunities (the pressure of natural selection). The result is evolution, but the fact that many steps at both ends of the chains are connected makes isolating any one portion a difficult and certainly an artificial task. The life cycle, in fact, is the unit of evolution (it is the organism!), the unit of innovation and elimination (as indicated in Bonner's model shown in Figure 6-14), and it is for this reason more than any other that the life cycle has a central position in the structure of biology." Bonner concludes that it is the life cycle that evolves as a whole but with each infinitesimal step being acted upon, by natural selection. In a

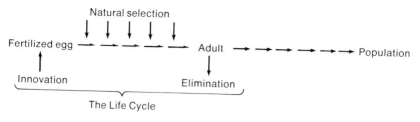

Figure 6-14. Bonner's model of the life cycle.

sense the life cycle is therefore an arbitrary unit defined by time, the size of the organism, and the points of innovation and elimination in an endless series of interdependent molecular steps. Although each step can be viewed as being acted upon independently by external factors of the environment, actually according to this model each step must fit with its adjacent steps. Each step is the result of the preceding step and a cause for the next step and any modifications accumulate during the life cycle. There is an inevitability to this flow regardless of shifts to the side with the modifications as diagrammed in Figure 6–15.

SPECIATION

Unfortunately, no amount of understanding of the natural selection process per se can reveal how many genotype changes are required to produce a new species. Species difference, the inability to interbreed, is the only criterion of a significant evolutionary change that is acceptable. Darwin had come to believe that speciation occurred during and as a result of adaptive radiations; however, whether adaptive radiations were the cause or effect of speciation and whether speciation could occur without an adaptive radiation was quite unclear then and is still uncertain in many details. On the one hand there were the mutationists who considered that a single giant mutation could produce a new species, and on the other there were those who thought that speciation could result from a simple recombination of those alleles already present in the population.

The mutationists were encouraged when Karpechenko was able to produce a new species in 1927 by forming a new hybrid. He did so by first combining radish with cabbage, each of which contain nine chromosomes (N). This cross was inferior and most often sterile. He was able to

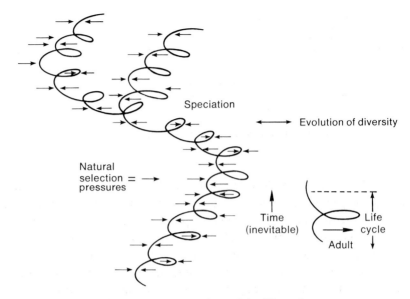

Figure 6-15. The flow of the life cycle.

induce polyploidy in this hybrid, however, resulting in a double hybrid with two sets of the original chromosomes as diagrammed in Figure 6–16. This resultant plant, which he named *Raphanobrassica*, was superior to either parent, a fertile vigorous plant that could not mate with either of its parents, i.e., a new species. This was considered quite significant as polyploidy is very common in plants.

It was clear, however, that this could not be the only and probably not the main source of new species. It was necessary to examine the possibility of speciation by forming new combinations of alleles. This must also include those alleles brought into the gene pool by gene mutation. Such speciation is a much more difficult process to assess. As more and more has become known, however, about the ease of diffusion of any allele throughout the gene pool and also that there is a large number of allele differences between any two species, it was generally accepted that speciation by recombinations of alleles is most difficult if not impossible. This eventually led to the conclusion that speciation could not occur unless the population became divided into reproductively isolated groups. And because of the ease of interchange of alleles the only way that this could occur would be if the subgroups were geographically isolated.

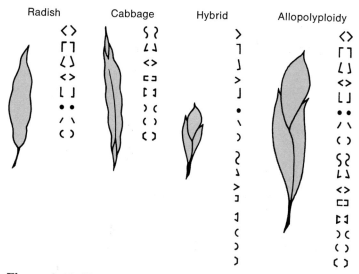

Figure 6-16. Karpechenko's synthesis of the species *Raphano-brassica.*

The main evidence for the requirement of geographic isolation was the finding of high endemicity associated with recently formed isolated geographic regions. Also, these endemic species were found to be closely related to a widely distributed form on the other side of the isolation barrier. This even holds true with regard to distance between subisolated regions as witnessed with the percentage of endemicity found for the various Galapagos Islands, shown in Figure 6–17.

The question of how many genotype differences might be required for speciation is made difficult because some traits seem to alter easily and have little effect on the production of a different species. For example, some behavior differences may appear that would reduce breeding between the different forms. On the other hand there may be differences such as those of habitat preference that would have less effect upon crossbreeding.

This model of speciation is as yet supported only by empirical evidence. It has been too complex to formalize theoretically and allow predictions and testing. From controlled experiments on microorganisms, however, it has been estimated, quite roughly, that about 100 gene changes would be required for speciation. This is considered a conservative estimate at that. And if it requires about 300 generations to replace an allele in a population

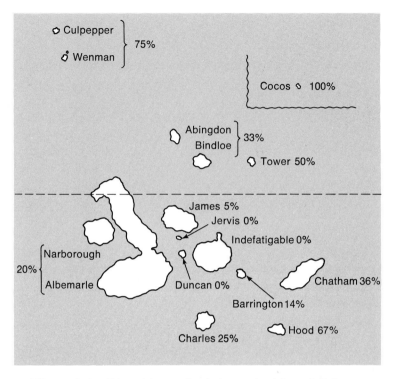

Figure 6-17. Disposition and relative endemicity of Galapagos finches.

by natural selection, then at least 30,000 generations would be required for speciation. If species differ by 1000 genes, which is considered very likely, then 300,000 generations would be necessary. Some very crude support for this estimation has been offered by noting that very recent islands like Great Britain, which has been separated from the continent since the last glaciation, or about 10,000 years, have experienced no speciation, while those islands with many endemic forms are several million years old, suggesting that somewhere between 10^4 and 10^6 years are required for the process.

MAN—SOME QUESTIONS ON
NATURAL SELECTION AND SPECIATION

The above presents a good case for natural selection and speciation, thus supporting the concept of evolution. It also reveals that there are great holes in the theory and many missing details about the mechanism natural selection. Perhaps no case demonstrates the uncertainties as does the question of whether man is evolving today. For there are many very respected evolutionists who are doubting that man is evolving today. If this were the case, we would have serious concerns about the basic nature of evolution and natural selection. One of the arguments against man's present evolution is that man has lost the opportunity to speciate by geographic isolation. Man is still one species although he had been on the way to diverse forms as evidenced by the subspecies or races of man that exist. But now with his expanding numbers and interbreeding of races on the increase, there can only remain one species and even the diversity of the race will diminish in time. That is unless other planets and stars offer man his Galapagos Archipelago, which seems quite remote at present, but something to remember.

Further arguments suggest that even progressive evolution of the species is slowing down because of the decrease of selection pressures to change the present genotypes, and man's adaptability to his environment. This decrease in selection is said to be the result of the increase in socialization of medicine and protective institutions that are retaining detrimental alleles in the population. In fact many feel that this reduction in selection pressure allows too much diversity or spread of genotypes and thus degeneration of the species, which will result in an ocean of mediocracy rather than of valuable specialization. Also man's technology has permitted him to become considerably less dependent on specific parts of his environment and, in fact, is having a greater effect on changing his environment than it is on him. Darwin's own great grandson feels that the population explosion of man has been so great and his technology so unwisely used that there is a loss of natural selection to check population and technological growth by encouraging degeneracy, famine, and wars. This will have the effect of not only slowing down his growth rate, but forcing man to return to more primitive conditions where the environment can again exert selection pressures. Perhaps this will act like a brake to allow man's intelligence and behavior to progress to the point where such drastic steps are not necessary. Regardless of views on the above there is no doubt that, in place of man's physical evolution, cultural evolution, the progress of ideas, has taken over and represents the leading edge of man's change. According to this, any new form of diversity in man will be in ideas, not physical structures. In other words, the brain of man is so complex that it does not need to change further to accommodate this endless expansion of ideas; witness history and the inherent capabilities in man as yet unrealized. Could cultural evolution be inhibiting physical evolution?

All the above must remain as conjecture until more is known. How many alleles are being preserved by socialization that would have been removed by natural selection? Are diseases like diabetes and some nervous disorders alleles that can exist in modern times? There are those who say it is impossible for degeneracy to result from a relaxation of natural selection and the increase of genotypes. Diversity has nothing to do with degeneracy. And as for eugenics programs, including legalized abortions and mercy killings, almost nothing is known about their effects on allele frequencies. There is evidence, such as changing blood types in Africa, to show that at least some allele frequencies are changing. But what does this mean? As Dobzhansky has written, "Theoretical discussions of these matters are plainly insufficient. Natural selection has been talked about for more than a century; we must now start measuring it, in man and in other organisms. As things stand at present, we have very few even approximate estimates of the Darwinian fitness of genetic variants in human populations. Such quantification is indispensable for understanding the status and the perspectives of the gene pool of our species. As it is now, all predictions and eugenical utopias have at best only the status of educated guesses or personal opinions. In particular, it is of crucial importance to learn what part of the human variability is due to recurrent mutation and what part is maintained by balancing natural selection. Indeed, while the mutational genetic load (the burden or unfavorable, unadaptive part of the genotype) should, as far as possible, be minimized, the balanced load is a 'load' only in the sense in which the expenditures which a community makes to bring up and to educate its younger members are a 'load' on that community." Likewise some better measures of the genetic homeostasis of a population must be found.

THOUGHTS

The search for explanations for the order of continuity has been very productive. There are, as inevitable, the remaining questions indicated above. The progression of answers to what it is that is continuing, i.e., cells to chromosomes to genes to gene pools, is not complete. This search has produced probably the greatest number of satisfactory syntheses as any, however. The cell theory emerged and could be synthesized with hereditary observations and heredity could be synthesized more clearly with evolution as the mechanism natural selection became clarified. Indeed, the search for the order of continuity permitted a synthesis of, or at least a common crossroads for, all the orders of life considered so far. It is also with this search that we come the closest to the ideal scientific process as judged in Newtonian terms. Here we could see that the perfection of a tool, the microscope, and related procedures allowed us to go from the seeable, the cell, to the unseeable abstractions below—panmerism and the gene. With Mendel and Morgan and his great students we find the beauty and productivity of simple quantitative experimentation and testing. The Bridges' case of the exception proving the rule is brilliant. The deductive process is at its greatest in the development of the gene, the crossover, the gene pool, etc.

vii

order: in matter and energy

It has this throwing backward on itself,
So that the fall of most of it is always
Raising a little, sending up a little.

West-Running Brook, ROBERT FROST

Life is a special and very complicated form of the motion of matter.

A. I. OPARIN

This order has resulted from a long search for the source of life. A definition of its source has also come to include a distinction between living and nonliving matter. There has been no consistent development of explanations for the source of life until recent times. As expected with no clear observations, earlier explanations were based on inaccurate observations and mainly basic assumptions. This search more than any of the others must be the most speculative. Historically, explanations jumped from transformations to autogeneration to creations and finally to the very long process of evolution of matter and energy. This chapter traces these assorted explanations historically. Two extreme semiartificial models of life, clocks and candles, are used throughout as reference guides. The order implied in this chapter was not really developed as a conception

234

until long after evolution had declared the existence of a times arrow and many indirect facts and observations had accumulated. It was then conceived that organic evolution was but a part of the cosmic evolution, the time pattern of change of all matter and energy. All areas of science had contributed facts about matter and energy and with this conception these facts could be integrated. The latter part of this chapter briefly traces the hypothetical explanations for the different steps of this conceived order as subunits of matter and energy combine into ever higher levels of organization.

EARLY IDEAS ON THE "SOUL" OF LIFE SUBSTANCE

The earliest myths from all parts of the earth suggest that the people of these times had no difficulty defining the living. Indeed there was no nonliving and the "soul" or life was in all substance as a continuum. The basis of this belief was the apparent ease of sudden transformations from stone to man and the reverse, to mention but one common theme. In later myths each stone and man had his own soul. Transformations in outer expressions of this soul, stone to man, man to bird, and others, were still common but they may take a generation to change or some special circumstance. It is important to realize that even the early ancient Greeks who spoke of the transformation of living organisms from the earth, slime, sea oozes, or what have you, considered that these latter were as alive as anything else in the universe. It was probably Anaxagorus (510–428 B.C.) who first offered a principle related to the "soul" stuff of substance when he considered that it was impossible to create or destroy the life element. Later Aristotle put forth an explanation of the transformation process that was well accepted. He declared the basic elements, earth, water, air, and fire, contained different amounts of soul—earth having considerably less than the other elements. The transformation process involved the combining of specific relative amounts of the different elements in the formation of a particular organism: the more earth, the more plant-like; the more water, the more aquatic-like; etc. Aristotle elaborated the recipes for the formation of a number of organisms. He considered that these "spontaneous generation" transformations were no more difficult than the type of tranformation from a parent organism to its offspring. In this latter process, the soul is passed onto the egg of the developing offspring. Although by no means restricted to this hereditary method of transformation, most higher animals including man were formed in this way.

CANDLES AND CLOCKS

Although transformations were still considered possible, there was by the seventeenth century various elaborate descriptions of the soul of

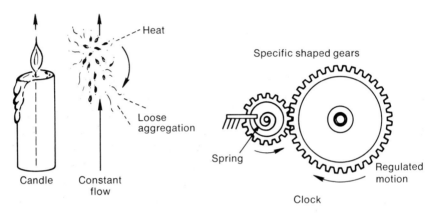

Figure 7-1. The candle versus the clock models of the element of life.

animals. Among these a popular model diagrammed in Figure 7-1, was that painted first by Leonardo da Vinci. He wrote, "The body of anything that takes nourishment constantly dies and is constantly renewed; because nourishment can only enter into places where the former nourishment has expired, and if it has expired it no longer has life. And if you do not supply nourishment equal to the nourishment which is gone, life will fail in vigor, and if you take away this nourishment, the life is entirely destroyed. But if you restore as much as it is destroyed day by day, then as much of life is renewed as is consumed, just as the *flame of the candle* is fed by the nourishment afforded by the liquid of this candle, which continually with a rapid supply restores to it from below as much as is consumed in dying above; and from a brilliant light is converted in dying into murky smoke; and this death is continuous as the smoke is continuous; and the continuance of the nourishment, and in the same instance all the flame is dead and all regenerated, simultaneously with the *movement of its own nourishment.*" This abstract model of the soul emphasizes that life is the flow of a substance or principle from without that does maintain a delicate and dynamic equilibrium of the parts of the organism whatever they may be. This model recognizes the instability as well as the aggregating power of the soul. It also emphasizes the absolute dependence on the environment of the organisms. There is inexactness, indiscreteness, and the necessity of the continuous vibrant motions of the substance that is living, including a substitution of parts.

An opposite model that emerged later, was that probably most aptly stated by René Descartes (1596–1650) when he wrote, "All the operations of the bodily mechanism, such as digestion of food, beating of the heart, pulsing in the arteries, feeding and growth of the members, breathing, waking and sleeping, vision, hearing, smell, taste, getting warm, and similar qualities revealed by organs through the external senses, impressions of ideas through common sense or imagination, retention of such ideas in the memory, internal impulses caused by desire and passion, and lastly, the movements of the limbs, which become so accurately adapted to the purposes put to them through the senses, come from the *arrangement of the organs* and from no other source whatsoever—just as the working of a *watch* or the action of an automaton depends on the gears and weights which cause it to go." This abstraction emphasizes the exactness of fit and interaction of the specifically arranged parts. There is a greater emphasis on the particularness and structure of the parts. There is a constancy, a stability, and a self-regulation property rending the organism independent from its environment.

TRANSFORMATIONS AND ORGANIC MATTER

The descriptions of the soul of organisms at this time involved the conditions of the substance that caused it to be living rather than the substance per se. There was no need to reconsider the ability for substance to be transformed from the nonliving to the living condition. In fact many experimentalists supplied additional recipes to Aristotle's transformation cookbook. Only one among many was Newton, who was convinced that plants could arise from the transformation of the emanations of comet tails. As time progressed, however, several observations began to suggest that the transformation of complex animals from nonliving matter may be much rarer than had hitherto been presumed. One of these was the discovery by Francesco Redi (1626–1697) that the white worms often found crawling from decaying meat were actually the larvae of flies. He went on to discover that if the meat was covered by muslin, the white worms were never found on the meat but rather on the muslin and that they were the result of eggs layed by the flies onto the muslin. Thus this was really a hereditary type of transformation being mistaken for a "spontaneous generation" type of transformation.

With further observations such as this, there developed the opinion that complex organisms may rarely if ever be transformed from stone-like matter. Rather there must be a unique substance, an "organic" substance, as well as a soul associated with organisms, and perhaps spontaneous generation could only occur when this unique substance was available and then restricted to the simpler kinds of organisms. This suspicion was elegantly demonstrated by many observations made possible by Leeuwenhoek's discovery of microorganisms. It was soon established that only where there was a fermentation process (the production of alcohols from crushed fruits and malts) or a putrefaction process of some dead organism could these microorganisms be seen to appear. Apparently the organic substance could be used again by their transformation into vital microorganisms. The actual transformation process, the sudden appearance of a microorganism, could be followed under the microscope. Experiments performed with a variety of materials confirmed the uniqueness of organic substance by its ability to serve as a precursor for the transformation to microorganisms.

With the establishment of organic matter there was a natural tendency to relate it to the soul of the organism. Buffon ingeniously hypothesized that the soul acted as a vital principle that held "organic particles" together in a variety of dynamic combinations, each one of which was a specific kind of organism. These particles were the substance of both hereditary and spontaneous transformations. In the former, the particles from an organism could multiply and recombine to produce a similar offspring. In the latter, the particles from decaying or fermenting organisms could break loose and recombine in different combinations yielding microorganisms.

This model was well accepted and enhanced by J. T. Needham (1713–1781), a contemporary of Buffon, who described the vital principle, the soul, in terms of "vital forces" that pulled and held the aggregate of organic particles together. Needham was concerned with the relationship between these vital forces and the recognized vital heat of an organism. He "took a quantity of mutton gravy hot from the fire heat and shut it up in a phial closed with a cork so well masticated that my precautions amounted to as much as if I had sealed my phial hermetically" and found that the gravy was full of microorganisms. From this and many similar experiments, he concluded that added heat could work with or was the same as the heat generated during fermentation or decay and indeed was the vital force that aggregated the particles together. To put it more strongly, if vital heat, i.e., vital force, was present, microorganisms would have to occur. The need for air to support combustion and vital heat had been well established long before these experiments, and thus a candle model of life was strongly supported.

Spallanzani had come to believe that an organism was far too unique and stable to be maintained by the dynamic fancies of heat, however. He thus set about in 1765 to subject many kinds of organic concoctions to long periods of heating and then to immediately seal them to prevent further air from reaching the heated substance. He found that microorganisms did not appear nor did decay take place. He thus concluded that there probably was no such thing as vital force and certainly if there was, it could not be related to heat. He even extended his conclusion to maintain that organic matter could not be transformed into microorganisms. This set off a series of debates and counter experiments between Needham and Spallanzani that attracted considerable attention and succeeded in placing the validity of spontaneous transformation into question. In the long run, however, Needham upheld his conclusion on the argument that the prolonged heating in Spallanzani's experiments had either been too intensive and destroyed the vital nature of the vital forces or that it had destroyed the vital factor of air. As Spallanzani found that whenever he allowed fresh air into his concoctions microorganisms appeared, his only recourse was to claim that the microorganisms had managed to come into his flasks with the fresh air—a very unconvincing argument.

What Spallanzani could not show with experiments on transformations he attempted to do by analysis. He became one of the outstanding workers to elucidate the nature of organic substance. His method was to collect and separate in their most natural state as many of the "immediate principles" that make up the bulk of plants and animals as possible. He collected juices and substance from the stomach, from squeezed plants, and from body excrement. By the end of the eighteenth century Spallanzani and other workers had managed to isolate what were later named tartaric, oxalic, malic, citric, lactic, uric, formic, gallic, benzoic, and several other acids; urea (1773), sugar of milk (lactose); cane sugar; cholesterine; and waxes. There was no longer any doubt that organic matter was distinct from inanimate matter. In spite of this wealth of information about organic matter Spallanzani could find nothing to deny the possibility of vital forces, only that the organism was indeed a very complex mixture of many "immediate principles."

What did bring Needham's vital force under discredit was the remarkable analysis of air that progressed through the latter part of the eighteenth century and culminated in the beautiful experiments of Lavoisier when he demonstrated that the organism was more like a candle than even Needham could accept. It was oxygen, a specific chemical element found in air, that was necessary and consumed during either the combusion of a candle or during the respiration of an animal, and a second gas, carbon dioxide, that was emitted during these processes. Respiration was also responsible for the production of the body's heat. The similarity was too close to allow a vital force heat to be of a nature any different from that produced by the simple burning of inanimate matter.

Lavoisier was also responsible for isolating a number of "immediate principles," and by means of careful measurements with combustion experiments concluded that these principles were usually made up of at least carbon, oxygen, hydrogen, and nitrogen. Spallanzani extended Lavoisier's experiments to many other organisms and was able to say that oxygen was absorbed, carbon dioxide given off, and heat generated in the same constant proportions in essentially all higher organisms. He also discovered that fermentation did not require oxygen although carbon dioxide was given off and heat was generated. He was unable to explain this discrepancy but did suggest that there might be different kinds of respiration in microorganisms.

The discrediting of vital force heat and the finding that the vital factor in air was oxygen, an element that could not be harmed upon heating, reopened the question of the ability of organic substance to be transformed into microorganisms. In spite of the lack of a more acceptable explanation than that of Needham's, L. Gay-Lussac (1778–1850) demonstrated by some elegant experiments in 1810 that indeed transformation does occur. He took a glass tube sealed at one end, filled it with mercury, and turned it upstanding in a dish of mercury, as shown in Figure 7-2.

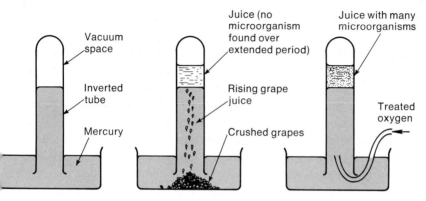

Figure 7-2. Gay-Lussac's experiment on transformation.

In this position the mercury would be pulled down by its own weight from the closed end of the tube, producing a perfect vacuum in a space above the column of mercury. He then crushed a grape under the mercury in such a way that the juices rose to the top of the column inside the tube. He noted that the juices could be maintained in the tube for extended periods of time and no decay took place and no microorganisms could be found in the juices. If he added even a single bubble of heated or treated oxygen, however, fermentation would immediately start and the juices would be full of microorganisms. Since oxygen could not be damaged, apparently Spallanzani had been destroying some kind of vital principle by overheating his concoctions. Also in contradiction to the results of Spallanzani apparently oxygen is necessary for fermentation and for the transformation process by whatever mechanism it might occur.

THE CONSERVATION OF MATTER

Lavoisier through his careful measurements of weights and volumes contributed profound new concepts regarding matter itself. He was able to show that oxygen had weight. He did this by showing that there was an increase in weight of quick silver when it and oxygen were heated together at moderate temperatures. He also demonstrated that oxygen was released if the combined form is heated at still higher temperatures. This process could be reversed many times without a decrease of the weight or the volume of oxygen involved. Thus Lavoisier demonstrated that oxygen is neither created nor destroyed and yet can apparently combine with other substances with a change of state, i.e., gas to solid and the reverse. He concluded that regardless of the state of matter, solid, liquid, or gas, oxygen was always conserved. This meant that during combustion or respiration, oxygen was combining with some substance and a different gas, carbon dixoide, was presumably coming off the same substance. By weighing a combining substance, which could be pure carbon, it was clear that combustion or respiration was a loss of this substance with the degraded matter being converted into a gaseous state.

By 1800 the conservation of matter hypothesis was supported by the experiments of others who demonstrated that hydrogen and oxygen could be made to combine to produce water and the reverse without loss of matter. Because of the simplicity of this reaction involving only two easily measured gases, John Dalton (1766–1844) was able to find that hydrogen and oxygen always combine in fixed ratios by weight, 1:8, respectively. From this data Dalton conceived that the only way that two gases could combine and form a liquid is if the gases are made up of small particles that can combine in a set manner forming a liquid made up of the composite particles. Dalton assumed that these particles, called *atoms*, differed from one element to the next by weight, the oxygen being eight times the weight of a hydrogen atom, so that only one atom of each combined to form water. He formulated these thoughts in his atomic theory of matter in 1808. It was (1) two or more elements combine to give a particular compound in constant and definite proportions by weight; (2) when two elements combine to form two or more different compounds, the amounts of the one element that combine in the different compounds with a given amount of the other are themselves in a simple ratio; and (3) when an element A forms compounds with either of two elements B and C, and when B and C combine with each other to give a third compound, the proportions in which B and C separately combine with a given amount of A will also be the proportions in which they combine together. Outside of being very small there was no idea of the size of an atom or, say, how many atoms there might be in a gram of hydrogen. Nor was there any idea of what held different atoms together. The particulate explanation for matter, at least inanimate matter, however, was the only logical model to explain the hydrogen and oxygen reaction, and it was considered that any model for other reactions would also have to be of a particulate nature.

INORGANIC(?) CATALYSIS

With the increased study of reactions during the early 1800's, certain very unusual reactions were discovered. It was found that a liquid later known as hydrogen peroxide would decompose into oxygen and hydrogen without the necessity of heat, a strict requirement for the earlier studied reactions involving these two gases. But what was even more surprising was the finding that if the metal platinum was present, hydrogen and oxygen could combine to form water at room temperature. It also was first shown in 1811 that the plant substance starch could be changed into sugar if boiled in acid and the reaction caused no alteration to the acid. Later, the plant product alcohol was found to give off carbon dioxide with the uptake of oxygen at room temperature if finely powdered platinum was present.

It was Jons Berzelius (1779–1848) who suggested that "in reactions of this nature there is present a chemical force quite distinct from any hitherto known." He defined these reactions in which an agent was necessary and yet not altered during the reaction, e.g., the platinum or acid, as "catalytic" reactions distinct from the more common "analytic" reactions wherein all the reactants changed.

A "glutinous component" of wheat was discovered in 1815 that could convert starch to sugars. These studies eventually led to the isolation of a substance from fermented barley in 1832 by A. Payen (1795–1871) and J. F. Persoz (1805–1868) that could convert the starch to sugars, apparently by a breakdown of the starch. They found that it was insoluble in alcohol, soluble in water, and when "heated to 65–75°C with starch, it has the remarkable property of promptly detaching the envelopes from the modified internal substance, the dextrin (a sugar), which readily dissolves in water, while the insoluble coatings either float to the top or are precipitated, according to the movements of the liquid. This singular separating property has prompted us to give the substance that possesses it the name of diastase (Gr. separation) which exactly expresses this fact."

Berzelius offered the reasonable explanation that basically all catalytic reactions worked on the same principle but that organic catalysts such as diastase derived their "force of reaction" from the particular complexity and organization of compounds peculiar to living substance. He considered that the similarity between inorganic and organic catalytic reactions could well mean that in addition to the obvious vital processes involved in organisms there were these diastase-like breakdown processes producing less complex products such as sugars. He wrote, "There are well justified reasons to suppose that in living animals and plants, thousands of catalytic processes take place between the organic fluids and tissues, ... It is likely that some time in the future we shall find out the cause . . . is the catalytic power of the tissues forming the organs of the living body." The regularity of tissue structure being discovered at this time suggested that a more clock-like organization of living substance could account for at least the nonvital reactions of the organisms.

Berzelius' conclusion was supported by the accidental discovery made by Frederich Wohler (1800–1882) in 1828 that a relatively simple body product could be formed by an analytic-type reaction. Wohler was attempting to form ammonium cyanate by combining lead cyanate and ammonium hydroxide but instead obtained a crystalline substance that resembled urea and had quite different properties from ammonium cyanate. If this was true, it would be the first time that an "intermediate principle" had been formed by an analytical reaction. Together with Justus von Liebig (1802–1873) he carried out extensive analysis of this reactant product and his own urea and came to the conclusion that they were one and the same substance. Berzelius' contention that organic catalytic and inorganic catalytic reactions were basically similar led them to test the hypothesis that analytic reactions involving organic compounds could be analyzed in the same way that inorganic ones are. As Liebig wrote, "The extraordinary and to some extent inexplicable production of urea without the assistance of vital functions, for which we are indebted to Wohler must be considered one of the discoveries which a new era in science has commenced."

There were some who argued that urea was not a typical organic compound, being much like the inert product carbon dioxide. Others pointed out that the lead cyanate was similar to urea, it being a body isolate, and that the reaction was probably a very simple one. Nevertheless, Liebig, especially, became more and more optimistic and wrote by 1838, "(The ability to separate and analyze) all organic compounds as long as they are not a part of an organism (only involved in the catalytic processes and not the vital ones) must be seen as not merely probable but as certain." He enthusiastically set up a large laboratory group to analyze every organic compound that could be found and announced his method over his laboratory doorway with the sign "God has ordered all His creation by weight and measure."

Liebig's optimism was contagious and he swept many contemporaries into the task. Together they came to the conclusion that the carbon atom could be assumed to act as any of the other elements but there was as ever great uncertainty as to how it interacted with other elements. They reaffirmed Lavoisier's preliminary investigations that all organic compounds were made up of carbon, nitrogen, oxygen, and hydrogen, but there were obviously far too many isolated compounds to assume that some simple element-to-element composite based on Dalton's laws could explain these compounds.

A diversity of models and ideas was suggested to explain the complexity of carbon compounds. All of these attempted to retain the elemental nature of carbon but there was no general agreement or way to test any of these models. Urea and ammonium cyanate proved to have the same relative proportion of elements $C:O:N:H$ as $1:1:2:4$ and yet different properties. Gay-Lussac suggested that the difference in these "isomers," as they were called, may be due to a difference in the organization of the elements. Liebig, considering Berzelius' conclusion, suggested that perhaps organic compounds could be identical with metals or other large atoms (by weight) and thus serve the role of, say, platinum in the catalytic reactions of the body. In spite of these many uncertainties, by the middle of the century these workers were able to conclude that living matter consists of extremely complex mixtures of only three major classes of organic compounds: (1) the hydrocarbons containing only carbon and hydrogen; (2) the carbohydrates containing carbon, hydrogen, and oxygen; and (3) proteins containing carbon, hydrogen, oxygen, and nitrogen.

CELLULAR CATALYSTS

In about 1836 Schwann and others discovered that a particular microorganism, yeast, was responsible for fermentation. He also isolated an organic catalyst from the digestive juices that could break down proteins. From these results he concluded that organic catalysts were probably of two types, the organic compound type such as diastase or his digestive isolate and a second type that required the complex organization of a cell such as the yeast for fermentation. He saw a great difference in the complexity of the reactions with the former type of catalyst not being able to carry out the more elaborate processes of fermentation.

Liebig disagreed. His analysis of compounds and his model of a single type of catalyst allowed him no justification to consider that fermentation was any different from any other breakdown process. He violently attacked any cellular catalyst model in his influential book *Die Thierchemie* (1842). Here he described his understanding of all organic catalytic reactions. "Different bodies, in the act of combining or decomposing (fermentation or putrefaction) are found to be in a position (even independent of the cell) to induce such processes, that is to cause changes in balance between the component atoms of the molecules, particularly of organic molecules which are highly complex, and causing these atoms to become arranged in other states of balance of greater stability, i.e., in the form of smaller moleculed bodies, in accordance with their natural affinities. Among the substances especially active in this respect must be reckoned in the first place those organic substances containing nitrogen (proteins), particularly the albumins."

The many discrepancies regarding fermentation were finally clarified to some extent when Louis Pasteur (1822–1895), starting in about 1854, concluded a survey of all the fermentation processes possible. By 1862 he could write, "I have found that all the fermentations—slimy, lactic, butyric, the fermentation of tartaric acid, of malic acid, of urine—were always related to the presence and the multiplication of (specific, e.g., tartaric acid is fermented by *Penicillium*, etc.) organized beings. And the fact of the organization of yeast, far from being a troublesome thing for the theory of fermentation, on the contrary placed it with the common rule and made it typical of all ferments." Pasteur's conclusion supported Schwann. They both agreed that there were both unorganized (diastase-like) catalysts and organized (cellular) catalysts. Again Liebig was unconvinced, and he attacked Pasteur's experiments as being inadequate and probably poorly done. Pasteur countered by a series of extensive experiments in which he convinced everyone except Liebig that the intact yeast cell was absolutely necessary for fermentation. In the process he demonstrated that yeast has two forms of metabolism (in this case catalytic processes). If the yeast is exposed to oxygen, fermentation does not occur, but rather respiration takes place with the uptake of oxygen and the emission of carbon dioxide. If no oxygen is available, however, the yeast produces alcohol and carbon dioxide; i.e., fermentation occurs. He also discovered that carbohydrates are catalyzed by both processes and essential for both processes to occur and that one of these processes must occur if the yeast cells are to continue to live. In other words, there is a specific carbon compound requirement for the maintenance of at least these microorganisms. In addition to carbohydrates, he found that less specific nitrogenous compounds and phosphate were required for the maintenance and growth of yeast. Thus even microorganisms were very complex and specific structures with unique requirements and could not fit the sweeping generalities of catalysis as Liebig was proposing.

AUTOGENERATION

Any detailed examination of and hypothesis regarding fermentation put Pasteur and others face to face with the question of spontaneous transformation of organic compounds into microorganisms. This phenomenon had again been brought into question with the rise of the cell theory, which required that all cells come from previously existing cells and that microorganisms were included in this category as they are often single cells. It had also been shown that hereditary transformation required the flow of cells, the egg and sperm. In 1836 both Schwann and Schulze, advocates of the cell theory, attempted to discredit the remaining transformation theory. Schwann found that if air was heated before it reached previously boiled meat, there was no transformation. If the meat was surrounded with liquids containing sugar, however, then microorganisms and fermentation resulted. Schultze found similar results by passing the air through concentrated sulfuric acid. Other doubters tried and failed to prevent the transformation process up through the middle of the century.

This led F. Pouchet (1800–1872) to propose in 1859 that indeed there was apparently an autogeneration process similar to the one proposed by Needham but not necessarily involving heat. As with the older theory, only organic compounds and probably the albuminous kinds could re-aggregate if accompanied by fermentation or putrefaction. Pouchet carried out a number of experiments that supported this theory. The subject had become so controversial by this time, however, that the French Academy of Sciences offered a prize to anyone who could explain the details of autogeneration, as it came to be called. Pasteur was necessarily involved because his results could be questioned if microorganisms could arise from organic compounds in his solutions. Fermentation, catalyzed by organic compounds as Liebig had proposed, may precede and cause the process of autogeneration as Pouchet had claimed. If this were possible, the cellular form of catalyst would not be necessary for fermentation. Liebig also claimed that "albuminous substances exposed to contact with air, undergo an alteration, a particular oxidation of an unknown nature, which gives them the characteristic of a ferment (fermenting catalyst)."

It was the general conclusion of those who opposed the autogeneration theory that the cells of the microorganisms were entering the experimental flasks by way of the air, but no one had been able to prove this. Pasteur undertook to do so by filtering "a measured volume of air through gun cotton which is soluble in a mixture of alcohol and ether." A microscopic examination of the solvent revealed a variable number of corpuscles, the form and structure of which indicated that they were organisms.

Pasteur then placed about 10 grams of sugar and about $\frac{1}{2}$ gram of albuminous material and minerals obtained from brewers' yeast for each 100 milliliters of water into a flask, as shown in Figure 7–3. He wrote,

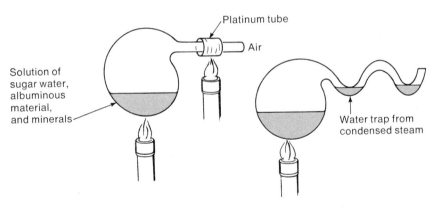

Figure 7-3. Pasteur's experiments on transformation.

"Sugared yeast water brought to a boil for two to three minutes and then exposed to air which has been heated does not alter in any degree (except for a darker color with time apparently from the direct oxidation of the albuminous matter), even after being kept eighteen months at a temperature of 25–30°C, while if one leaves it in ordinary air (or adds cotton containing the filtered organisms), after a day or two it is in the course of a manifest alteration and is found to be full of bacteria, vibrios, or covered with mucors." He obtained the same results if he heated a similar solution in a flask with a twisted neck. After the heating, the steam condensed in the lower dip of the neck and apparently trapped any microorganisms from entering the cooling flask. If later the neck was cut off, again the solution was soon full of microorganisms.

In these simple and elegant experiments published in 1862, Pasteur convinced all that the failure of previous experiments to give the same results was due to a lack of assuring that microorganisms in the air did not contaminate the sterile organic substance. He repeated Gay-Lussac's experiments and showed that microorganisms were present on the mercury or glass surface. He also demonstrated that the mold, *Penicillium*, could grow equally well on rock sugar, ammonium tartate, and the ashes of brewers' yeast as well as albuminoid substance in place of the ammonium tartate. Thus growth of microorganisms did not require an albuminoid (protein) as had been claimed by many. These experiments firmly closed the question of autogeneration as being quite impossible and emphasized that the vital part of the cell was the result of its complex organization or internal structure and not contained within any single organic compound. Pasteur won the prize.

AUTOGENERATION AND EVOLUTION

Of especial interest was the effect of Pasteur's publication on the various impressions of the stability of life. Before his publication, life seemed to be a more flexible entity, perhaps even changing state from the inorganic to organic to the complex organic and back again. But after the publication, life became associated with a very unique complex structure, the cell, which was stable and perhaps endlessly so. It is significant that just previous to this publication, Darwin had shaken the scientific world with convincing arguments that life was evolving. If at that time one had stretched his imagination, he might well have considered that all present living organisms were derived from a single cell and he would have also had to ask where that came from. Very few, except Darwin, had had the time to think about this aspect of evolution. He wrote about this topic in ambiguous and wise ways. On the one hand he wrote, "It is often said that all the conditions for the first production of a living organism are now present which could ever have been present. But if (and oh what a big if) we could conceive in some warm little pond, with all sorts of ammonia and phosphate salts, light, heat, electricity, etc., present that a protein compound was chemically formed ready to undergo still more complex changes, at the present day, such a matter would be instantly devoured or absorbed, which would not have been the case before living creatures were formed." He also wrote, "It is mere rubbish to talk about the origin of life, one might as well talk about the origin of matter."

It is perhaps amazing how Pasteur's paper had such a one-sided effect to override any influence of the origin of life aspects of evolution. Apparently *The Origin* had more of an effect of extending time rather than of endless change. Even Liebig would agree with Pasteur as the former wrote, "It is sufficient to admit that life is as old and as eternal as matter itself, and the entire argument about the origin of life loses apparently all sense by this simple admission. And really, why can we not imagine that organic life is just as much without beginning as is carbon and its combinations, or as is all uncreated and indestructible matter and the forces which are eternally bound up with the movement of matter in universal space?"

There was, however, one difficulty in assuming that life was eternal, mainly that according to contemporary hypotheses, the earth had arisen from a molten mass that could not have supported life. Pasteur's experiments demonstrating the large number of organisms in air helped Liebig and many others to conclude that "the atmosphere of celestial bodies can be regarded as the timeless sanctuary of animate forms, the eternal plantations of organic germs." Life is never really created but only transferred from one planet to the next. Meteorites were considered a very likely means of dispersing tiny microorganisms. Another possible method of travel was suggested by Svante Arrhenius (1859–1927) at the turn of the last century. From his ingenious calculations the pressure due to light was strong enough to propel bacteria and spores through space at high rates of speed and at temperatures no greater than 100°C. He estimated that it would take only 14 months for spores to travel from the sun to the earth and only 9000 years from the nearest star to our planetary system. Thus it was generally assumed that although life-forms may evolve, the single cell was eternal.

ORGANIC MOLECULES—EARLY CONCEPTS

The momentum generated by Liebig to analyze organic compounds never let up. Friedrich Kekule (1829–1896) reaffirmed Liebig's working hypothesis that carbon could be considered like any other atom and by analyzing enough reactions he concluded in 1857–1858 that carbon was tetravalent; i.e., it could combine with four other atoms. Where Liebig had stressed content of the organic compounds, Kekule stressed structure. He conceived that carbon atoms may often be found as a chain of atoms, which eventually led him to conclude that the hydrocarbon benzene was made up of a ring of six carbon atoms. The story goes that the idea came to him as the result of a dream in which a snake bit his own tail.

Pasteur aided in this search for structure when he discovered by accident in 1858 that during fermentation in *Penicillium* only half of the tartartic acid nutrient of the medium was used. He found that the remaining tartaric acid medium rotated the plane of vibration of polarized light to the left, while the original medium did not rotate the light at all. Although it was not clear how a solution of compounds actually affects light, it was presumed that the structure of the organic compounds is involved. Pasteur concluded that there must be two opposite isomers of the tartaric acid present in the original medium and that each isomer rotated the light in opposite directions. The *Penicillium* apparently could only use the isomer that rotates the polarized light to the right.

Pasteur extended his studies and found that indeed for any carbohydrate nutrient source that he used with any microorganism only the one isomer, which apparently rotates light to the right, is consumed. Also upon the examination of many of the carbohydrates extracted from plants, he found that these are all of the one isomer that can be utilized by other organisms. These studies reinforced Pasteur's interpretation of the unique and highly organized structure of the reaction machinery of the cell extending as it were to the specific structure of the compounds with which the machinery dealt. "This property (specific isomerism) may be the only sharp differentiation between the chemistry of dead and living matter which can be made at present."

The fact that isomer differences were found to be associated only with carbon compounds also suggested that there must be some unique structural relationship between carbon and its adjacent atoms. Rotational light studies on many carbon molecules led to the conclusion that the four atoms adjacent to carbon are directed toward the four corners of a tetrahedron. The tetrahedron restriction on the configuration of atoms about the carbon atom encouraged others to seek the three-dimensional shape of more complex carbon compounds. Before the end of the century the shape of many of the intermediate-sized compounds such as glucose and other sugars were elaborated, especially by the ingenious methods of Emil Fisher (1852–1919).

In 1870 Mendeleev had managed to construct his periodic table in which all the known elements could be shown to fall into eight different categories according to their ability to combine with one another or with water. In other words there were set rules that determine which atoms could join with which other atoms to form defined molecules. This highly ordered classification of the atoms not only led eventually to models of the internal structure of the atoms and the bonds that hold atoms together, but it clearly demonstrated once and for all that the atom carbon was a normal member of the family of elements with no unusual qualities to ever again allow one to consider that it was alive. The climate of optimism that resulted from these discoveries encouraged many to feel that someday a highly ordered clock-like structure of life would be found. The procedure was to start from the atoms and build up and thus construct the gears of the clock. Once the gears were understood, the clock could be explained.

THE CONSERVATION OF ENERGY

While the clock builders sought the element of life through the structure of its matter, however, there were others who felt that this approach alone could never explain the essence of life. Life was not static structure but involved motion and change. Besides, the organic compounds that were being clarified were presumed to be intermediate products of a series of catalytic reactions, and no one assumed that these reactions were the vital processes that everyone was trying to understand. In fact just what these breakdown processes had to do with life was unknown. Heat was given off it is true but, as Lavoisier had shown, this heat is the same as that produced by the combustion of inanimate matter. He had also dispelled the idea that heat could act as some sort of vital force, so there did not seem to be any purpose for it.

Before any insight into these questions could be found, much more about heat alone had to be explained. Heat of course is sensed by a change in temperature. As Lavoisier had shown, a hot body melts more ice than a cold one and thus the amount of water resulting can serve as a measure of heat content. When combustion takes place, heat is generated proportional to the amount of material burned; hence a precursor of heat seems to be present in the material. It was also noted early that heat can be generated by scraping two objects together. This was especially obvious when a cannon is bored out of a metal block.

During the early part of the nineteenth century, many machines were being developed for various deeds and it became necessary to standardize their accomplishments. Newtonian mechanics had derived a term called *work*. This abstraction could serve as a unit of accomplishment for any machine. The problem was to measure the work of a machine. The work of an average horse versus the machine was used as a reference at first, but this proved unsatisfactory or impossible with some machines.

Count Rumford (1753–1814) considered the idea of using the heat generated in the cannon block as the machine under question drives the borer. He found that the increase in heat as measured by the change in temperature of the block was proportional to amount of work done by the machine. This led Julius Mayer (1814–1887) to write in 1841, "We must bring ourselves to think of two such dissimilar items as mechanical work and heat as two forms of the same thing," one being able to be converted into the other. He also reiterated Lavoisier's assumption that the heat generated by living systems is no different in nature from the heat generated by nonliving systems. With the current understanding of heat and work, however, the idea seemed too strange to be acceptable. Nevertheless, James Joule (1818–1889), thinking along similar lines, in about 1843, measured the amount of heat generated by different kinds of machines performing the same amount of work. He measured the temperature change in a water bath produced by an electric heater coil through which electric current from a generator is sent, by the flow of water forced through small pipes by a pump, by compressing gas with a pump, and by turning paddle wheels in the water by an engine. He found that regardless of the system, the work done by the prime movers was equal to one another when the same amount of heat was produced.

These findings led Herman von Helmholtz (1821–1894) to amplify Mayer's idea in his *Uber die Erhaltung der Kraft* (1847). Helmholtz used the term *energy* to represent all the forms of work and heat and put forth the postulate that as matter is neither created nor destroyed so energy is conserved regardless of the change of form. He also proposed that there was no reason to consider that the complex processes of the living organism did not abide by this law of conservation of energy. The concept of energy, which has best been understood as the potential to do work, was difficult to accept, but then it is probably the most abstract concept ever conceived by man.

ENERGY AND VITAL PROCESSES

Stating that all forms of energy are conserved in a defined system such as an organism is one thing but applying it toward an understanding of vital processes is another. There seemed to be no relationship between the heat released in the catalytic reactions and the vital processes. On the face of it, all that could be said is that presumably an equivalent amount of energy was contained within the nutrient molecules, mainly the carbohydrates, as released as heat during oxidation (combined with oxygen).

It is primarily due to the studies of Claude Bernard starting from about 1850 that some insight into the energetics of vital activities was found. Bernard discovered that "sugar always persists in the liver and the blood of carnivorous animals whose food contains no sugar." Also as "the blood which enters the liver by the portal vein contains no sugar, whereas the same blood which leaves by hepatic veins always contains appreciable amounts ... it was impossible to avoid the conclusion that the sugar originates in the liver," and is carried by the blood to the rest of the organism. He also discovered that sugar production in the isolated liver is prevented by boiling. He isolated a substance, which he called *glycogen*, from the liver that could be demonstrated to be the precursor of the sugar.

As glycogen is not taken into the organism, the only conclusion is that glycogen is synthesized by the organism. Glycogen turned out to be similar to the plant substance starch, and thus apparently the organism can synthesize carbohydrates as well as break them down. This was a completely new notion, as it was considered before this time that all carbohydrates were intermediate catalytic breakdown products.

Bernard also showed that the juices of the pancreas contained catalysts that broke down the food in the digestive tract. Starches were broken down to sugars, large hydrocarbons were broken down to smaller ones, and proteins were broken down to amino acids, and these smaller molecules are absorbed into the blood stream where they are made available to all parts of the body. These processes were accomplished by the unorganized type of catalysts. Bernard assumed that this process was reversible depending on concentration and that glycogen is synthesized from sugar by means of organized catalysts when the sugar concentration is high.

Pfluger (1829–1910) showed in 1872 that oxidation of carbohydrates takes place in the tissues of an organism rather than in the blood. Soon afterward Bernard showed that glucose from the blood was the carbohydrate being consumed by these tissues, and others confirmed Bernard's hypothesis that the liver could convert glucose into glycogen.

Thus it was that by 1875 Bernard, supported by Pasteur, could write with confidence that carbohydrate oxidation in the cells of the organism was supplying the energy for syntheses of other molecules within these cells. Presumably these synthetic processes produced the structural parts of indeed the whole cell. All the processes of the organism must be driven by the energy supplied by the cellular oxidation of carbohydrates or, in Bernard's famous statement, "Life, that is function, is death. And it is worthy of mention that we delude ourselves habitually in such form that when we want to point out phenomena of life, what we really talk about are phenomena of death (oxidation, putrefaction, fermentation). Vital phenomena properly so-called do not attract our attention. Organizatory synthesis works silently within the body, its actual occurrence well hidden away from sight, gathering up secretly the materials which will afterwards be consumed .... The existence of all animals and plants is maintained by these two types of necessary and inseparable processes: organization and disorganization. Our science has to fix the terms and conditions which guarantee the correspondence of these two types of phenomenon." Even the response of an organism to some stimulus was now considered to be a disorganization-type process that had to be reorganized from a general supply of energy in the organism. Thus once energy was recognized as the potential to do work, then the energy released by disorganization processes (now called catabolism) flows to and produces the organization processes (anabolism) that are the structures and functions of life.

Bernard had thus as diagrammed in Figure 7–4 applied the law of conservation of energy to the organism. It took almost 100 years to show that the heat that was produced by the respiration process in Lavoisier's experiments was not that produced directly by the breakdown of carbohydrates. Instead it was actually the heat yielded from the net breakdown of all sorts of molecules after the synthetic and other vital processes of the organism had taken place. There was a loss of energy from the organism system equal to the energy being supplied by the carbohydrate nutrients. Bernard reinstated the candle model of life by adding a loop in the flow of heat of the candle.

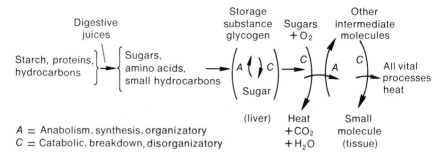

Figure 7-4. Bernard's concept of catabolism and anabolism.

SUBCELLULAR ELEMENTS OF LIFE

Many were concerned that the flow of energy must be dictated by structure and this could only be understood by determining more about the structure of organic molecules. It occurred to some like Pfluger that there may be living particles responsible for these vital (synthetic) processes. He proposed a model for such a particle made up of proteins, carbohydrates, and hydrocarbons complexed as rays extending at right angles like coordinate axes. These rays were constantly breaking down and reforming. Oxidation with the release of carbon dioxide and other products was considered to be the breakdown process, while the replacement of these parts was accomplished by the assimilation of nutrient molecules. These complex particles, later called *biogens*, were considered as the basic building blocks of the protoplasm of cells. There was believed to be a large number of these biogens in any one cell. There was disagreement as to whether such particles were the entire clock or whether they worked together in some fashion as gears of a large clock. In either case it was the "organization of molecules forming the protoplasm and in their manner of association one with the other (wherein) lies ... the fountain-head of the whole range of elementary phenomena of life." It was this structure and organization that determine "the direction which ... energy must take ... which preserve the cell." This was supported when Thomas Graham (1805–1869) discovered that proteins such as albumins were suspendible in solutions and could aggregate to form supermolecules or colloids with properties similar to cellular protoplasm. There was also an attempt by many to construct model cells by mixing various molecules together. Some of these behaved like certain cell types, such as amebae, but they did not add any new insight into the possible structure of the cell.

Inherent in this approach was the assumption that the basic element of life lay at the subcellular level, but there was no direct evidence of this until 1897 when Edward Buchner (1860–1917) accidentally discovered that fermentation could take place without the yeast cells being intact. He had been testing various fractions of yeast extracts and was attempting to preserve them. One method of preservation was to add large quantities of glucose. Whenever he did this, however, fermentation took place with the evolution of carbon dioxide and the production of alcohol. He was able to isolate an active material which turned out to be an albuminoid which was water soluble and heat labile. He designated it zymase. Buchner wrote, "The question now remains whether zymase is to be considered as one of the long well-known enzymes (the general name used for the diastase-like particle catalysts)." Bernard had shown that enzymes split large molecules into smaller ones at regular points within the compounds, yielding smaller molecules of regular size. Zymase-like catalysts, on the other hand, are involved in much more complex processes including the ability to break carbon bonds.

In spite of this uncertainty, Buchner had clearly demonstrated that, contrary to Pasteur's conclusion, the cell is not only unnecessary in its intact state for fermentation, but that the cell can be taken apart and its pieces can be examined independently. This finding added an impressive support to the contention of the clock builders that the structure of the gears can be understood and probably be put together again, thus explaining the living particles and the cell. It could also be concluded that the size of the element of life must lie somewhere between that of a large compound and some complex of particles within the cell.

THE MOLECULAR FITNESS OF THE ENVIRONMENT

Whether clock or candle, the twentieth century has added a new dimension and meaning to the element of life. This has, in effect, been an integration of all the matter and energy of life through time. One of the most important contributors to this new concept was L. J. Henderson (1860–1917) when he expressed his ideas in his book, *The Fitness of the*

Environment in 1913. Henderson made a thorough study of the "atmosphere of solid bodies," the molecules that make up the environment of organisms. He concluded, "The fitness of the environment is one part of a reciprocal relationship of which the fitness of the organism is the other .... (The fitness) is not less frequently evident in the characteristics of water, carbonic acid, and the compounds of carbon, hydrogen, and oxygen than is fitness from adaptations in the characteristics of the organism. Given matter, energy, and the resulting necessity that life shall be a mechanism, the conclusion follows that the atmosphere of solid bodies does actually provide the best of all possible environments for life."

Of all the molecules in the environment, water appeared to be the most fit for life to Henderson. It is a liquid allowing the necessary redispositions of matter during the vital processes. Life as it exists on earth could only operate at temperatures from a little above the freezing point of water to about 50 or 60°C. Water is the only substance that easily remains liquid in this range and does so up to high enough temperatures that a great deal of the surface of the earth will have its water in a liquid state. It also serves as a buffer against rapid temperature changes by having an extremely high specific heat. The very unusual property of ice to be less dense than water and thus float allows it to melt sooner with increasing temperatures than if it were more dense than water and rested at the bottom of lakes, for example. Water also participates as a whole or its parts in many of the more significant reactions of the vital processes. Of most significance is the fact that water serves as the most capable solvent for all inorganic salts and most organic molecules, allowing them to interact in a random and dynamic fashion. Water also has an extremely high surface tension that can influence the formation and shapes of membranes and cause the movement of itself by capillarity. Finally, of significance is the ability of oxygen and carbon dioxide gases to dissolve in water in adequate amounts to be available for reactions involving these gases in solution.

It was such arguments as these applied to all the molecules of the environment of living organisms that led Henderson to feel that "the properties of matter ... are now seen to be intimately related to the structure of the living being and to its activities; they become, therefore, far more important in biology than has been previously suspected." Henderson applied Darwin's adaptation to the molecular level and discovered that not only was the environment suited for life, but the same molecules make up life. The environment is basically no different from the organism except in the organization of the matter. It would appear that the reciprocal relationship of fitness must have always had to be true for life to exist. He thus concluded "that ... peculiar and unsuspected relationships exist between the properties of matter and the phenomena of life; that the process of cosmic evolution is indissolubly linked with the fundamental characteristics of the organism; that logically, in some obscure manner, cosmic and biological evolution are one."

This was not the first time that the gradual evolution of life from inanimate matter had been suggested as a possibility. Haeckel with his thoughts about ontogeny and Weismann with his biophores had considered that at some distant time under certain special conditions, life must have emerged from the inanimate. These were considered as deductions from further thoughts about evolution. However, Henderson supplied the first logical argument based on observations, circumstantial as they were, that a gradual transformation of matter into a new pattern of organization may have yielded life. He served as a new Lyell and expanded time and space for life to be able to begin, not spontaneously as Pasteur had denied, but as an ever so slow emergence.

THE IMPROBABILITY OF LIFE
TO ARISE FROM INANIMATE MATTER

There followed some speculations as to how life may have originated. It seemed quite logical that, as Henderson had pointed out, there would have to be some method of assimilating carbon atoms, and as the simplest common form of carbon was as CO_2, then this molecule must have served as the source of carbon. This was also quite reasonable since a system for assimilating CO_2, with the only other requirement being water and light, was well-known and common in plants and protista, mainly photosynthesis. As simple as this reaction may appear, however, very little was known about the mechanism and the more that was learned about photosynthesis and the cell in general, the more complex and baffling life became. It was generally agreed that any thoughts about its origin must remain highly speculative and certainly not open to experimental verification, hence not of a significant scientific value.

There were also some strong doubts about the ability of life to arise on the basis of energy changes that would be required. This was indirectly due to the ingenious efforts of Willard Gibbs (1832–1903) who first formulated the cause of molecular reactions. He derived a relationship between free energy, heat flow, and entropy. And with molecular reactions, there must be some randomization of the participating reactants and thus an increase in entropy. Another way of stating this was that the most probable state of matter was a randomized one. With every reaction of molecules some potential energy would be lost to a less useful form of energy, heat. It was soon stated that the useful energy of the universe is being constantly degraded as matter and energy become dispersed; there is an inevitable "entropic doom."

When these relationships were derived, it was soon suggested that evolution was a violation of the law of increasing entropy change. Evolution is clearly an aggregating process that would constitute a decrease in entropy. Many assumed that this could only be explained if there were unique molecular reactions involved with vital processes that not only violated the general law of reactions but probably could never be understood in terms of physical laws. The same arguments would apply to the origin of life, this also involving an aggregation of matter. This difficulty was finally explained by pointing out that if catabolic reactions were coupled to anabolic reactions in the organism and if the net increase of entropy associated with the catabolic reactions was greater than the net decrease of entropy of the anabolic reactions, then the organism as a whole system would not be violating the law of entropy change. A diagramatic model of how this might be accomplished is indicated in Figure 7-5. It was also pointed out that inorganic salts can form crystals, coming

Gibbs' relation: $\Delta G = \Delta H - T\Delta S$, ΔH; change in internal energy $\Delta H = 0$
$\therefore$ Free energy, $\Delta G = -T\Delta S$ (Absolute T × entropy change)

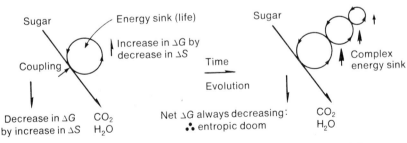

Figure 7-5. Energy coupling, changes in entropy, energy sinks, origin, and evolution of life.

out of solution and aggregating together without violating any physical laws. There is still a net increase in entropy in the system in terms of loss of useful energy. In either system, crystal growth or the organism, the aggregation process could only be accomplished with a large use of free energy that would have to come from outside the system.

There remains, however, the rather uncomfortable problem of explaining how local "energy sinks" exist in the universe where coupled reactions can compound the dispersive tendency of other reactions and establish an ordered structure of matter, especially when the system is a dynamic one like the cell. It suggests that there may be other physical laws to be discovered to account for these local fountains of energy. These would not be violations of any present laws, but additional ones. To find such a law would be to find the law of life. It would describe the candle model of the origin, development, and evolution of life.

Damaging argument against the possibility of the origin of life was that regardless of the ability of a coupled system to produce more complex life-forms, how did the coupling system come into being? By all understanding, this system itself is a complex one involving a complex of specific proteins. This was also the time that the gene theory of heredity was reminding all that any life element must be able to reproduce itself in an exact image; one more property early forms of life would need, and probably would require, a complex mechanism. This suggested that there would have had to have been a chance aggregate of specific kinds of matter. By all calculations, the likeliness of such a chance occurrence proved to be exceedingly small, so small that most considered it impossible and thus turned to the meteorite source of life on earth as the answer. A few others suggested with little enthusiasm that after all even a very, very small probability was not zero and that it might have happened just once in the universe and that was here on earth.

A MODEL FOR THE ORIGIN OF LIFE

The sum total of these arguments had the effect of quietly putting Henderson's hypothesis to sleep. The idea that life may have originated from inanimate matter had occurred to A. Oparin, however, quite independently from the logic used by Henderson. Oparin published a brief pamphlet on his preliminary thoughts in 1923, but at this time there was little to support his hypothesis. Nevertheless, he carefully sought every new finding for evidence and slowly worked out a very logical argument for the pattern and mechanisms of the origin of life. He published this argument in one of the most influential books of the present century, *The Origin of Life* (1936).

Oparin was interested in the origin of the planetary system. According to the common theory at this time, a "near collision" of another star with the sun about 3 billion years ago caused part of the sun to spin out like the tail of a comet and collect as balls of matter that slowly cooled and formed the planets. This took considerable time with long periods of high temperatures followed by gradual cooling. Oparin contributed to this explanation by calculating the abundance of natural elements in the atmospheres of the stars and planets on the basis of analyses of light coming from these bodies. The light could also be used as an indication of the surface temperatures of the stars. These were found to range from about 30,000 to 2000°C for the different stars. The sun being a yellowish-white star was estimated to have a surface temperature of about 6000–8000°C. It could also be concluded that although carbon is an isolated atom in the atmospheres of the hotter stars, with the cooler stars such as the sun there are carbon-carbon, carbon-hydrogen, and carbon-nitrogen bonds, and a fair number of hydrocarbons.

Oparin noted that from light reflecting from the surfaces of planets there appears to be water vapor and carbon dioxide on Venus, water vapor and a little oxygen on Mars, methane and ammonia on Saturn and Jupiter, and at least methane on Uranus and Neptune. Also an analysis of meteorites revealed that they were of two general types, so-called iron meteorites mainly made up of iron oxides and so-called stone meteorites mainly made up of aluminum and silicon oxides. Both of these contain small amounts of carbon in the form of hydrocarbons. From this data, Oparin concluded that "carbon is found either in its elementary form, or in the form of compounds with nitrogen, but more often with hydrogen . . . only the earth and Venus, the planet closest to it" show carbon in the form of carbon dioxide in their atmospheres. Also because of the kinds of carbon bonds that occur with the cooler stars, he added, "Carbon, at least in part, first appeared on the earth's surface in the reduced form (combined with hydrogen such as methane, CH_4), particularly in the form of hydrocarbons."

This in turn led Oparin to hypothesize that the carbon dioxide and oxygen were of secondary origin. Oxygen is one of the most reactive atoms. It would quickly combine with any hydrogen and metals to form oxides, and only if there was a great excess of oxygen would it combine with carbon or exist as an oxygen molecule in any appreciable amount. As these are not found in the stars or majority of planets, it can be assumed that the atmosphere of the primitive stages of the earth had an excess of hydrogen atoms, i.e., was a reducing atmosphere. Likewise free nitrogen as exists in the present atmosphere of the earth must have been of secondary origin, nitrogen originally being found as ammonia, NH_3.

Thus according to Oparin the early atmosphere of the earth must have consisted of hydrogen, water, vapor, ammonia, and methane. And, something drastic must have occurred to change this atmosphere. Oparin also concluded that as the earth began to cool with temperatures similar to those on the surface of the sun, hydrocarbons containing more than one carbon atom would form. With further cooling and the eventual forming of the oceans, the hydrocarbons could be modified and/or combined with the aid of heat and metal catalysts available at that time. Such processes were known to be possible from the many laboratory studies of the synthesis of various organic molecules. Thus it was perfectly logical to trace the formation of complex organic molecules from simple inorganic ones and even the atoms themselves. This he assumed was quite possible under reducing conditions.

Oparin reasoned that if oxygen was not present on the primitive earth, then aerobic organisms could not have existed. Likewise plants could not have existed because they require carbon dioxide for the assimilation of carbon by photosynthesis. He was encouraged in his search for presently existing organisms which could have existed on primitive earth when he found that many microorganisms did not need oxygen or carbon dioxide. These were simple organisms that were able to assimilate large complex organic molecules as their source of nutrients and building material. Most of these were fermentating microorganisms that produced alcohol or a variety of organic acids and often CO_2.

Even though these were simple organisms, their metabolic apparatus consisted of a fair number of steps that could not have existed in the earliest form of life. Here again he was encouraged to find that "at least the first stages of all these fermentations, even if not identical, are quite similar." From this survey he concluded that "the primitive metabolism of energy was entirely anaerobic and dependent on the interaction of organic substances with molecules of water." With this data he concluded that the first organisms were anaerobic heterotrophs and produced among other things carbon dioxide.

Some of these organisms may have evolved pigment systems to absorb light to help catabolize nutrient molecules. These latter could have evolved a pigment system to absorb light as a source of energy, which in turn could be used to assimilate the carbon dioxide then present in the atmosphere and in so doing release free oxygen. Other microorganisms may have taken in ammonia and released nitrogen with the aid of light energy.

The resulting autotrophic photosynthesizer would in time add enough oxygen to the atmosphere to permit the complex aerobic metabolism of present animals to evolve. It would be to their advantage to be secondary heterotrophs once there were sufficient complex molecules produced by photosynthetic organisms. A summary of Oparin's model for the evolution of metabolic systems is diagrammed in Figure 7–6.

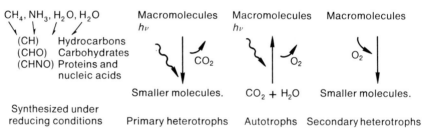

Figure 7-6. Oparin's evolution of metabolic systems.

Although this hypothesis of the origin of life appeared to follow a nice sequence, there was a big gap between complex organic molecules and even the simplest of "organisms" that could only be speculated upon. There were no data to help him fill in this gap. His knowledge about the various metabolic systems led him to propose that the earliest "cells" were aggregates of diverse proteins which could assimilate nutrient molecules from the environment and carry out the simplest of catabolism and anabolism which would cause the growth of the aggregate and its subsequent division once a critical size was reached. At this time there was still a common understanding that proteins aggregated and were held together as colloids. These "coacervates" of proteins could easily form with little energy of adhesion necessary.

Oparin assumed that from the many coacervates that would occur naturally in the early oceans, many would have the right combination of molecules to be able to carry out some kind of catabolism and anabolism, however inefficient it might be. Then the coacervate with the most efficient utilization of the nutrient organic molecules of its environment would grow more rapidly than the others and would become more numerous. Thus there would be a natural selection for the more efficient "cell" types. At the same time there would be a selection for the coacervates which happen to be synthesizing relatively similar proteins which could carry on the more efficient metabolism in additional cells. In time this process would be perfected enough that the coacervate could reproduce its own parts by genes. This model of the origin of the cell depends on the ability of many kinds of proteins to act as catalysts, however poor they may be at this endeavor, and the gradual accumulation of intermolecular regulation and diversity of specialized function of the proteins.

Oparin realized he was merely charting the course that must be studied as he wrote, "We are faced with a colossal problem of investigating each separate stage of the evolution process . . . . The road ahead is hard and long, but without doubt it leads to the ultimate knowledge of the nature of life. The artificial building and synthesis of living things is a very remote, but not an unattainable goal along this road." He was convinced that not only would an understanding of life be incomplete, but it could never progress until there was knowledge as to its origin. He criticized the methods of the clock builders, as most of his contemporaries were, because "in principle this amounts to their wanting crudely speaking, to take the living body apart into its component screws and wheels like a watch and then try to put it together again . . . (when) an understanding of the nature of life is impossible without a knowledge of the history of its origin."

In the years that immediately followed the publication of Oparin's book, it had very little influence. For one thing the book was not well-known. Most of those who did read it considered it but one more attempt to talk about a futile subject. Those who happened to be interested in the origin of life criticized his hypothesis on many grounds. It was impossible to accept the idea that the atmosphere of the earth had ever been a reducing one containing a surplus of hydrogen when there is so little free hydrogen today and so much nitrogen and definitely oxygen. Also it was soon convincingly shown that the origin of the planets was not from the sun but had resulted from the aggregation of cosmic particles, each aggregate being independent of one another and the sun. This not only put less credence on elements of the primitive earth resembling those of the sun, but it also suggested that the surface of the earth had probably always been at relatively low temperatures—below those that could have aided in the synthesis of complex organic molecules. Also these complex molecules could probably never have remained long enough for a cell to form. Proteins for example are naturally unstable molecules and the amount of ultraviolet radiation present at that time would certainly help break them down. There was also criticism of the coacervate hypothesis of cell origin. It was realized at about this time that proteins do not form colloids. Also it was generally thought that there must be a stable self-reproducing entity as an earliest step to assure exact continuity and permit natural selection to operate and thus evolution to occur. Even Oparin's colleagues and friends who agreed with his hypothesis considered that it was not worth much consideration as it was out of the question to ever study the origin of life in the laboratory.

Rather, for those who enjoyed speculating about life's origin, much more interest was generated by Edwin Schrodinger's *What Is Life?* (1945). Schrodinger considered that the gene is the key to life for it "is so long-lasting and constant as to border on the miraculous. (There is a) resemblance between a clockwork and an organism ... the latter also hinges upon a solid ... the aperiodic crystal forming the hereditary substance, largely withdrawn from the disorder of heat motion." And life would have to have evolved from such a self-reproducing solid. Also, there was one attempt in 1950 to demonstrate that CO_2 and water under an intense radiation (carried out in a cyclotron) could form organic compounds. The result was that such a small yield of even any simple organic molecules was produced for the tremendous amount of energy used that it was concluded that not only were such experiments impractical but that the origin of life was indeed a most improbable occurrence.

There was, however, one individual, Harold Urey, who had not only read Oparin's book but had come to the same conclusion regarding the reducing atmosphere of primitive earth from his own studies of the planets. Urey assumed that when the planets were originally condensing, they must have consisted of the same elements but that the size of the planet and its distance from the sun would determine the eventual atmospheric makeup. The larger the planet, the greater the abundance of lighter gases that could be retained by the larger force of gravity. Also the farther from the sun, the lower the temperature and thus the more likelihood there was of the planet retaining the lighter gases. From theoretical calculations he estimated the gas contents on the different planets and found general agreement where it was possible to experimentally make an estimate. This is impossible with some planets because of vapors or clouds. Nevertheless from these calculations he concluded that although the earth was relatively small and comparatively close to the sun, for a considerable part of the history of the earth, the atmosphere would have been highly reducing and would have contained hydrogen, water vapor, ammonia, and methane as Oparin had suggested.

With this conclusion, Urey and his graduate student Stanley Miller reexamined Oparin's hypothesis that organic molecules could be synthesized under the reducing conditions of the primitive earth. The cyclotron experiment had been carried out under oxidizing conditions. Miller carried out a very simple experiment in 1953 to test Oparin's hypothesis. He placed the estimated proportions of hydrogen, methane, and ammonia into a closed reflux system as pictured in Figure 7–7. He also added water,

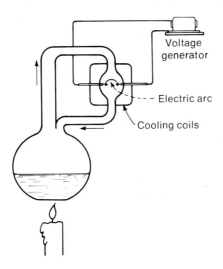

Figure 7-7. Miller's experiment.

which could be boiled and refluxed throughout the system so that all the gases would flow past an electric arc. The electric arc was considered a likely source of energy for the synthesis, conceivably produced by electrical storms. After refluxing this sterilized system for about 15 hours, he found a wide variety of organic compounds accounting for 15% of the carbon originally introduced as methane. Most significant was the finding that one-quarter of this aggregated carbon was in the form of either alanine or glycine, the two simplest amino acids.

The simplicity and elegance of this experiment were overwhelming. Many secondary school children repeated it. There followed a total collapse to the inhibition of studying the origin of life. From the position of ridicule and impossibility, such studies immediately became something of value and necessity. Never before has there been such an explosion of interest as the result of one single simple experiment.

In the pursuing years, this optimism has been justified. It was soon found that almost any source of energy was adequate including ultraviolet, radioactivity, X ray, or any other kind of radiation, ultrasonics, gentle heating, or in many cases no source of energy if at room temperatures. It was also found that the proportions of gases were unimportant as long as the conditions were reducing. Even carbon dioxide can be present and incorporated into organic compounds. If any free oxygen is present at all, however, the synthesis cannot take place. It has been found that not only are amino acids formed but also sugars, a large assortment of hydrocarbons including ring compounds, and in fact all the kinds of molecules that are considered significant to the vital processes.

Of especial interest has been the ability of abiogenetic experiments to synthesize proteins. This has been accomplished with a number of different initial conditions including temperatures ranging from room temperature to about $1000°C$ as long as there was a reducing atmosphere. Sidney Fox found that heating amino acids at about 100 to $150°C$ yielded protein-like structures which in turn would aggregate together in sheets and these would continue to grow and form membranes which stabilized as hollow spheres, "microspheres," of fairly constant size (1 to 3 microns in diameter) which is about the size of many bacteria. These microspheres could act as catalysts to catabolize glucose in the presence of oxygen with CO_2 given off, but the heat generated was all dissipated with no anabolism or growth resulting. Nevertheless, these experiments suggest the possibility of aggregating protein systems analogous to Oparin's coacervates. It can certainly not be said that life has been created by these experiments, but the ease and success of these early adventures have bettered anyone's wildest expectations.

SUPPORT FOR THE MODEL

Thus on the whole Oparin's hypotheses have been more than supported. Where his facts were wrong, they were of less support than more recent data. Also additional facts have been in his favor. It is now fairly well agreed that the earth, as estimated by radioactive dating, is close to 4.7×10^9 years of age with the oldest known rocks from about 3.3 to 3.8×10^9 years. There is universal agreement that the early atmosphere was reducing and that water vapor, ammonia, and methane were present. There is disagreement as to whether there was free hydrogen and other gases such as free nitrogen, carbon monoxide, or possibly carbon dioxide. Some feel that by the nature of the carbon compounds in meteorites they must have formed from carbon monoxide. Others claim carbon dioxide is present on celestial bodies that have atmospheres similar to that of primitive earth. Fortunately Oparin was only using an apparent lack of carbon dioxide as an indication of a reduced atmosphere. It has been found that it makes little difference what the form of carbon is in the abiogenetic experiments and free hydrogen does not have to present. There is additional evidence of hydrocarbons present in meteorites and in crude petroleum deposits, most of which were probably of abiogenetic origin.

One of the most crucial questions yet to be answered is the age of the origin of life. Fossil microorganisms have been dated as old as 3.2×10^9 years, but it can hardly be expected to find fossils of the earliest forms of life. It has been assumed that the vital processes would only be able to operate if molecules of certain specific characteristics were available and able to be synthesized by early "cells." This would in effect tend to produce a relatively greater number of these essential molecules than the many other complex organic kinds that would occur abiogenetically. It has been suggested that the best markers of origin of life may be certain naturally occurring long hydrocarbon chains that have characteristic groups of atoms at period intervals along the chain. If formed abiogenetically, there would be no reason for the groups to occur at the set locations along the chain. As yet these have not been found. The ease of natural synthesis of organic compounds adds support to the theory that macromolecules arose abiogenetically but adds hardships for analyses of the origin of life.

Very much related with the time for origin of life is that of the form of metabolism of the earliest "cell." The oldest known fossils are bacteria and probably blue green-like algae dated at 3.2×10^9 years and definitely blue green algae at 2.7×10^9 years. The fact that algae are found so early and also that CO_2 may have been present in the primitive atmosphere has led some to champion the notion that photosynthesis was the first form of metabolism. It has been estimated, however, that there is about 2.4×10^{23} moles of carbon on the whole earth. If all the free oxygen that is available was combined with carbon, this would make up only a small fraction of the total carbon and consist of about 4×10^{19} moles of CO_2. There is at present only 0.5×10^{16} moles of CO_2 in the atmosphere with a constant use and generation of 1.5×10^{16} moles of CO_2 per year. If CO_2 had been produced at anywhere near the present rate by fermentation processes, all the oxygen now in gaseous state would have been combined with carbon as CO_2 by 3000 years! Obviously the rate of these processes was considerably less extensive than at present, but this does indicate that there would have had to have been a process reconverting the oxygen in the CO_2 form back to a free state relatively early in the origin of metabolic systems. Although photosynthesis is not the only system for assimilating CO_2, it is one of the most effective and it does return the oxygen to its most available condition, the free molecular state. But here again photosynthesis could not have taken place for long without aerobic metabolism being present or else all the oxygen would have been put in a free form and thus unavailable for photosynthesis. There must have had to have been a relatively close balance between CO_2 uptake and generation once free oxygen became available; thus, photosynthesis would have had to have been closely balanced by respiration. The best estimates are that the atmosphere of the earth was a reducing one until at least 2.0×10^9 years when the photosynthetic processes began producing enough oxygen to convert the atmosphere to an oxidizing one. Oxygen is believed to have been in abundance by 1×10^9 years with aerobic systems by 0.8×10^9 years ago or a little later. No doubt these dates are going to have to be corrected many times, but they do show that at least some estimates of metabolic systems can be made and that Oparin's sequence of systems is still the most favored.

It is interesting that it is the totally speculative part of the origin of life, the origin of the "cell," that brings into focus the division in approach used in attempting to understand the element of life. Oparin and Fox consider that metabolism, the accumulation of matter and energy into a compact unit, is the most important issue. They imagine an aggregation of diverse proteins. This model is obviously of the candle type. The far more popular model is one similar to that proposed by Miller in which a stable molecule serves as a crystal seed. This seed accumulates smaller molecules that collide with it synthesizing these together to form other stable molecules very close to if not identical with the original seed molecule. Thus the emphasis is on reproduction and exactness of a specific fit of molecules, a clock model. But whether candle or clock, both would have to somehow acquire the emphasized property of the other. There are great difficulties with either model. Only to mention one is the fact that both require considerable aggregation of organic molecules. Even if all the carbon of the earth was present in the oceans in the form of organic molecules, it would be far from the "organic soup" in concentration that it was once considered to be. It has to be kept in mind that the total biomass constitutes only one-billionth of the total earth mass.

THE INEVITABILITY OF LIFE

Regardless of the differences in approach or the difficulties ahead, the exploration is underway with a sweeping optimism that has brought with it a completely new view of life. Where before Oparin's hypothesis and Miller's experiment the origin of life was at best an extremely remote possibility, now after this one experiment it is now believed that, given the right conditions, *life must occur.* The origin of life is now considered to be a set of many but natural and inevitable molecular reactions. This assumption has led to the conclusion that life must exist in other regions of the universe. One estimate has been based on there being at least 10^{20} stars in our universe. If such is the case and if only 1 in 1000 of these has a planetary system, then there would be 10^{17} such systems. If 1 in 1000 of these has planets in the correct temperature range, there would be 10^{14} in number. If 1 in 1000 of these were the right size to hold the optimum amount of atmosphere, there would be 10^{11} such planets. Then if for other reasons only 1 in 1000 of these could support life, there would be 10^8 planets like the earth. This is even considered a conservative number by many. But regardless of the exact number, life must exist on many other planets of the universe, and if the figure above is used, there would be at least 10^5 civilizations more highly advanced than ours!

The ease of the origin of life reminds us that some day man will no doubt "create" life in the laboratory. Of course man will not create life; he will discover the conditions under which the reaction of life must occur.

EVOLUTION AND THE ORIGIN OF LIFE

There is a curious similarity in the development of the idea of the origin of life and that of evolution. Both ideas were considered long before they were accepted but relegated to the ridiculous, mainly on the basis of lack of evidence and breadth of imagination. Both occurred to their most successful proponents by the direct observations of geological evidence over a new expansion of space and time. Darwin found his on the other side of the world with the changes of landmasses through the millions of years. Oparin found his in the stars and planets with the atmospheric gas changes through billions of years. Both accumulated many facts and sought for mechanisms to support their hypotheses over a very long period of time before publishing their thoughts in one book. Even the titles are similar. Where Darwin was able to unify all forms of life throughout the space and time of the earth, Oparin unified all matter and energy throughout the space and time of the universe. Where Darwin reminded man that he was a cousin to the ape, the sea urchin, and the oak, Oparin has reminded him that he is also the cousin of the shale and the stone. Where Darwin has supplied the optimism for man to help regulate his natural surroundings and produce new species, Oparin has supplied the optimism that basic processes of life can be amended and that new forms of life can be "created."

Progress associated with evolution has been explained as a result of the creative nature of natural selection. As the environment changes, the species must adapt with new structures or be eliminated. As single structures can take on diverse functions, complexity will increase as long as diversities can be sustained. Diverse organisms and structures have seemed to increase the opportunities for more diversity and thus even more complexity. The cause of this progress goes back to the inevitable change in total environment. This would be analogous to the turning of a tail to one side as a source of moving a fish forward: the more swishing, the farther the fish moves.

There has always existed a certain incompleteness with this explanation of the driving force of evolution. It does not explain why life pushes out into the opportunities once change is forced upon it. The abiogenetic experiments have revealed the natural tendency for matter to enlarge and diversify. There inevitably must be local energy sinks of "living" matter wherein the entropy decreases. This idea is still too new to be formulated but the belief in this inherent ability of matter is strong. Perhaps one can think of evolution as a hand moving into a glove. The glove, as diagrammed in Figure 7-8, is equivalent to the environment with its fingers serving as diverse opportunities or niches. The spreading of the fingers as they move into the glove is the diversification and change made possible by natural selection and the environment glove. The progress

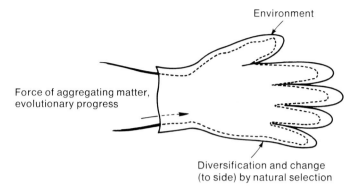

Environment

Force of aggregating matter,
evolutionary progress

Diversification and change
(to side) by natural selection

Figure 7-8. Hand and glove.

of evolution is represented by the movement of the hand and fingers forward and is due to the inherent ability of matter to interact in ever more ordered patterns with time. The diverse ways of evolution are adequately explained by natural selection, but the force or push of life that makes diversification possible must lie within the forces acting on matter.

THE LEVELS OF ORGANIZATION OF MATTER

It is with an intent of explaining how this inherent property of matter can account for the process of evolution and all the diverse functions of its product forms that the remaining part of this book is designed. In order to do this, it is important to remind ourselves that matter is not homogeneous but apparently consists of a series of subunit aggregates of matter at different levels depending on their size, as diagrammed in Figure 7-9. Apparently the unit of matter on each of these levels is organized by its own "laws" depending on the characteristics of its subunits. There are then different kinds of "forces" operating to cause the aggregation and thus one can think of different systems of energy on each level. The search into definitions of life in terms of its ability to transform or arise from inorganic matter has revealed that the matter of life is the same as that of nonlife. Also the lower levels of organization of matter found in living substances are identical with those of the nonliving. What we shall be concerned with in the later chapters are the unique laws operating at the higher levels of organization of living matter. This

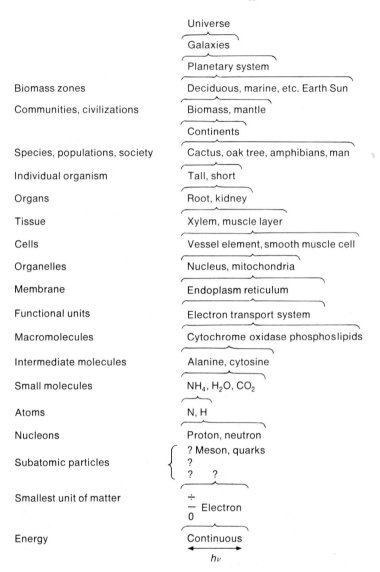

Figure 7-9. Levels of organization of matter.

search has also suggested that organic evolution is but a part of the cosmic evolution, and thus these higher levels of living matter have resulted from natural aggregations of the lower nonliving units of matter. There is apparently an order of aggregation of subunits building up higher levels of matter with time. Indeed this order is the reason for this chapter.

What has also been concluded from studies of the even more remote parts of the universe is that the various levels of inorganic matter have evolved by aggregations of subunits. This expansion of our spatial senses has yielded various explanations as to a beginning of the universe, the most popular being that it all began about 6 to 7 billion years ago with an explosion of a "cosmic egg" made up of nucleons and electrons. It is true that the nucleon "onion" is being peeled and sublevels of mesons and quarks are being proposed, but it is thought that the forces that hold a nucleon together are greater than the disruptive ones due to the cosmic explosion with temperatures estimated at 10^{11} to $10^{12}C°$.

The story goes that with expansion the temperatures would drop below $10^9C°$ allowing nuclear forces to hold together the colliding nucleons, thus generating nuclei and atoms. There are apparently certain laws which determine the relative combinations of neutrons and protons which can form a stable nuclei, the conforming to which cause nuclear energy yielding reactions of the suns and nuclear weapons. Estimates of the kinds of nuclei that have been formed in the present universe by this aggregating process show that greater than 99% of these are hydrogen or helium nuclei. Then where the temperature has dropped below $10^4C°$, electrostatic forces can hold the electrons and nuclei together and the atoms form. These can exist at the surface of most stars and of course on planets. An estimate of the relative abundance of atoms on earth by weight show O_2 at 55.19, Si at 16.08, H at 15.40, with the metals being next highest and with C at 0.119 and N at 0.038%. This indicates that there are certain laws operating to cause a chemical selection of certain atom-atom combinations in the formation of molecules that make up the planets. These chemical laws are well-known and would suggest that the most stable bonds and those which would form first would be between the small atoms H, N, C, and O. Although some of these molecules could grow naturally under reducing conditions, most of the molecules formed would not adhere to one another and would be lost to space. On the other hand, the metals, especially silica and aluminum in oxide forms, would grow as crystals as long as the oxides were available. In this way the relative abundance and disposition of the atoms of the earth can be accounted for as diagrammed in Figure 7–10. The less abundant organic molecules could grow to macromolecular size by the schemes devised by Oparin and others and thus the building blocks of life could come about. In order to try to understand why only certain organic molecules are selected for and aggregate within living processes, we must know more about the "laws" of a cell. This we do in the next chapter.

The above is a scheme in which aggregation of subunits of matter could occur with lowering temperatures. These all depend on attractive forces between subunits. Perhaps the best way to imagine how new sets of weaker forces could demonstrate their effect with the lowering of temperature is through energy diagrams as shown in Figure 7–11. Each

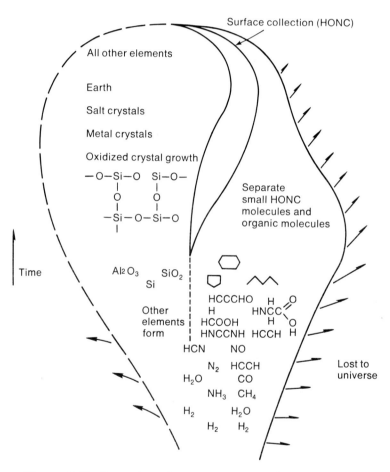

Figure 7-10. Separation of matter during the formation of the earth.

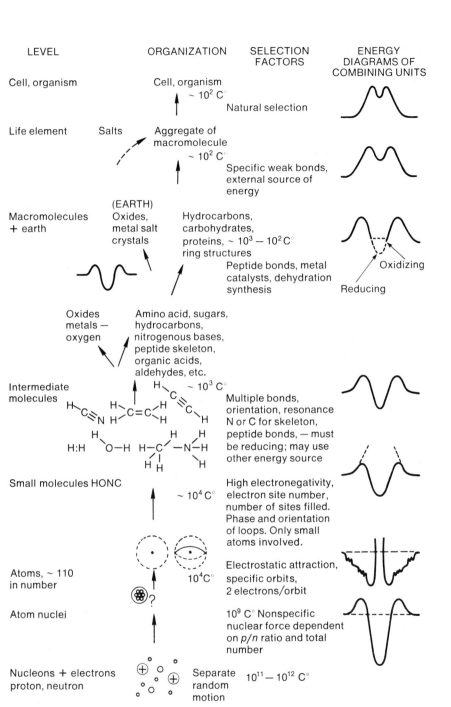

LEVEL	ORGANIZATION	SELECTION FACTORS	ENERGY DIAGRAMS OF COMBINING UNITS

Cell, organism — Cell, organism ~ 10^2 C°

Natural selection

Life element — Salts — Aggregate of macromolecule ~ 10^2 C°

Specific weak bonds, external source of energy

(EARTH)

Macromolecules + earth — Oxides, metal salt crystals — Hydrocarbons, carbohydrates, proteins, ~ $10^3 - 10^2$ C° ring structures

Peptide bonds, metal catalysts, dehydration synthesis

Oxidizing

Reducing

Oxides metals — oxygen — Amino acid, sugars, hydrocarbons, nitrogenous bases, peptide skeleton, organic acids, aldehydes, etc.

Intermediate molecules — H—C≡C—H ~ 10^3 C°

H—C≡N H—C=C—H

H:H O—H H—C—N—H

Multiple bonds, orientation, resonance N or C for skeleton, peptide bonds, — must be reducing; may use other energy source

Small molecules HONC ~ 10^4 C°

High electronegativity, electron site number, number of sites filled. Phase and orientation of loops. Only small atoms involved.

Atoms, ~ 110 in number — 10^4 C°

Electrostatic attraction, specific orbits, 2 electrons/orbit

Atom nuclei

10^9 C° Nonspecific nuclear force dependent on p/n ratio and total number

Nucleons + electrons proton, neutron — Separate random motion — $10^{11} - 10^{12}$ C°

Figure 7-11. The evolution of matter.

diagram describes the forces acting upon two subunits of that particular level. These subunits are as rolling balls coming from opposite directions moving at rates equivalent to their kinetic energy of motion. As they approach, various forces of interaction cause bumps or valleys, depending on whether they are repulsive or attractive, respectively. For aggregation to occur there would have to be a valley wherein they both could be held together. The balls may well have to roll up over a bump to reach this valley, this requiring sufficient kinetic energy. If the subunit balls can remain together at a level below the height of the rolling plane when they are apart, then potential energy is given off. If they end up in a valley with a height above that of the rolling plane, then energy is taken into the system and this must come from the kinetic energy of the balls. The higher the temperature, the higher the kinetic energy of the "rolling" subunits. This would mean that as far as any of the bumps in these diagrams, if the temperature was $10^9 C°$ or more, these bumps would easily be crossed. At these temperatures, however, the balls could also just as easily roll back out of the valleys and no aggregation would result with the one exception being that at the nuclei level. Thus at lower temperatures the valleys at succeeding levels become effective in trapping the subunits. The interesting point is that with increasing levels there is not only a decrease in the scale of potential energy valleys involved, but these valleys shift gradually from deep troughs to little dips on the top of high hills. This means a shift from exothermic or endothermic reactions with less stability for the resulting aggregate. Cellular macromolecules are stable aggregates of subparts under the reducing conditions of primitive earth but become unstable in the oxidizing atmosphere of present earth.

THOUGHTS

In this section we could see that asking some age-old questions regarding the nature of life revealed only comparatively recently that life is made of the same stuff as any other substance. It is the organization of this stuff which adds properties which we can distinguish as living. We also discovered an order of progression during the flow of time with aggregations of matter and energy resulting. This could be called the order of emergencies, where time has proceeded to unfold in a definite sequence many levels of organization of matter most of which have well-developed descriptions. The job of describing these various levels has been relegated to the different sciences. Physicists care about atomic levels on down and planetary systems on up; chemists care about molecules, small to macro; geologists care about crystals of metals on up; and biologists care about cells and on up. This order of cosmic evolution integrates these efforts and hopefully blurs their distinction for the benefit of each other.

Explanations of this order are of even greater uncertainty and circumstantial than those for the evolution of life. This order is but a wilder scheme and abiogenetic experiments or studies of present existing planets cannot prove that life has arisen in this way. They do allow and encourage us to think about it. Because of the many areas of science involved, perhaps any style of thinking is helpful. Regardless of the level of interest of the scientist, however, there are certain peculiar characteristics of the search that may demand a certain attitude of the searcher. The searcher must seek a multitude of facts derived from any source including diverse exploratory experiments. He must be able to correlate all these data to fortify his assumption that the basic hypothesis is a correct one. He must strain his logic and imagination in order to do this, and he must feel comfortable about working with the most unseeable and abstract of ideas. Above all he must be bold and be a bit resistant to criticism. He will often feel he is working in a vacuum and will need great confidence with little assurance. There are certainly giant gaps in the hypothesis, and heroic search for explanations and data will be necessary. But this will make the rewards that much greater for those who work the path.

viii

order: in composition

When we discover the existence of an intraprotoplasmic enzyme or other substance on which life depends, we are at the same time faced with the question how this particular substance is present at the right time and place, and reacts to the right amount to fulfill its normal functions.

J. S. HALDANE

It is as if two men, the one blind from birth and the other a cripple, were desirous of going traveling, and the crippled man were to mount on the shoulders of the blind man and were to direct him. . . .

Sayings of Buddha

This order is a recently derived order resulting from analyses of the cell and metabolism, their early histories already reviewed in the last two chapters, respectively. It is the order of the parts of the cell. It is the cell. The search here is for the laws governing the organization of matter and energy in the most basic unit of life. The examination of this order is sometimes referred to as modern biology, as more new explanations and concepts have been generated in this search over recent years than there has been from all the other searches put together. It is wrong to call it modern in the sense that the searches for explanations of other orders are obsolete. Rather the order of composition (cell) has come ripe with explanations over the past 50 years. The others have either not seen their day yet or are pausing for a next blossom of ideas.

There will be no attempt to even survey the tremendous amount of data that have accumulated from experiments carried out by searchers. Rather this section follows the development of various strategies used for understanding the cell. Each strategy is delineated on the basis of the experimental questions it asks and the experimental techniques available. The chapter is organized semihistorically around the flow of these various strategies. The first, biochemistry, carrying over from the last century is concerned about the isolation and characterization of the many enzymes and substrates of the cell. The purpose is to reconstruct the pathways of the flow of matter by piecing these dissected parts into a logical scheme. Integrating easily with this approach is that of the physical biochemist who cares about the energetics of this flow. The models of these are greatly aided by the strategy of the biophysicists who can reveal details of protein-environmental interactions. This permits us to survey examples of these protein reactions from simple to complex. Models of complex protein reactions and associations are in turn greatly stimulated by the results of cell physiologists and cytologists as they unfold ultrastructure and function of the cell. There is then the remarkable synthesis of the results of the structuralists with those conceived by the geneticists. The synergistic effect of this synthesis also effects the remarkable intracellular regulatory reactions conceived by the molecular biologists. This gives a model of the cell that needs some further thought as to its basic nature. These syntheses of strategies also bring with them a controversy as to respective roles of the principles of physics, chemistry, and biology in describing life. The products of this massive search are also applicable to the problem of the origin of the cell.

PATHWAYS OF MATTER

With the turn into the present century, the legacy of Liebig, Pasteur, and Buchner was continued with stronger than ever desires to understand the nature of the organic catalysts, the enzymes, and the metabolites that these enzymes react with. Enzymes were considered to be associated in some way with proteinaceous material but very little was known about these mysterious proteins. The great Emil Fischer (1852–1919) had proposed in 1894 that enzymes operate on their metabolites in a lock and key fashion. He derived this model on the basis of the specificity of different kinds of fermentation reactions in separate organisms with different enzymes. He did extensive analytical investigations into proteins and concluded that they were chains of amino acids, but this did not reveal any catalytic properties per se. Inorganic catalysts were metals and it remained a good working hypothesis that an enzyme, the catalyst, whatever its nature, was attached to a protein. Some support was given to this hypothesis when Buchner's method of removing small factors from cell exudates revealed by 1905 that an organic heat-stable agent was necessary for fermentation.

Although the exact nature of an enzyme remained controversial, the ensuing years revealed much about enzyme action and the complexity of enzyme systems. Again the procedures of Buchner made this possible. In vitro experimentation was carried out wherein part of the cell exudate could be placed in a test tube and controlled types and amounts of ingredients could be added with resultant reactions noted. In this way it became possible to examine the conditions under which enzyme action proceeded. By 1910 it was found that enzymes operated at optimum rates at some fixed temperature, hydrogen ion concentration, and salt concentration; i.e., enzymes operated best under specified environmental conditions. It was also found that the rate of enzyme reaction was a function of metabolite concentration. For a given amount of cell exudate from which all metabolites had been removed or used up, the rate of catalyzing additional metabolite increased proportionally to the concentration of this added metabolite up to a certain level after which there was no further increase in rate, as shown in Figure 8–1. Fischer's lock and key model

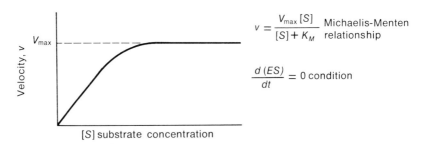

$$v = \frac{V_{max}\,[S]}{[S] + K_M} \quad \text{Michaelis-Menten relationship}$$

$$\frac{d\,(ES)}{dt} = 0 \text{ condition}$$

Figure 8-1. Velocity of enzyme reactions versus substrate concentration.

served as an explanation for this result by suggesting that for a given number of enzymes there were a corresponding fixed number of active sites for catalysis and as the number of metabolite molecules, or substrate molecules as they became called, for a given enzyme increased, they would saturate the site and prevent others from being acted upon at the same time. By 1913 reaction rates of most typical enzyme reactions were formalized by assuming that the rate was dependent on the enzyme-substrate complex concentration. This resulted in the so-called Michaelis-Menten relationship: $v = V_{max} \, S/(S + K_M)$, where v is the velocity of reaction, V_{max} is the maximum velocity, S is substrate concentration, and K_M is the so-called Michaelis-Menten constant related somehow inversely to the affinity of enzyme and substrate molecules. Thus by measuring the rate of change of substrate or product upon the addition of a range of known amounts of substrate, a characteristic constant K_M, unique for each enzyme-substrate relationship, could be found. This made it possible to know when one was looking at a single enzyme from a mixture, assuming that no other enzyme was affecting the substrate or product. The meaning of K_M has changed since the relationship above was devised. At that time K_M was considered equal to the dissociation constant of the enzyme-substrate complex back toward enzyme and substrate. It became apparent later that in some reactions the complex dissociated with the formation of the product and enzyme even more easily than it did toward substrate and enzyme. This condition was taken into account by considering that with an excess of substrate the substrate flows in producing a "steady state," constant amount, of enzyme-substrate complex with product leaving at the same rate. The resulting relationship is identical to the one above, but K_M has a less obvious meaning.

Regardless of this complexity, the procedure of measuring rates of reactions with added substrates in in vitro experiments became the standard bioassay method for tracking a specific enzyme during an enzyme isolation. Once a substrate or product was known, the procedure above was limited only by the ability to measure its rate of change. Spectrophotometers and manometers became the most useful tools to follow reactions. At the same time improved methods of separating proteins from one another were devised, especially those dependent on differential solubility, e.g., varying salt or nonpolar solvent concentration.

With these procedures and techniques in hand, many enzymes were being discovered. Enzymes involved in the oxidation of fatty acids, the formation of lactic acid, the dehydrogenation of organic acids, and several dehydration steps were discovered. Otto Warburg, by measuring the uptake of O_2 with manometers he designed, indicated the existence of a respiratory enzyme in sea urchin eggs. He even devised a model of this enzyme based on its similarity with response to cyanide and carbon monoxide inhibition on both normal respiration and the oxidation of an iron-nitrogen charcoal he had made. But most of the enzymes being uncovered were those associated with fermentation processes. These were discovered by finding that different halogens blocked fermentation but had different effects on the reactant solutions. If fluoride is present, 3-phosphoglycerate and 2-phosphoglycerate accumulate; while if iodo-acetate is present, then fructose 1,6 diphosphate also accumulates. This suggested that at least two different enzymes are involved in respiration and that these substances are intermediates being products of one enzyme and substrate for another. Thus a metabolic pathway involving several steps of intermediates was conceived and proposed for fermentation. By the early 1930's Gustav Embden and Otto Meyerhof together with Warburg had worked out the intermediates and isolated the 11 enzymes involved in fermentation.

It was also during this period that James Sumner was able to purify the first enzyme, urease, which hydrolyzes urea to CO_2 and NH_3, and show that it was pure protein. All studies since have allowed the conclusion that all enzymes are proteins although they may have tightly or weakly bound cofactors necessary for their action.

By the late 1930's evidence had accumulated that aerobic metabolism involved the dehydrogenation of carboxylic acids with the electrons of these hydrogen atoms flowing along several cytochrome enzymes that are oxidized and reduced sequentially and with the electrons eventually reducing the oxygen molecule to water. H. A. Krebs discovered that malonate inhibition caused the accumulation of succinate regardless of which of the other carboxylic acids were added including those that appear to follow succinate. This suggested to Krebs that these acids acted in a cycle. He confirmed this by noting that fumarate, malate, or oxaloacetate caused oxaloacetate to combine with pyruvate in equimolar ratios and that this synthesis was necessary for the accumulation of succinate. Also under these inhibitory conditions he found for each mole of oxaloacetate and pyruvate consumed, 2 moles of oxygen were consumed, and 1 mole of succinate, 1 mole of water, and 3 moles of CO_2 were produced. Thus it was that the Krebs citric acid cycle was supported.

This method of using inhibitors worked well but there were a number of questions about pathways that could not be clarified until radioactive tracers became available in the late 1940's. The substitution of radioactive atoms for normal specific atoms in substrates or intermediates is the ideal way to determine the steps in a pathway. The radioactive atom has a minor to negligible effect on the reactions and can be detected in small amounts. Thus tagged specific molecules can be thrown into the flow of matter without making a ripple and followed at various points downstream. Enzymes often catalyze reactions either direction depending on the concentrations of substrates or products. Thus studies while in vitro may not necessarily indicate the normal operation of an enzyme in the cell. With tracers the normal flow becomes apparent. Also tracers can be used to measure the quantity of intermediates, as pools of available molecules, under normal operations. Another major advantage in this method is that with the search for the radioactivity in reaction mixtures unknown, pathways and branches can be discovered. It is then a matter of analytical chemistry to determine what the tagged molecules are. This method was so successful, especially when combined with chromotography methods of analysis, that by mid-century most of the pathways of both catabolism and anabolism had been elucidated.

Radioactive tracer studies did more than reveal pathways, they dramatically demonstrated a whole new notion about the cell. This was that every part of the cell is within this network of flow of matter. All the enzymes and even the structural parts of the cell are constantly being replaced. Like a brick wall from which randomly picked bricks are pulled out and replaced, the wall remains at any one time but over a period it is a completely different wall in terms of bricks. The rates of these turnovers is quite astounding. Proteins will commonly turnover in seconds. The whirlpool of matter that Bernard added to da Vinci's flame includes the house of the cell as well as its parts.

In addition to the above there were other contributions from these strategies used by the biochemist. One nice surprise finding was that all living cells have at least some variation of the Embden-Meyerhof pathway and aerobic cells have similar pathways to the Krebs cycle and cytochrome system. Life has a uniformity. Another pleasantry is that all catabolic pathways have connecting points regardless of the source of metabolite and that anabolic pathways parallel their catabolic counterpart, as shown in Figure 8–2. Thus in most cases, any one kind of metabolite can serve as a part of most any building block of the cell. In spite of the diligence of this strategy only a fraction of the many enzymes in any cell have been isolated and characterized. The pathways of matter are well-known and the types of changes are probably fairly accurate but the enzymes themselves may be difficult to isolate. From the cell about which most is known,

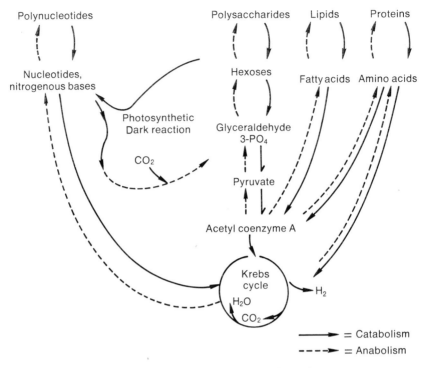

Figure 8-2. General anabolic-catabolic pathways.

the bacterium *Escherichia coli,* only about one-third. of the 3000 odd
different kinds of proteins have been characterized. By far most of these
are enzymes. This may seem alarming when we realize that the *E. coli*
is only about one five-hundredth the size of an average animal cell with
a corresponding number of protein types. Nevertheless, the biochemist
is fairly certain, especially from radioactive tracer studies, that he has
revealed those pathways involved with the major amount of matter flow.
The search into proteins has proved its worth.

THE FLOW OF ENERGY

Lavoisier and Bernard made it clear that where there was flow of matter
there was also flow of energy and that energy released during catabolic
reactions must somehow drive anabolic processes. The chemical energy

moving with the metabolites drains off and some of this flows back into the matter pathways of synthesis. Apparently according to this model this ability determines which molecules can serve as metabolites.

Clues as to how this transfer of energy could occur were long in coming and only after much less related data were collected could some model be conceived. By 1905 it was known that inorganic phosphate was required for fermentation. In 1929 ATP was first isolated from muscle and thought to be involved in energy transfer, but the relationship was not understood until 1937 when Herman Kalckar found that inorganic phosphate disappeared from around oxidizing tissue slices with the formation of ATP, glucose-6-PO_4, and fructose-6-PO_4. If this tissue preparation was poisoned by cyanide, aerobic metabolism was blocked, no phosphate was transferred, and ADP was found in place of ATP. It was thus assumed that ADP is phosphorylated to ATP during respiration.

About the same time Warburg showed that glyceraldehyde-3-PO_4 could be oxidized by a dehydrogenation of the aldehyde with inorganic phosphate taken up forming 1,3-diphosphoglycerate without a release of energy as expected by the oxidation. This suggested that the phosphate group's presence preserves the loss of oxidation energy. It was Fritz Lipmann who analyzed this ability of phosphates in various organic compounds to render the intact molecule a higher energy of hydrolysis than most simple inorganic phosphates. From his many experiments Lipmann concluded that all preservation of energy released during catabolism involved the phosphorylation of ADP to ATP soon, if not immediately, after an oxidation of the catabolite. It was later shown that ATP was directly or indirectly required for all synthetic steps and cellular processes like contraction and active transport. Lipmann proposed that it was ATP that was the common carrier and therefore the agent of energy flow as diagrammed in Figure 8–3. ATP is considered so significant to the life process that many insist that it is a key substance to indicate the presence of life on some extraterrestrial body.

But to understand that ATP is the common carrier did not reveal the mechanism of phosphorylation associated with aerobic metabolism. Wherein with the Emdem-Meyerhof type of phosphorylation, oxidation occurred to a substrate molecule already containing a phosphate group that could then be transferred to ADP, no such phosphate compounds exist among the Krebs cycle carboxylic acids. Instead when these acids are oxidized, their electrons in the form of a hydrogen atom are transferred to electron acceptor molecules, NAD, NADP, or FAD. There is a delay in the

Catabolic pathway of matter

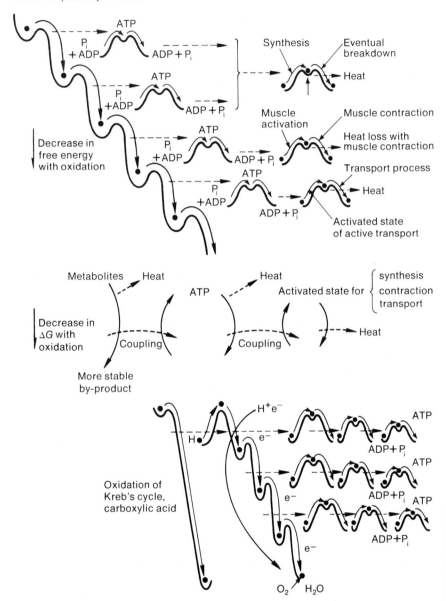

Figure 8-3. Representative pathways of energy flow.

phosphorylation process. Thus the flow of energy, as far as aerobic metabolism is concerned, involves other forms of matter than just ATP. These acceptor molecules seem to be even more efficient at preserving this energy of oxidation than does the direct phosphorylation process. Again these acceptor molecules are universal where ever indirect phosphorylation is concerned and could serve as an indication of this kind of energy coupling being present on some other planet.

Insight into this delayed phosphorylation pathway was given when Kalckar showed in 1940 that at least two phosphate molecules were taken up by ADP for each pair of electrons that reduced a single oxygen atom to water. Later measurements showed that this P/O ratio was 3. This indicated that with the flow of electrons from the Krebs cycle acids via the electron carriers to oxygen three side reactions to the final phosphorylation step existed. By the end of the 1940's Lipmann and others had shown that this side reaction involved at least two steps on the basis of "uncoupling agents" that would allow the electrons to reduce oxygen without phosphorylation also occurring. Meanwhile continued study had revealed that the electron transport system consisted of at least six electron acceptor-donor-type molecules including four cytochromes. The sequence of these molecules was determined by spectrophotometric measurements of the cytochromes with the presence of different kinds of specific inhibitors that could block electron flow at different points along the pathway. Also the reduction potentials of these molecules, when they could be isolated, supported this sequence, the potential increasing in a positive direction toward oxygen. Finally by 1950 Albert Lehninger isolated this electron transport system and side phosphorylation system and could show that with the addition of reduced NAD, he could generate three ATP molecules and take up one oxygen atom. He also proposed a model of where along the electron transport system the side phosphorylating steps were located along with a scheme for these latter reactions.

There have been several interesting questions and conclusions regarding the use of ATP as the universal single energy carrier. There is no question as to the value of having a single carrier in order to make a pool of energy available to a variety of uses. However, why ATP? One argument goes that it has the optimum free energy of hydrolysis both to accept energy from small organic oxidation reactions as well as to supply energy for synthesis steps. The assumption is that within the limitation of the intermediate molecules the metabolite is broken down by an optimum number of stages to generate the maximum number of ATP's. If the steps were larger, so the argument goes, then more energy would be lost as heat. If there are too many breakdown steps, then either more phosphate transfers would have to occur or the amount of energy transferred would be too small to make synthesis feasible. There must be some optimum and this would be a compromise as to size of the energy package transferred to gain the maximum amount of energy preservation and use to the cell. According to this argument, evolution has probably shaped the pathway of matter flow to fit ATP's characteristics. Supporting this argument is the high efficiency of energy coupling, especially found with energy preservation processes. It is difficult to measure this efficiency exactly because of unknown microscopic environmental conditions surrounding the phosphorylation sites. Both pH and reactant concentration affect the actual free energy change so much that it could vary anywhere between 7.0 to 13.0 kilocalories per mole of ATP. Regardless of this uncertainty, however, the cell is by far the most efficient energy converter known.

The other side of the argument is why ATP per se? It is true that the use of phosphate groups for transfer is ideal as they can seem to transfer a maximum amount of energy without suffering an oxidation-reduction reaction that would mean a greater loss of energy, but why the AT part of the molecule? Again one can find that the structure of ATP allows resonant states and charge repulsion that lend it its high free energy change of hydrolysis but so could other molecules. In fact almost any of the commonly occurring nucleotides, e.g., CTP, UTP, etc., have the same energy changes. Then the question becomes, why adenine? The only suggestion that has been offered is that of the nitrogenous bases studied adenine is the most stable and may have formed most easily abiogenically in the primitive oceans and thus have been available. In any respect, more thought about the reciprocity of ATP nature per se versus the optimum design of metabolic pathways and how these could have evolved is called for. Also this reminds us that this flow of energy plays a significant role in helping to shape the whirlpool of life matter and its organization in resisting the entropic doom, not just as an efficient energy sink but as a clear pathway-pathway coupler.

PROTEIN STRUCTURE

Studies on the flow of matter and energy demanded an explanation of the nature of enzymes. Their tremendous number in kind each with such diverse specific functions boggled the mind. How could one kind of molecule, the protein, have such exacting and yet diverse specificity? Since candlemakers were satisfied with the flow studies above, it was the structuralists, the clock builders, who felt most challenged to answer this question. They were convinced that the only way to understand protein structure was that employed for simpler organic molecules in the tradition of Liebig, Fischer, and others. This assumption was that no matter how complicated the molecule, it must be understood by building up from the subunits, atoms, or small molecules. They were concerned most with location and relative position of the various atoms. Know these and the properties of the protein will become clear.

This strategy of building the protein was frustrated by the difficulty in analyzing its subunits, the amino acids. There seemed to be different kinds, and separation methods during the early part of the century were not adequate. While time was allowing better techniques to be developed, the biophysicist was poking at the protein for other clues. Molecular weight is always an important parameter, and by the 1930's osmotic pressure measurements and the useful ultracentrifuge was indicating that proteins varied considerably in molecular weight, from the thousands to millions. This indicated that there were many amino acids in even the smallest proteins. Also proteins were found to differ in charge as indicated by the electrophoresis methods developing at this time. Diffusion and light-scattering experiments were also giving some data on overall shape. X-ray scatter pictures were even attempted on naturally occurring proteins, mainly by W. T. Ashbury, about 1930 on, with the surprising result that regularity or recurrence of subunits was demonstrated. Even though many of these techniques helped characterize different proteins and served as isolating devices, however, they added no further knowledge as to the substructure of a protein.

It was not until the relatively simple technique of paper chromatography was developed during the early 1940's that it became possible to separate amino acids and determine their separate structures. Also in the following years, it became possible to determine the composition of a specific protein in terms of the number of the different kinds of amino acids. The first protein to be so characterized was lactoglobulin in 1947. As each kind of protein had a relatively different composition of amino acids, it was most commonly assumed that the uniqueness of a protein and its specific action was somehow due to this composition. It was also assumed for example that if the relative number of amino acids were 1:2:4, leucine:alanine:glycine, then these would be distributed along the chain with a sequence lgaggaglgaggagl, etc. There was only one heroic dissenter who claimed this model was too simple. This was Frederick Sanger who set out in 1945 to examine the sequence of amino acids, the primary structure, in the 51 amino acid protein, insulin. Sanger eventually developed the method of partial hydrolysis and amino group labeling that revealed in 1953 that the amino acids are not arranged according to their relative abundance but as a very unique sequence. Since then improved methods have determined the primary structure of many proteins and Sanger's conviction has been upheld.

Sanger's result was heralded as a brilliant success for the structuralists after the fact, and they were quick to assume without question that a protein's function was determined by its specific amino acid sequence and that it would be but a short time before a knowledge of this sequence could be used to construct the protein's overall structure. Using the strategy of the organic chemist, much effort was put into synthesizing proteins by combining the normal or substitute amino acids in order to learn the laws of protein structure. This was done first for the 9 amino acid peptide hormones, oxytocin and vasopressin, in 1953. Over 50 analogues were synthesized for these similar hormones and their biological activity was measured. Although all nine sites must be filled, it was discovered that substituted amino acids at some sites could still support biological activity, one of these actually producing more activity than the natural occurring peptide. There are also natural analogues for some sites depending on the animal source of hormone. Since then similar studies on larger proteins have revealed that as the protein becomes larger even more sites can withstand multiple substitutions and again the biological activity may even increase. Also with these larger proteins, sections of amino acids may be omitted without effect. Although this strategy is still being employed, ambiguous results such as these have lessened the likelihood that exact laws of protein structure depending on amino acid sequence should be expected. This also heightens our curiosity in the as yet unanswered question as to why there are 20 and the particular 20 amino acids used in proteins.

The strategy of the structuralists that revealed most about protein structure was that of the X-ray crystallographer and model builders. Especially rewarding to their cause were the results of Linus Pauling who started his search in the late 1940's. Pauling was convinced that the secret of protein structure must involve restrictions imposed on the orientation of adjacent amino acids by the peptide bond that held them together. He synthesized dipeptides of glycine and carried out extensive X-ray analysis of their crystals. He showed that the amino and carboxyl ends of two adjacent amino acids are held together by a stronger C—N bond than most such bonds as judged by its shorter distance. He also found that the atoms adjacent to this carbon and nitrogen were all (six in total with the C and N) found within one plane. These results indicated that resonance existed between the carbon and nitrogen and the carbon's oxygen causing a restriction on rotation around the C—N bond axis. Thus the only flexible joints along the protein chain are at the alpha carbon of each amino acid. This is the kind of restriction every good structuralist needs.

Pauling then went on to repeat Ashbury's studies on natural proteins that are found in comparatively pure forms such as in hair or silk. He also found regularity and agreed with Ashbury's conclusion that much of the protein chain was in a spiral shape. He even repeated experiments with salts that cause the chains to lose much of their regularity and, as had Ashbury, he concluded that hydrogen bonds must be responsible for the spiral. Using the knowledge that bends in the chain must occur at alpha carbons and assuming that every carboxyl oxygen and amino nitrogen are bonded by straight hydrogen bonds, Pauling constructed a mechanical model of the spiral chain that is now known as the *alpha helix*. This model first proposed in 1951 was highly acclaimed and its existence as part of most proteins has been confirmed since then. It is significant, as it adds a secondary structure restriction that should make protein structure prediction easier. It is also a very rigid structure that holds the amino acids into a tight configuration with their side groups extending away from its main axis, thus holding them in a fixed available position. The question then is why are some of the amino acids in the alpha helix while others are not?

When more proteins were analyzed, it was found that although alpha helix secondary structure is common, most of the amino acids of many proteins are not found in this configuration. Pauling discovered while examining certain preparations of hair and silk that the chains were not in spiral but agreed with pleated plane models where chains of proteins were cross-bonded with hydrogen bonds with the parallel chains folding up and down together. Another type of secondary structure found later was that for the protein collagen in which three protein chains are cross-bonded with hydrogen bonds and form a triple helix as they wind around one another. Collagen-type secondary structure is presumably restricted to long tough fibrous proteins that make up linings such as skin and gut.

There are also many amino acids in any given protein that do not seem to be in any secondary configuration. Nevertheless, looking at the types of amino acids that are found in these various secondary structures, certain generalities can be made. Pleated plane structures require a high content of glycine or alanine, the smallest amino acids. The alpha helix is only suited for intermediate-sized side groups. Proline and hydroxyproline break up either a pleated plane or alpha helix and yet are required for the collagen structure. This still leaves a great deal of flexibility and there are exceptions, expecially where amino acids that could form an alpha helix are not found in this state. The secondary structure cannot be assumed from the primary structure.

Another point of frustration was that these studies above were all done with fibrous long proteins rather than the common globular proteins that make up the bulk of the cells working proteins, including most if not all the enzymes. Globular proteins are usually large and extremely delicate, making their crystalization difficult and an interpretation of their X-ray scatter pictures extremely uncertain. Nevertheless, one very courageous investigator, Max Perutz, began an X-ray analysis of hemoglobin as early as 1937. Amazingly enough he found some regularity, enough to encourage him, but not enough to allow interpretation as to its cause. It was not until Pauling had defined the scattering patterns of an alpha helix that Perutz was able to make any further headway. He could then say that hemoglobin did indeed contain some alpha helix formations that should have helped him analyze the remaining spots; however, this was not enough help for such a large protein. Finally in 1953 he conceived the idea of attaching large atoms, e.g., Ag, Zn, Hg, etc., to specific parts of the protein chain. As these reflect X rays readily and remain fixed in location, it was possible to determine some overall configuration of the protein chains. Again, the complexity of his protein denied Perutz an unambiguous interpretation of his reflecting spots. However, a colleague in the same laboratory, John Kendrew, had been working since 1947 on myoglobin, a muscle protein one-quarter the size of hemoglobin with a similar function as hemoglobin, and he tried the heavy metal technique on his protein. With great effort and the use of computers Kendrew was able to formulate a rough overall configuration, or tertiary structure, by 1957 and could pinpoint most of the major atoms by 1959. With help from this interpretation, Perutz soon afterward could define the configuration of hemoglobin, some 20 years after he started.

There were some striking surprises revealed by these studies, results that not only agreed between the two proteins but for all others analyzed since. First of all there are four globular units that make up hemoglobin, two identical pairs of two protein chains, alpha and beta, each of which is similar in tertiary structure to one another and to myoglobin. Also all of these units contain a heme group, which can trap oxygen molecules, attached in a very similar location. Each unit has from about 140 to 150 amino acids. The single chain in each unit is made up of a number of stiff alpha helix sections but the tertiary structure is so compact that not a single water molecule could fit inside. The most amazing finding was that all the side groups of amino acids (except one used in binding the heme group) that are found inside the globular unit are nonpolar in character. This means that all the polar side groups extend from the surface of the units. The conclusion from this is that there are, as shown in Figure 8–4,

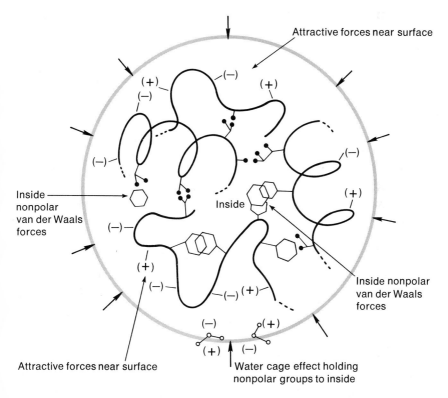

Figure 8-4. The forces of protein structure (only parts of protein shown).

three types of weak bonds involved in determining the tertiary structure of proteins. The first is the van der Waal forces of attraction between nonpolar atoms. The second is a possible electrostatic attraction between the polar groups on the surface acting as a surface tension. Finally and probably the one of most significance is the force due to the water cage repulsion of the hydrophobic nonpolar groups, collapsing them together toward the inside. All of these are weak bonds that could form spontaneously under normal cellular temperatures.

There was some elation from these results. Weak bonds could explain the ability of a heated protein to become denatured and lose its specificity or functions. Apparently the chain unfolded. Also if this heating was done very slowly with the temperatures returned to normal over periods of hours, a denatured protein could often recover its functions, the chain apparently refolding into the same pattern. This ability to reverse is considered strong evidence that there is a single best tertiary structure for a given sequence of amino acids. The overall structure would depend on the composite and location of polar and nonpolar side groups. Random attempts at folding would eventually find this best fit. Also analogue computers attached to consoles that can display a picture of the model have had some success in analyzing for the most stable state of a growing protein chain. As the chain grows and the various degrees of freedom are surveyed, it is possible for a known protein to see it fold with secondary structures appearing first and finally the correct tertiary one. The lowest free energy state and thus the most stable condition agree with the known tertiary structure.

In spite of this success there are severe problems. The myoglobin-hemoglobin results indicated the major one. It was found that the tertiary structure of myoglobin and the hemoglobin alpha and the hemoglobin beta subunits have essentially identical tertiary structures. This is amazing when one notes that only about 1 out of 10 of the myoglobin amino acids are the same as those for the hemoglobin units. This would mean that either different amino acids must be able to substitute for one another and yet produce the same tertiary structure and/or else only a few key amino acids determine this structure. Neither answer is satisfactory to a structuralist. There appears to be too much redundancy regardless and it begs an answer to the question of why the 20 amino acids. The analogue computer supports this flexibility by finding that substitutions can be made and still the same best fit tertiary structure emerges. Also there is far from any satisfactory formal description of this tertiary structure. A structure depending on a composite of weak forces placed at specific places is not as yet able to be formally described. A most stable energy state is not a unique or very useful parameter. Protein structure appears to be far out of the descriptive range of both quantum mechanics and classical organic structure analysis.

ENZYME MECHANISMS

Many structuralists supplemented the approach above by studies of enzyme mechanisms. Enzyme reactions are considered the most productive to study when interested in protein function as they are easy to follow technically and involve a highly specific active site. The assumption is that if one knew enough about the catalytic reaction in terms of change to the substrate, then it may be possible to hypothesize reaction steps and the reactive groups that make up the active site of the enzyme. There are considered to be about six major classes of enzymes based on type of reaction, e.g., transferases, which transfer a group; isomerases, which convert one isomer to another; etc. Most of these types of reactions are thought to involve at least several sequential enzyme-substrate complexes that may allow more than one hypothetical mechanism to produce the same resultant product. There are other types of reactions for which no mechanism has even been proposed.

Probably the most acceptable class of mechanisms are those for hydro-lases where a long-chain molecule is split at a subunit linkage position with the addition of a water molecule; an OH group is added to one end of the fracture, while the H group is added to the other. For this type of reaction it has been proposed that a side group of an amino acid donates an electron in the form of a hydrogen to one of the atoms at the bond that is broken, while a second amino acid's side group that is already dissociated serves to either accept an electron in the formation of a covalent bond or else stabilizes the substrate atom on the other side of the bond to be ruptured. This push-pull mechanism has been supported by dissociation-characteristic changes in enzyme reaction with pH variation and by finding that poisons to the reaction often adhere to amino acids with these types of side groups. It was not, however, until 1965 that the tertiary structure of an enzyme was determined by X-ray analysis. This was accomplished by David Phillips and was done with the hydrolytic enzyme lysozyme. Phillips managed to determine the structure both with and without a substrate equivalent. From this analysis he was able to verify the push-pull hypothesis, as diagrammed in Figure 8–5. In lysozyme this

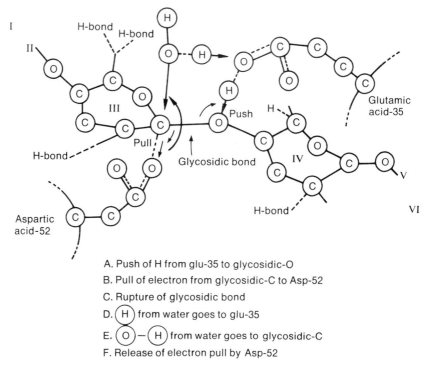

A. Push of H from glu-35 to glycosidic-O
B. Pull of electron from glycosidic-C to Asp-52
C. Rupture of glycosidic bond
D. (H) from water goes to glu-35
E. (O)–(H) from water goes to glycosidic-C
F. Release of electron pull by Asp-52

Figure 8-5. Proposed mechanism of enzyme action of lysozyme (proposed by David Phillips).

is apparently accomplished by a glutamic and an aspartic acid side group. It is believed that catalytic sites for hydrolytic reactions may contain any dissociable group, e.g., histidine, lysine, cystine, serine, etc.

Of especial interest was the finding that in addition to the catalytic groups, the active site must also be considered to have holding groups, side groups that hold the substrate in some exact position for the catalytic groups to be effective. Lysozyme happens to cleave a polysaccharide and it holds six glycosidic ring units in a rather deep cleft. The substrate is held by at least six hydrogen bonds and van der Waal bonds with a number of nonpolar side groups. Thus in this enzyme at least some nonpolar groups are present at the surface and are involved in the active site. Another surprise was the finding that the cleft twists, when filled with the substrate, a distance of about 0.75 angstroms. This twist is in such a way that the polysaccharide chain is bent with the third ring structure distorted, thus weakening some of its bonds. D. Koshland had proposed in the late 1950's that such a change in tertiary structure was necessary for enzyme reactions but this was the first direct support. This change in tertiary structure in the enzyme-substrate complex, called the *induced fit* by Koshland, was not considered a general phenomenon by many at this time. Since then, however, every enzyme that has been able to be analyzed by X rays with and without a substrate demonstrate the same change. One enzyme, carboxypeptidase A, shifts about 8 angstroms. Presumably the substrate alters the most stable configuration of the protein and it shifts to this state. The amount of this shift must also be taken into account in the design of the protein's amino acid sequence. This is a very powerful reminder of just how specific a protein must and can be. At the same time it bothers us even more that enzymes can operate with some amino acid substitutions or even parts of their chains removed.

Another similar result has been found for hemoglobin. In this case when one of the subunits takes up an oxygen molecule, that unit registers this effect somehow to its couplet subunit, which is then modified to have a stronger affinity for an oxygen molecule than it did beforehand. Each alpha unit is coupled to a beta unit and the modification is believed to work in either direction. This modified couplet then shifts with respect to the other couplet about 0.7 angstroms on the one side. This in turn apparently induces the unoxygenated subunits to have a higher affinity for oxygen. Thus there is a cooperative effect of enhancing the rate of reaction once a single unit has become oxygenated. Any shifts in protein structure resulting from its combining with a factor are called *allosteric changes*. This type of interprotein interaction reminds us that allosteric response can be initiated by sites on the protein other than the active site. One cannot but wonder whether allosteric responses are common for all proteins.

A survey of the various enzyme reactions also tells us that a considerable part of the surface of a protein could be involved in the active site. We must remember that in addition to a single substrate molecule being accommodated by the active site many enzymes must have specific holding and in some cases reactive groups for other substrate molecules, a variety of cofactors including metals, and energy or electron carriers, all in adjacent regions. How all these groups can be so accurately spatially arranged by a tertiary structure and possibly rearranged by systematic allosteric changes is a challenge that baffles our imagination and suggests that there is a great deal more that we need to know about protein structure.

PROTEIN COMPLEXES

Proteins form complexes with other kinds of molecules and with one another in ways that give confidence to the structuralists who assume that life and the cell can someday be explained in terms of close fits of their parts. Enzymes by their function form complexes, some strongly as exampled with the heme group in myoglobin and others weakly as with substrates, cofactors, salts, etc. Proteins may be made up of several polypeptide chains intertwined about one another as is the case for collagen. Others like hemoglobin may be made up of a set number and composition of globular protein subunits that happen to be held together by disulfide bridges and/or surface groups forming weak bonds. In fact most of the enzymes are made up of two or more subunits, their organization referred to as the quaternary structure. In some of these, when the subunits are separated, each part can still carry out catalytic reactions. In others this enzymatic property is lost. Each alpha-beta couplet of hemoglobin can take on oxygen without the other. The subunits may be identical or, like hemoglobin, have several types. It was also found that, even from the same cell, proteins with the same enzymatic activity, isozymes, differed structurally due to their composition of subunits, e.g., lactic dehydrogenase may be found in the five possible combinations of two kinds of subunits taken four at a time—AAAA, AAAB, AABB, ABBB, and BBBB. It would seem that proteins have experimented with all sorts of arrangements and complexes.

There are also multienzyme complexes that must remain intact for their action such as the one used for the oxidative decarboxylation of pyruvate to acetyl CoA and CO_2. This pyruvate dehydrogenase complex from *E. coli* has a molecular weight of about 4 million. It contains a total of 37 enzymes of 3 different kinds and involves about 96 coenzymes at any one time. Each of the enzymes has bound coenzymes: each of 24 pyruvate dehydrogenases binding a thiamin pyrophosphate, the dihydrolipoyltransactylase has 24 polypeptide chains each containing a bound lipoic acid, and each of the 12 dihydrolipoyl dehydrogenases has a bound FAD. In addition there are 24 CoA, one for each lipoic acid, and 12

NAD'S, one being reduced by each FAD that does not appear to be bound to the rest of the complex. It has been hypothesized that complex allosteric changes are required for this complex system to operate. This complex can be dissociated by lowering the salt concentration around it. The bound enzyme-coenzyme units can be isolated. As such they are ineffective. If these units are mixed together again in the proper salt concentration, however, the active complex reforms spontaneously. Apparently there are complementary sites in the proteins that recognize and bind one another into the proper organization. Like the folding of a protein chain into a particular tertiary structure, presumably proteins can be designed to self-assemble into a certain pattern without a source of energy. This emergence of a higher order at room temperatures would only require a sufficient number of accurately placed surface groups that could form weak bonds with one another. The ability of dissociated subunits to self-assemble and regain a lost function is proving to be one of the most powerful analytical strategies of the structural molecular biologist. One feels one is watching a natural process at work.

There are many protein complexes that are too large and complicated to be able to be analyzed by X-ray analysis. These most important complexes were not known or appreciated until the combination of thin film electron microscopy with differential centrifugation was developed by the cytologists about 1950. The techniques for using the EM on biological material evolved considerably and its prospects continue. When first used, as it was on eucaryotic cells, it revealed a whole new world of subcellular organization. The incredible array of membranes and dense bodies blasted down the generally accepted model of the cell as a bag of diffusible enzymes. When a cell is ruptured and differentially centrifuged, a wide range of protein complexes can be separated. These can be analyzed for specific functions and viewed separately under the electron microscope where they can be correlated with structures found in electron micrographs of the intact cell. Electron micrographs were also used to verify that the reassociated units of the pyruvate dehydrogenase complex were organized the same as its undissociated form. New techniques and improved resolution have led many to feel that it is only a matter of time before electron microscopy will replace X-ray analysis for protein structure determination. Even if this is never realized, it can be said that the EM even more than the X-ray scattering apparatus has upheld the structuralists' dogmas.

When one looks at an electron micrograph of a typical multicellular cell, it is clear that the cell is made up of many kinds of characteristically structured parts. The extensive endoplasmic reticulum that folds around most of the plasmic spaces has located throughout it many membranous organelles, microtubules, dense particles, and dense stroma, all of which involve protein complexes. Even the procaryotic cells usually have specialized infoldings of the plasma membrane, dense particles and stroma restricted to certain regions, and sometimes small membranous organelles and microtubules. When this substructure was discovered, it was immediately obvious that the biochemist's concept of the cell as a bag full of diffusible enzymes and the cell physiologist's concept of a cell as a single compartment with a single selectively permeable membrane controlling its content were gross oversimplifications. There must be an organization of parts and a great deal of subcompartmentation even within the simplest of cells.

Membranes per se have formed the basis of extensive studies and have indicated additional properties of protein complexes. First of all, there is assumed to be a general plan to the organization of molecules in what could be called the supporting structure of the membranes. Electron micrograph and physical optic studies have suggested that this supporting structure tends to be uniform from one membrane to another, being more dense on the surfaces than in the middle, somewhat like a sandwich. The dense areas are mostly protein, while the middle is mostly lipid. Likewise the dimensions across all membranes tend to be from about 90 to 120 angstroms. What is rather amazing is that lipids are almost always associated with membranes, these lipids having both polar groups at one end and nonpolar groups at the other. It is considered that the greatest difference between membranes is in the relative composition of lipid types, there generally being about four different kinds in any one membrane. Models as shown in Figure 8–6 have suggested that nonpolar parts of the lipids attract one another, while their polar ends hold the proteins. It is true that this complex is held together by weak bonds, but the specificity of the lipid to protein bonds is far from clear. It is possible to dissolve most of the lipids out of the membrane without losing its integrity. On the other hand a rare membrane is made up of essentially all lipids. Stability is still offered as the explanation for the lipid protein combinations although it is known that some of the lipids serve a functional role at least in certain membranes.

Although stable, membranes are flexible. Membranes break and re-form readily. They bud off as small vesicles from larger vesicles, or two

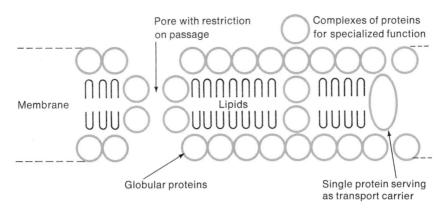

Figure 8-6. Model of membrane structure.

vesicles can fuse into one. When their lipids or some of their proteins have been bleached out of the membrane, they can be reconstituted by re-assembling processes to their original relative complement. Artificial membranes can be produced by natural means without a supply of energy once a polar-nonpolar interface exists. In this case lipids must be added first and then the proteins. Membranes can also grow endlessly by inter-collating the proteins and lipids into preexisting membranes. It is for these many properties that some feel that all the membranes of a cell grow out of one of the organelles, probably the nuclear membrane. If this were the case, the membranes would have to differentiate by substitutions into different entities related to their functions as their lipid and protein complements are different from one membrane to the next within the same cell.

In addition to serving as compartmental barriers these membranes hold other proteins and protein complexes on their surfaces and their protein content is at least partially made up of functioning proteins and protein complexes. The most studied example is that of the electron transport phosphorylating complex, a functioning unit that is found in the inner layer of the plasma membrane of aerobic bacteria and in the inner membrane of the mitochondria in eurcaryotic cells. Apparently these are found as intact units, each capable of carrying out complete aerobic phosphorylation, regularly spaced within and accounting for as high as one-quarter of the membrane mass. Photosynthetic light reactions and phosphorylations are also found in units and always as part of a membrane. In either case these units can be removed by varying the surrounding salt concentrations, a reversible process. It is of interest that at least for the aerobic units of the mitochondria, when removed, the integrity of the remaining smaller membrane is the same. This latter is important because some feel that all the membrane is made up of functional units organized in some way to give the appearance of a structural entity. This idea is not disproved by reconstitution experiments as it is known that functions other than aerobic phosphorylation are associated with these membranes. It does mean that the remaining parts of the membrane can coalesce without the aerobic units and this ability will have to be answered for.

Another most interesting finding is that these functioning units actually involve other phenomena that operate in coordinated fashion. For example the aerobic phosphorylating complex from liver mitochondria has a molecular weight of about 1.4 million and is made up of four subunits. As diagramed in Figure 8–7, these are the succinate electron transport system, the NAD-FAD electron transport plus its coupling and phosphorylating system, the cytochrome b and c electron transport system plus its coupling and phosphorylating system, and the cytochrome a and a_3 electron transport system and its coupling and phosphorylating system. Each of these can operate weakly when isolated from one another or can be reassociated for better complete aerobic function. If the lipids are removed from any of these subunits, electron transport cannot occur. This function is regained with the addition of the lipids that had been removed. Likewise the electron transport part of each subunit can be separated from its phosphorylating system by the presence of high concentrations of urea, which is known to break weak electrostatic or hydrogen bonds or by uncoupling agents. This separation is also reversible. It can be shown that the degree of affinity of the coupling and phosphorylating systems to one another depends on the degree of reduction of the electron transport agents involved, e.g., cytochrome c. When coupling is weaker, the electrons flow faster and the electron transport agents are in a less reduced condition.

Little is known about the details of interaction of these parts within this complex unit but several other observations must be taken into account. When the system is highly coupled and phosphorylation is high, the membranes containing this unit are in a contracted state. In opposite fashion, when electron transport is high and phosphorylation low, their membranes swell or stretch. Obviously changes in the attachment arrangements or allosteric changes occur to the subunits of this complex. This stretched state causes a four fold increase in permeability in general.

(a) Slow electron transport, highly coupled state

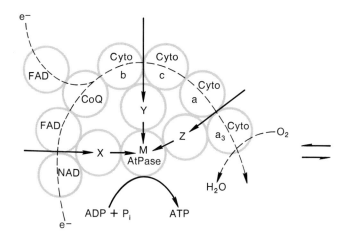

(b) Fast electron transport, low coupled state

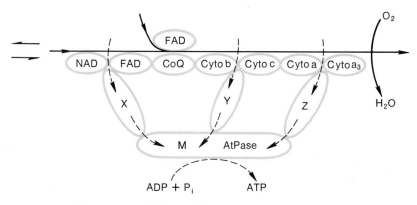

Figure 8-7. Models of electron transport system.

Also during the contracted state ions are actively moved across the membranes. This movement can be shown to depend on the coupled state of the electron transport-phosphorylating systems. The kinds of ions that are actively pumped depend on which are present. Divalent ions such as Ca^+, Ca^{++} are pumped inside liver mitochondria, while K^+, Na^+, or H^+ are pumped out, as shown in Figure 8–8. Likewise, in order for phosphorylation to occur, some ion transport must occur.

Thus a protein complex can serve a multifunctional role, and probably these functions are reciprocal. An analogous complex is found for the

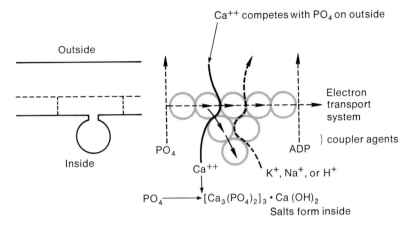

Figure 8-8. Ca^{++} pump in mitochondria association with phosphorylation.

electron transport-phosphorylating system of photosynthesis. Interestingly similar ions are pumped in directions opposite to that in mitochondria. It would seem that the complex is designed for a particular set of reciprocal activities rather than being restricted to one combination. Presumably the design of the proteins and allosteric changes are involved; however, these multifunctional units are, as well as is known, restricted to membrane structures. It has even been suggested that one of the reciprocal functions of the unit may be as a structural device for the membrane integrity. Regardless, one can see why it would be efficient to have the metabolic unit also directly affect its own environment by altering the ion and perhaps substrate concentrations on different sides of the unit. Moreover, the unit could be a useful device for translocating ions where energy is required. Such reciprocating units could add governance or balance to the clockworks of the structuralists' cell.

PROTEINS AND GENES

In hindsight, it is rather amazing that more concern was not given to proteins from a hereditary point of view. The absolutely essential role of proteins for all living processes should have suggested that they were influenced by heredity. Likewise the rapid turnover of proteins, requiring about 90% of the energy carried by ATP, might have led one to think that protein synthesis was related to gene action. This is what we can say in hindsight. If much was said, however, it was not deeply considered. Genetics did not fit into the mental constructs of the biochemists or biophysicists above. Genetics had its own language and did not seem to involve energetics or organic structure in any obvious way. Max Delbruck, a leading biophysicist, wrote in 1935 "that genetics is autonomous and must not be mixed up with physico-chemical conceptions. . . . (Fruit fly studies have led to estimates of) gene sizes which are comparable to those of the largest known molecules endowed with a specific structure (i.e., proteins). This result has led many investigators to consider that the genes are nothing else than a particular kind of molecule, . . . (but) one must remember there exists here a significant departure from the chemical definition of the molecule."

It was rather the geneticist in search of the meaning of a mutant organism that first brought genes and proteins into relationship with one another. Although similar notions were expressed earlier, their true significance was not felt until detailed analysis of the ability of restoring normal growth by adding specific intermediates to the media of a micro-organism was discovered. This was first done for *Neurospora* by George Beadle and E. Tatum in 1940 (and later for *E. coli* by Joshua Lederberg). They found that for three particular mutants it required arginine, arginine or citrulline, or arginine or citrulline or ornithine to be added to their media to restore normal growth. This pattern suggested to them that for each mutant an enzyme was nonfunctioning in this metabolic pathway. They checked this by noting a buildup of intermediate before the enzyme step that they deduced was inoperative on the basis of the experiments above. As these malfunctioning enzymes were related to a particular gene that could be mapped, and followed Mendelian genetics, Beadle and Tatum concluded that each enzyme is dependent on a single gene: the 1-gene 1-protein hypothesis, as this soon became known.

This gene-metabolism synthesis had a resounding effect on models for the gene. It was only natural to conclude that genes are also proteins and some considered that for each kind of protein found in a cell there was a prototype present in the chromosome that could replicate itself manyfold to produce the active proteins of the cell. These chromosomal protein genes could also replicate themselves during cell division and retain their identity from generation to generation. It was also conceived at this time that each protein could serve as a template to align identical corresponding amino acids into a set pattern. This was before Sanger had found out about the specificity of the amino acid sequence but the chain of amino acids idea was not new. Although this was an interesting idea, few considered it strongly and no models of how a protein could operate to replicate itself were put forth.

Instead, another discovery that turned out to be a bombshell and caused a refocusing of efforts happened in 1941, too soon for the models above to really be developed anyway. This was the follow-up of a strange result first observed by F. Griffith in about 1928. Griffith was studying the cause of death by *Pneumococcus* and discovered a mutant strain that had lost this ability. It was called a rough strain as it did not have a shiny smooth mucopolysaccharide coat surrounding the rough inner core. By accident he discovered that if heat killed normal strains and these roughs were inject into an animal, the animal died and normal smooth-coated bacteria were present. Since neither heat-killed normals nor roughs injected separately into the animal caused death, he could only conclude that some "transforming principle" was leaving the heat-killed smooths and transforming the roughs into normals. It was O. Avery and his colleagues who methodically dissected out the various kinds of molecules present in the heat-killed cells and added them with the rough strain that discovered that the principle was one and only one kind of molecule, desoxyribonucleic acid (DNA).

This was shortly followed up by a further resounding implication that DNA plays some hereditary role by the bacteriophage studies of Alfred Hershey and Martha Chase. Delbruck and others had been studying the ability of bacteriophage to enter a bacteria and self-replicate itself several hundredfold. It had been a basic assumption that it was a protein from the virus that entered the host and served as the genetic material for the new phage. Hershey and Chase were able to test this assumption when radioactive tracers became available. In separate experiments they labeled the protein coats with S^{35} and the DNA core with P^{32} and discovered that it was, except for a trace of protein, the DNA that entered the host and was not only responsible for replicating itself but also was responsible for the viral-type protein that made up its coat.

All eyes turned toward DNA as the answer of the gene molecule. Their look was encouraged and guided by the writings of Erwin Schrodinger, especially his *What Is Life?* written in 1945. Schrodinger stressed that "the obvious inability of present-day physics and chemistry to account (for life and genetics) is no reason at all for doubting that they can be accounted for by those sciences." He suggested that genes preserve their structure because the chromosome that carries them is an aperiodic crystal. He considered that this large aperiodic crystal was composed of a succession of a small number of isomeric units with the succession acting as a hereditary code. In this way a great deal of information could be preserved that could dictate the cells' composition.

By 1947 it was clear that all multicellular organisms contain DNA and that its nitrogenous bases were found in such relative concentrations that adenine plus guanine equaled thymine plus cytosine although its $(A + T)/(C + G)$ ratio varied considerably from one organism to the next. It was also found that the DNA per cell was consistent with the haploid number, doubling when the chromosomes double during interphase, etc. It was also known that there was a second nucleic acid, ribonucleic acid (RNA) that was found to be in place of DNA in some viruses. RNA was also found in the cytoplasm of cells.

It was T. Caspersson who *first* suggested an explanation for the presence of the two kinds of nucleic acids in a cell and their role in protein synthesis. He had perfected a method of using a microbeam of ultraviolet light with which he could scan large cells and noted that the DNA of the cell was restricted to the chromosomes. He found RNA in the nucleus and even more in the cytoplasm near the nucleus. It was at this latter region that radioautographs had shown that proteins are synthesized. From this data Caspersson proposed that DNA was somehow responsible for the synthesis of RNA, which in turn moved out of the nucleus to the cytoplasm where it was somehow responsible for the synthesis of proteins. This scheme, which was later called "the central dogma," has served as a guiding basis for experiments ever since.

DNA structure was not an easy problem. Ashbury had seen regularity in DNA with X-ray analysis in 1945 but could make little of it. In 1950 Maurice Wilkins repeated this observation and under the influence of Pauling and his alpha helix concluded that there may have been several sugar-phosphate backbones of the nucleic acids twisting around a central axis. Francis Crick derived the mathematical relations that showed that Wilkin's X-ray pictures could have arisen from helical chains, but neither the pictures nor the theory could tell how many chains were involved. The greatest other uncertainty was the arrangement of the nitrogenous bases. It seemed that they were perpendicular to the main axis but whether toward the center or away was impossible to ascertain. By 1952 Wilkins and Rosalind Franklin could show that there was a repeat structure along the axis every 34 angstroms and that the diameter of the cylinder the helical chains generated was about 20 angstroms. They also felt that the nitrogenous bases faced toward the axis. These data especially led Crick and James Watson to take up the game of mechanical model building in the tradition of Pauling. After a number of attempts Watson tried various combinations of nitrogenous base pairing by hydrogen bonds, again clearly under the influence of Pauling. Being a geneticist he was inclined to think in terms of two chains, as an even integer of these would be required for division somewhat as in the case for chromatids during mitosis. In this case one base from each chain could pair with a base from the other chain. To his surprise only certain combinations of bases could pair up, the A with the T and C with G. This could account for their consistency in relative concentrations. Also if these pairs are inserted into the model, the diameter of the overall double helix so formed is the 20-angstrom value. Watson built a model so that it did twist with a repeat every 34 angstroms. Crick showed mathematically that this model could produce the obtained X ray pictures, and the Watson-Crick model of of DNA was born in 1953.

The analogous structure of DNA with that of proteins must be noted. They both have a sequence of subunits along a chain-like structure. The secondary structure of each is influenced strongly by hydrogen bonds, weak bonds so that they can form spontaneously without the need of an external source of energy. DNA secondary structure, like protein tertiary structure, also depends on van der Waal forces—between nonpolar side groups in proteins and between adjacent pairs of bases in DNA. DNA must also have a tertiary structure within the chromosomes of eucaryotic cells but as yet it is not known.

Considered as a sign of the importance and success of the DNA model was its ability to satisfy any requirements placed upon it and to suggest further models of how it acts—a rare and delightful event in science. DNA must be able to replicate itself and the model made this easy by suggesting that it merely unzippered up the middle with the resulting open chains serving as templates upon which the random nucleotides must form as a complementary sequence, thus synthesizing two identical double helices. Only an error in this complementation process would cause a loss in the unique sequencing over endless generations. An error, in other words, would be the explanation of a gene mutation. Likewise a model of gene action along the central dogma theme could be imagined. One of the DNA chains could serve as a template for the formation of the RNA which could somehow determine the sequencing of the amino acids of the proteins that each gene is responsible for. There would presumably have to be two different enzymes that could recognize the two kinds of ribose sugars between DNA and RNA and thus select which class of nucleotides would be accepted. Thus both self-preservation and gene action can easily be imagined.

Perhaps of most significance is that the DNA model represents a synthesis of genetics with molecular structuralists. This is remarkable enough, but there was an even more interesting result. This was that a completely new concept was superimposed upon the basic assumptions of structuralists, mainly that in the case of molecules involved in genetics these molecules have an informational content. They are not just structures per se, but there is a particular pattern of structure, as the word template implies, that gives the molecule additional meaning. This property was conceived by Schrodinger but not really developed or understood until the DNA model was born. Information vocabulary and theory as formerly applied to communication processes were quickly adapted to the central dogma. As RNA was similar in pattern to the DNA from which it was generated, this synthesis has been called a *transcription process*. As there are 20 kinds of amino acids to be sequenced by only 4 kinds of nucleotides, there would have to be at least 3 bases in some pattern determining each different kind of amino acid—this pattern designated the genetic code. The process of synthesizing proteins from the RNA is therefore called a *translation process*. The RNA specifically used for a given polypeptide is referred to as that protein's messenger RNA. Likewise the genetic code has nonsense and missense combinations, and the messenger RNA is started and stopped in its readout process. These vocabularies emphasize the complexity and extent of the processes involved.

Information theory as applied to genetics also implies that the DNA acts as a blueprint with an order of data that is transferred to and is responsible for the order of the organism within which it is found. The bases along the DNA are not in a random sequence but arranged in order to code out the proper sequence of amino acids in all the proteins of the organism. A highly specified one-dimensional order is transferred to the three-dimensional order of the protein, functional units, membranes, the cell, the organism, etc. In this context DNA allows us to think about spatial organization in a simpler probabilistic manner. Now the informational capacity of DNA can be defined as the number of possible arrangements of the bases. The informational content of a given DNA molecule refers to the probability of that one combination out of all possible combinations existing. For convenience and use of the digital computer, information is defined in terms of bits of information (H) equal to $\log_2 n$, where n is the number of possible arrangements. Thus the number of bits of information related with a given sequence of bases in the DNA would be the minimum number of either-or, yes-no, questions that would have to be answered in order to establish the unique sequence of bases.

If as mentioned above we assume that 3 bases are required for each amino acid, then it would require 300 specifically sequenced bases in the DNA responsible for a 100 amino acid protein. As there are 4 kinds of bases possible in any sequence in the DNA, this specific sequence would contain $\log_2 4^{300}$ or 600 bits of information. The protein being programmed by this DNA would have to receive information from the DNA. The amount of information in the finished protein would be $\log_2 20^{100}$ as there are 20 different possible kinds of amino acids that could be arranged in any combination of 100. This equals 440 bits of information. As should be there is less information than in its source, the DNA. One must expect this as there is a redundancy in that more than one code triplet of bases can call for the same amino acid. As the DNA of *E. coli* is estimated to have about 1.5×10^6 nucleotide pairs, there would be $\log_2 4^{1.5 \times 10^6}$ or 3×10^6 bits of information in this molecule and presumably less than this in the intact *E. coli* cell.

Informational analysis has been related to thermodynamics, noting that entropy is a measure of order. Conceptually it is possible to equate bits H as equal to $S/k \ln 2$, where k is a conversion constant. Entropy depends on a stable end state of some reaction, however, while informational analysis stresses the ordering process per se; i.e., informational content refers to the number of steps required to reach the order. In this way it is different and may be useful. Unfortunately to date informational analysis has not revealed anything that was not or could not be understood by simple probability calculations. It continues to work in one dimension and tells us nothing about the unique kinds of three-dimensional organization that different DNA's with the same informational content could produce. It is also but a potential of what information may be used. The idea is interesting but will have to be developed considerably before it adds that kind of insight and description we so badly need for explaining organization.

Regardless of the difficulties with information theory, its basic notion that a molecule like DNA has a great deal of uniqueness is another necessity for the DNA model. DNA must be able to account for all the organisms that have ever and will ever exist. This is a big order. Yet if we allow that each form of DNA describes a different form of organism, then there could be $(4)^{1.5 \times 10^6}$ kinds of *E. coli* considering the 4 possible bases in any sequence. Likewise with man or most vertebrates, there is at least 1000 times as much DNA or at least $(4)^{10^9}$ such organisms, an incredibly large number. Thus DNA more than satisfies this need as a diversifiable agent.

Studies such as these on DNA and the central dogma have introduced a tremendous number of facts produced by very clever experimentation on the part of a large number of individuals. These are well-known and need not be covered here. One outcome has been a model of the ribosome-*m*RNA-enzyme-cofactor complex that somehow shifts with allosteric changes as the *m*RNA moves with respect to the ribosome, and a polypeptide chain is generated. The key to understanding this complex is the adapter molecule, the *t*RNA, and the ribosome itself. The base sequences of some of the *t*RNA's have been determined and their structures have been proposed. The ribosome could contain as many as 50 different proteins and 3 different kinds of RNA. It can reassemble once disassembled. As yet this complex is very little understood and represents one of the main thrusts of the molecular geneticist. Even before the model of this complex is complete, it is clear that it will be a success for the structuralists. The many mechanical-like movements of its closely associated parts remind one of a delicate clock.

We need mention but several more points to show how the geneticist has brought the DNA molecule into focus as a gene in the classical sense. It was the remarkable techniques of Seymour Benzer that finally showed by about 1961 that, at least in viruses, gene mutations and the equivalent of crossover could occur anywhere along the DNA and that what could best be defined as a gene is that length of DNA that was responsible for a single polypeptide. Then by the mid 1960's it was possible to show that the linear mapping of either these mutations in bacteriophage or others in *E. coli* are in direct colinear relationship with the amino acid substitutions produced in corresponding polypeptides, as diagrammed in Figure 8–9. Here again is a beautiful synthesis of the methods of the classical geneticist and the molecular structuralist. By 1968 Marshall Nirenberg and his colleagues had completed the difficult and meaningful task of breaking the genetic code. The base sequences of the DNA from viruses and bacteria specific genes are now being determined and soon artificial genes will be manufactured and used in in vitro experi-

E. *coli* tryptophan synthetase (A chain section) Gene map

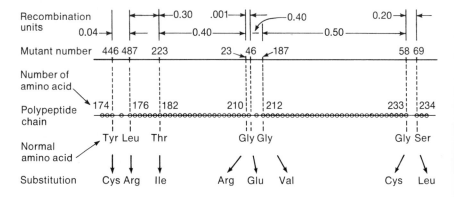

If cross mutant 23 (gly→Avg) with mutant 46
(gly→glu) can get normal gly in 210 position.

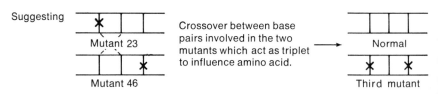

Figure 8-9. Colinearity of gene mutations and amino acid substitutions.

ments to produce *m*RNA's and proteins in order to explore the degrees of freedom or redundancy permitted in DNA as has been done with proteins and their biological function.

PROTEINS AND INTERDEPENDENCY

But perhaps the greatest contribution that the discovery of DNA had, and one its founder could not have anticipated, was the revealing of a new order in life (the order of this chapter) and the establishment of the field of molecular biology dedicated to the explanation of that order. DNA was more than an explanation of the gene and its action, it completed the circle of processes most basic to the cell. Now ATP is more than an energy carrier. It is involved in more than a flow of energy. It is a basic building block, directly and indirectly, for the nucleic acids, which in turn play a vital role in anabolic processes. It is as if some of the energy was stored in the form of a matter that had a template ordering role as suggested in Figure 8-10. As such, this template helps heighten, or

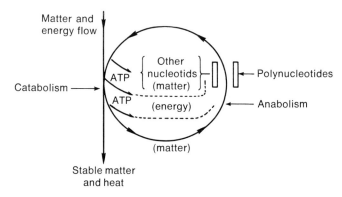

Figure 8-10. ATP helping to shape the loop of matter.

prolong, the whirlpool of matter we call *life*. At the same time this picture is incomplete, for it does not include the proteins that shape the pathways of flow of this matter and energy. For this consideration we can imagine a reflux system of matter being contained with restricting tubes of proteins, as diagrammed in Figure 8-11.

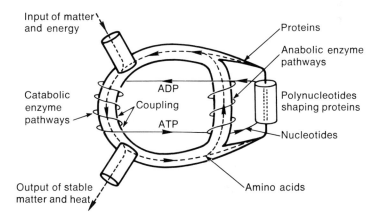

Figure 8-11. Protein reflux system.

Our diagrams have become too complicated, however, and we have almost lost the most basic beauty of all; there is a complete and total interdependency between the functions of proteins and polynucleic acids, as shown most simply in Figure 8–12. The catalytic center includes all

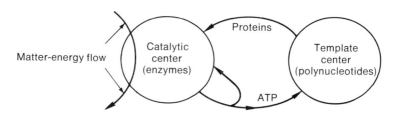

Figure 8-12. Basic "centers" of a cell.

enzymes whether anabolic or catabolic, and the template center includes all polynucleotides. As indicated, the polynucleotides could not exist if it were not for the coupling processes of specific enzymes creating ATP and putting it and its derivatives, the other nucleotides, together. Also the catalytic center requires its own product ATP for its operation directly as in Embden-Meyerhof catabolism and indirectly in the synthesis of its many organic cofactors. Then of course the proteins and their described specificities could not exist if it were not for the polynucleotide template center. Finally this synergesis is not complete without reminding ourselves that there are membranes and other organelles in even the simplest of cells that are totally dependent on both of these other centers and yet necessary for each of these centers to maintain their organization and interactions with their environment. As a cell elaborates, it is this last organelle activity that becomes most evident and plays an increasing role in the efficiency and the carrying out of the activities of the catalytic and template centers—indeed it is the visible cell. Thus we add one more diagram, Figure 8–13.

The beauty is that life can be explained as a synergesis of two kinds of macromolecules, and this synergesis exists because of the particular uniqueness of each of the macromolecules. The molecular biologist cares less about the properties of these macromolecules than he does about their interactions and the systems these interactions produce.

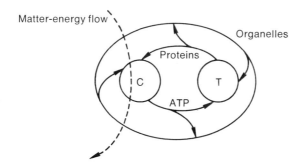

Figure 8-13. The cell system in terms of centers and organelles.

PROTEINS AND REGULATION

The most recent discoveries that have added so much more to this order we call the cell are its many systems of self-regulation. The incredible number of these systems operating so delicately at so many levels has made the models of centers above seem quite crude. The molecular biologist is being both frustrated and delighted with these finds— his models are getting so complex, he fears he can never explain this order; yet he feels more intensely than ever the need for his effort and the beauty of his object of study.

It should be no surprise that as proteins, the enzymes, determine the pathways of flow of matter and energy, so they are the sensitive receptor sites for regulation. There are at least four different classes of regulation all involving proteins in some way.

The first could be considered those involving natural processes of the active sites of enzymes per se. Each enzyme has a natural affinity for its substrate that influences the turnover rate. This rate must also be influenced by the concentration of substrate and product as a mass action effect. And if the enzyme happens to be able to catalyze more than one substrate or if natural inhibitors are present, then competition will influence the rate. If allosteric changes are part of the reaction, these too must have a most natural rate limiting the overall catalytic rate. Finally we must consider the system of enzymes, as the effect of any one enzyme in a pathway will influence the rate of others indirectly by affecting their substrate or product concentrations. The organization of these systems could affect the rate if, for example, diffusion is necessary to allow a product of one enzyme step to reach the next enzyme. With the complexity of the number of steps, it had been assumed that these inherent properties set the pace and concentration of intermediates within the cell.

A second type of regulation, however, that is hard to evaluate in terms of importance but known to operate is that of controlling the accessibility of various factors involved. Enzyme reactions are not only sensitive to the substrate or product concentrations but to many factors in their local environment. With the discovery of the subcompartmentation within the cell and active pumps, it is certain that hydrogen and salt ions, cofactors, as well as substrates and products may be limited or concentrated as a means of control. This may be accomplished by coordinated protein complexes such as the electron transport-phosphorylating complex in addition to pumps activated directly by ATP. These kinds of control are difficult to detect and measure, but more and more are being discovered as microtechniques improve.

Very closely related to this latter type of control is that influencing "pool" size. Any factors that are used over again such as building blocks or energy carriers exist as a pool. The size of this pool depends both on the rate of synthesis using these factors and the rate of degradation of this synthesate resupplying the factor. Where several kinds of building blocks are used, such as the amino acids or nucleotides, the relative kinds present as well as the pool size are important. Control works very effectively on both the input and output of the pool as the turnover rate is often considerable. For example it has been estimated that the pool size of ATP, ADP, and AMP is only about 1 to 2 million molecules and yet the rate of turnover from ADP to ATP or back is about 2.5 million per second. One can imagine how rapidly phosphorylation would be blocked if the ATP was not used very rapidly.

A third major type of regulation is that in which some factor directly acts upon an enzyme at some site other than its active site and influences the enzyme's operation. This factor may even be another enzyme or protein. For example, glycogen phosphorylase has an active *a* form and a less active *b* form, the difference being that the *a* form is made up of four subunits, while the *b* form has two. An enzyme, phosphorylase phosphatase, hydrolyzes a phosphate group off the subunits of the *a* form and the subunits dissociate. A second enzyme, phosphorylase kinase, uses ATP to rephosphorylate the dissociated subunits causing the reformation of the *b* form. Similarly the electron transport system operates differently if it is complexed with "structural proteins" than it does when isolated. Another example is when alpha and beta chains of hemoglobin are being synthesized. If there is an excess of one of these chains, the protein apparatus synthesizing the excess is inhibited by direct action of the excess chain.

Most of the cases of this control are due to energy carrier or building blocks acting directly on an enzyme at some site other than the active site. A general theory is that these enzymes have an even number of subunits and pairs cross react in response to the factor analogous to the cooperative effect in hemoglobin. These factors act either negatively or positively, apparently by allosteric changes to the enzymes, influencing the active site accordingly. What is amazing is the efficient choice of enzymes upon which these factors operate. This is best explained by the diagrams in Figures 8–14 and 8–15. Notice that where ATP is necessary

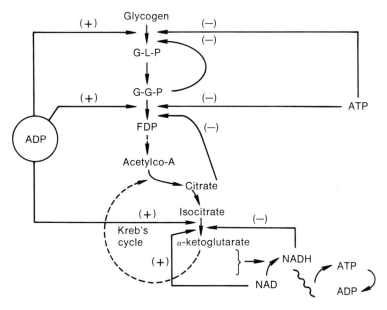

Figure 8-14. Catabolic control by key product and energy carriers.

ADP can enhance, while ATP or some end product inhibits. Moreover, feedback inhibition may act on an early common precursor enzyme and the first enzyme after a branch in a pathway. Of especial interest is the enzyme glutamine synthetase from *E. coli*, which apparently has eight separate sites receptive to different end product inhibition, all affecting

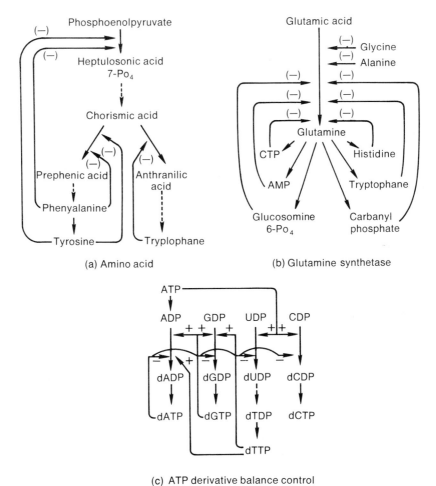

(a) Amino acid

(b) Glutamine synthetase

(c) ATP derivative balance control

Figure 8-15. Multiple system of regulation.

its active site. How complex can an enzyme be?

In Figure 8–15 is an example of cross-reactions wherein the products of one pathway can inhibit or activate a second pathway to assure a balance of related building blocks.

The final type of control operates on protein synthesis. Protein synthesis can be affected at the ribosome, thus controlling the rate of *m*RNA read out, or by influencing the rate of *m*RNA synthesis. The former seems to operate by way of sensitive centers on the ribosome and comparatively little is known about it. The latter has been another example where clever deductions of the behavior of select mutants at the hands of F. Jacob and J. Monod produced the interesting model bearing their name on gene regulation. They deduced that a gene they called the *regulator gene* produces a product called *repressor* which acts directly at a region of the DNA known as the operator site which is next to the so-called lac operon region of *E. coli*. This lac operon region of DNA happens to be three genes in length. When the repressor is attached to the operator site, no operon *m*RNA synthesis can occur; i.e., the operon is "turned off." If the *E. coli* is surrounded with lactose, however, this substance enters the cell and combines with the repressor preventing the latter from adhering to the operator site. When this happens, the operon *m*RNA is synthesized; i.e., the operon is "turned on." Lactose or its equivalent in an analogous system is called the *inducer* and the process is called *protein induction*. It operates by releasing a normal state of the inhibition of protein synthesis (of a particular operon). Long after this was deduced W. Gilbert isolated the repressor of the lac operon, found it was a protein made up of four subunits, and showed that it does indeed combine with DNA as the Jacob-Monod hypothesis had claimed. When it combines with inducer, it goes through an allosteric change that apparently prevents it from attaching to the DNA. The lac operon of *E. coli* was also the first single gene to be isolated.

In addition, protein synthesis can be repressed by the presence of a small molecule. It can be shown that an RNA complexes with this molecule, the combination known as the *corepressor*, which in turn combines with a regulator gene protein and together they attach to an operator site preventing the adjacent operon from serving as an *m*RNA template. Again the action is negative but the repressor molecule must be present before the otherwise dormant regulator gene product becomes active. Finally there is positive form of gene regulation wherein the regulator gene combines with an operator site and "turns on" the gene.

Again it is best to see a diagram as shown in Figure 8–16 to note the efficient locations of enzymes affected by this type of control. Protein synthesis is extremely costly and the more regulation, the better.

These many controls often operate on the same system of enzymes for even finer regulation. We must wonder if there are not others. Even with these we find that the cell is a complex network of interactions with little left for chance. Even the adaptation patterns seem to be set.

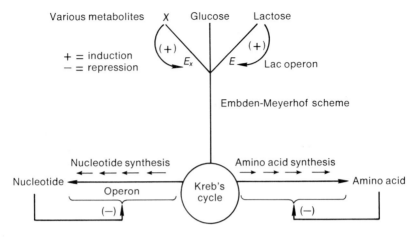

Figure 8-16. Distribution of enzyme controlled by regulator genes.

CELL MODELS AND CONTROVERSY

With this unfolding of discoveries, searches, and syntheses there is no doubt about the order of constituents that make up the cell, but we must ask just what is the cell. If cannot be described by any equation. No simple abstract of the cell exists. Rather we must resort to a spectrum of models to give our present impressions, impressions that will no doubt change. Early models of the cell as a bagful of enzymes that diffuse about one another allowing the flow of matter and energy served its purpose then. The dynamic change of all its parts including those that are structural as demonstrated so well with isotopic tracers made the cell become a candle beyond da Vinci's wildest dream. But then X-ray and electron microscope studies found structure upon structure so extensive that only mechanical-like motions as in a clock could account for cell processes. Finally poly-nucleotide-protein interdependency and especially the many circulating streams of regulating agents added complexity and logic for both clock and candle models. On the one hand these agents act so swiftly and precisely that the interlocking of parts of the clock are even more assured. On the other hand, this regulation depends on a dynamic flow of diffusible agents allowing the parts to be less mechanically connected for their precision of operation. In short, the cell is neither and both clock and candle by our present dim view. We must also recognize that there is a huge gap in our knowledge and thus concepts because of the inability to follow structural (molecular complex, membrane) movements with time to indicate rhythms of activity and regulation. The electron microscope is but a still picture of a possibly well-developed and rapidly moving plot.

This same ambiguity or dualism results from any attempt to apply physical theory to the cell. On the one hand it would appear that laws of energetics apply to simple enzyme systems especially when in vitro. Likewise, structural models of macromolecules changing shape and fitting together made sense within the confines of chemical bonds and physical mechanics. It is when we try to apply thermodynamics to the cell that we find the most difficulty. This statement will intimidate some because thermodynamics is the physical science that has succeeded to explain so much already about the cell. It has allowed us to see that the cell extracts free energy from its environment in the most efficient manner an isothermal device can and use that energy to place matter into a minimal entropy state. This is possible because as nonequilibrium thermodynamics can show these efficencies are maximum and the entropy is minimal only because the cell is an open system with matter and energy flowing through (and around) it. Open systems operating at steady-state conditions are the most optimum at obtaining these characteristics. Thus, of course, life does not violate any law of thermodynamics. The problem is that this law of open nonequilibrium isothermal systems is an overall statement that cannot reveal what the cell's order will be or describe its many operations. It can serve only as a general description after the facts are known rather than give us a description that we can use for prediction, testing, or growth of ideas. Another more basic difficulty is that the size of the cell may make thermodynamics inapplicable. Thermodynamics involves statistical treatments of diffusible reactants. It has been estimated that *E. coli* has only about 300 H^+ ions, 2 or 3 of several of its enzymes, only 2 or so repressor proteins from each regulator gene, and only the single DNA molecule. Much of the water is thought to be in a bound state and certainly there are many nondiffusible reactions involving complexed molecules. The cell is in the awkward state between a diffusible many-membered system and a machine with engaged gears—again somewhere between candle and clock.

Presumably because there is such a rich blending of physics, chemistry, and biology in these models of the cell, there is also great controversy as to which area is most successful at explaining it. Many, but not all, physicists and chemists believe that the explanations of life processes can be reduced to the laws of physics and chemistry. Many, but not all, biologists and philosophers of science believe that life phenomena are irreducible to the explanations of physics and chemistry. This is obviously a very subjective matter, with some physicists feeling that their craft is most basic and thus superior and some biologists feeling smug with their moving field and being defensive about their position. Regardless, the real difficulty seems to be with the definition of physics or physical laws. This difficulty might best be understood by stating a range of views from that of the physics absolutist to that of the biologists' equivalent.

The extreme reductionists' view is to say that even now it should be clear that life is but a complicated assortment of atoms all doing what any other atom does, and that our present laws of physics explain life quite adequately. A bit less extreme is the position that although we do not have the complete explanation of life now, history is on their side, vitalism had to give way, and it is only a matter of having more facts and it will be clear how well the laws of physics suffice.

A more moderate position of the physicist is that the laws of physics are sufficient but that the problem is too complex to obtain all the data and demonstrate how they, the laws, apply. A similar argument is that life processes may obey physical laws but that there are uncertainty principles at work, as with physics, and either the instruments of observation or the statistical nature of the processes is such that exact behavior cannot be measured and thus tested to see if physical laws apply.

A fairly neutral position is that although some laws of physics apply, many are not applicable because of insufficient conditions, e.g., the inappropriateness of classical thermodynamics because of the non-statistical nature of many cellular reactions, etc. The laws are all right and can serve as guidelines but they are not sufficient descriptions.

A moderate irreductionist's position is that the present laws of physics are inadequate but that from observations at the "cell level" new laws of physics will be deduced, they being extensions of old ones. One can imagine a fourth and fifth law of thermodynamics to explain the energy and matter sinks of life. These will apply to life processes only and thus be biological laws. These persons tend to emphasize that different laws apply to all the different levels of matter. The whole is more than the sum of its parts.

The final extreme irreductionists' position is an extension of the levels statements above, plus the emphasis that those laws that are found to apply at an upper level are conceptually different from those on the other levels. They may like to remind the physicist that the laws applicable to the nucleus of an atom have no relationship to the laws of chemical bonding. They see the statistical laws of genetics never explained by the statistical laws of quantum mechanics.

CELL ORIGIN

One must be reminded that the models of the cell are based on our present simplest form. We must remember that life has a time dimension and it has been stated before that until this parameter is included within our description of the cell it will remain incomplete. On the other hand our present gaze at the cell may help us in the development of explanations of the cell's origin. Reciprocity should occur.

This search into explanations of the order of composition concentrated on proteins and nucleic acids with success. This success reaffirms the basic tenet that life could arise from the inorganic by emphasizing that the properties of these two classes of molecules (plus ATP, really a part of the one) explain life. And in order to have these properties, the molecules needed to be made of amino acids and nitrogenous bases, which needed to be made of H O N C and no other atoms would do. Even if these macromolecules were not as they are today, their functions, catalytic and template, must have been present in the primordial cell.

Our survey of the pathways of matter and energy and the protein complexes involved are also in agreement with the logic of Oparin. The simplest, least efficient energy preserving, most common and least organized metabolic system is the Embden-Meyerhof scheme, which could well have generated the CO_2 from a variety of metabolites. Even simple photochemical reactions could have been present, however, with a complex light reaction or aerobic metabolism protein complexes and probably membranes were necessary, again in agreement with Oparin's scheme. In short the studies of the cell that have been done since Oparin's hypothesis was presented are on the whole in support of his broad hypothesis.

But by far the most rewarding and interesting discovery has been that something can be said about the evolution of proteins and nucleic acids with what appear to be new insights into the origin of the cell and its evolution. Protein evolution was first discovered by Sanger when he found that 3 out of the 51 amino acids of insulin varied depending on the animal source. At first he assumed that the patterns of the 3 amino acids would be more similar the closer the phylogenetic relationship of the animal. Why these 3 might vary and not affect insulin function was and still is an uncertainty. After enough animals had been compared, there was no direct relationship between these 3 amino acids and phylogeny. These studies were all done with mammals, however, and few amino acids were involved. Since then a number of proteins have shown remarkable correlation between differences in amino acids and phylogenetic relationship of their source organism. One of the best of these studies has been that on cytochrome c, a small protein with a little over 100 amino acids. Many mammals, chicken, tuna, a moth, and yeast were compared, with remarkable agreement with differences in amino acids and extimated age of divergency of these major groups. When the data are corrected for expected double substitutions at the same amino acid site, a linear relationship can be found with an amino acid substitution occurring every 26.2×10^6 years. This not only suggests that analysis of protein composition can aid in relating major groups of organisms but it also indicates that there has been a fairly constant rate of amino acid substitutions or effective mutations. This is most surprising when one notes that in this case of cytochrome c about 45% of the amino acids are different between yeast and mammals. Obviously this rate could not have continued over a longer history without affecting the active site unless the substitutions are equivalent with their predecessor amino acids in some way. This change could also reflect change in allosteric potential and the addition of regulatory response sites. But why is the rate so constant, and if there are equivalent amino acid substitutions, why are they made?

This rate of change is not uniform from one protein to the next. For example, in hemoglobin the rate of a single amino acid substitution in the alpha chain is 9.1×10^6 years, while in the beta chain it is 2.4×10^6 years. This wide range in rate is too extreme to assume that some single "evolutionary clock" is acting on all proteins in identical manner. These rates can be related with age of divergence from one another when they come from a common ancestral protein. For example, analysis of amino acid differences in the hemoglobin chains and myoglobin from the same organism suggest that they all arose from a common ancestor similar to myglobin as pictured in Figure 8–17. According to this analysis the slower evolving alpha chain diverged from the ancestor of the other hemoglobin chains about 380 million years ago, while the beta did not emerge and diverge from the delta chain until about 70 million years ago. Similarily the slowly evolving cytochrome c ancestor of mammals

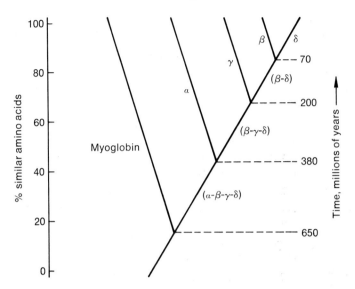

Figure 8-17. Phylogenetic tree of myoglobin-hemoglobin chains as derived from amino acid differences.

diverged from the cytochrome c of yeast at least 1.5 billion years ago. Curiously enough this phylogenetic relationship of proteins agrees with the developmental appearance of the different hemoglobin chains; i.e., the alpha diverged from and appears earlier in development than the beta chain.

More detailed analysis of protein evolution has been made possible by translating the proteins amino acids back into their equivalent triplet nucleotides by means of the genetic code and noting nucleotide differences. In this way it is possible to indicate the number of gene mutations that have actually been realized to produce the various proteins. When this finer method is applied, it has been surprising to find that proteins with different tertiary structures and with different holding and catalytic sites (different functions) must have had common ancestors. For example, lysozyme and RNAase, quite different enzymes, can be shown to have 10% of their nucleotides in common (a 10% molecular fossil) and an estimated age of divergence of about 2 billion years. From studies such as these it has been estimated that all proteins existing in the cell may have evolved from as few as about a dozen primordial proteins.

Protein chains have apparently elongated by duplicating their sequence and experiencing different amino acid substitutions in the two sections.

This can be shown by analyzing common occurrences of amino acids along a chain with repeated cycles of some fixed sequence of amino acids. This duplication process may occur many times. For example, it can be shown that the ferrodoxin of bacteria are about half the size of plant ferrodoxin and that an obvious duplication of the former has occurred. This bacteria protein, having 55 amino acids, can in turn be shown to have been the product of several previous duplications from an original sequence of the 4 amino acids: alanine, aspartic acid or proline, serine, and glycine, in that order. Duplication of regions of DNA by chromosome mutations could easily account for both protein elongation and divergency. Using this kind of logic one does not have to rely on large macromolecules to carry out the basic properties of life in the earliest cells. Very simple complexes may have done the job of catalytic action and template ordering, inefficient though they may have been.

The results of studies on gene action and protein evolution have caused many of the molecular biologists to disagree with classical evolutionists who maintain that natural selection operates only on the organismic level. The former point out that the primary phenotype of a gene is the protein and that in most cases proteins from a variety of genes must act in such a complex manner to produce the organismic structures that selection pressure operating only at this latter level could never be felt on any specific primary protein. Also there is thought to be too many amino acid substitutions, many of which seem to have no value to the operation of the proteins. This has suggested that a fair number of mutations are occurring causing many amino acid substitutions, some of which are clearly detrimental to the balance of the "molecular ecology" of the cell. Some few other mutations would fit with an advantage to the cell, but the selection would occur at the molecular levels. This process could be considered a constant source of variation and with this "internal natural selection" there would be a progressive change toward the most efficient system of molecules. Presumably the younger proteins would occupy positions of greater sensitivity. The accumulated effect of this could affect the organism's phenotype and also be under a natural selection pressure. There would thus be at least two natural selection processes at work. The classical evolutionist disagrees and feels that any molecular change must also be influenced by the organism's natural selection pressures because of the tight interdependency of all the cell's proteins: thus selection at the "outer level" is sufficient. At present there is no resolving of this conflict.

Our knowledge of polynucleotide evolution is less complete. Only recently have nucleotide sequences been determined. An interesting source of polynucleotides is the *t*RNA's. They are small and associated with specific amino acids. With the results so far, the same kind of *t*RNA's from different organisms have different and similar nucleotides that suggest phylogenetic relationships. DNA being a double helix allows hybrid studies between the DNA's of different organisms, and estimates of regions of similar nucleotide sequences can be made. For example, it has been calculated that 25% of *E. coli* DNA is present in man. It must be remembered that it has at least 1000 times as much DNA; therefore the fossil record of *E. coli* still present in man is only 0.025% of man's DNA.

Perhaps one of the most important studies will be that of determining the origin of the genetic code. The code is universal and it is most important to realise that there must have been some very clear reason for the particular choices of triplet nucleotides for each amino acid. Some assumptions and correlations that have been found are supplying a strategy of experimentation that could yield exciting results. First, because most of the redundancy in the code involves the last nucleotide, it is now assumed that the original code was a doublet, the first two nucleotides of the present code. This would mean that there were probably 15 primordial amino acids with asparagine, glutamine, methionine, tyrosine, and tryptophane added later. It has also been noted that all codes with U as the second letter call for nonpolar amino acids (Leu, Ileu, Phe, and Val). The doublet for alanine and glycine is CG. That for aspartic and glutamic acid is GA. Also experimentation with dinucleotides suggests some specific binding to amino acids. The amino acid is believed to be held between the planes of the two nitrogenous bases somewhat like a sandwich. If these strategies work out, they will also suggest ways that building blocks of a nucleoprotein complex having both catalytic and template properties may have been selected. Perhaps these many approaches will not only add new insights into the origin of the cell but reveal other significant processes at work in all present cells.

THOUGHTS

The last few sections summarize the results of this rich and exciting search for explanations into the order of the cell's composition. As exciting as these models are, there could well be more to come. No order has demanded and received so much experimentation. The searcher must know many techniques, be persistent, enjoy many rapid-fire experiments, and above all be careful to use a variety of approaches for any problem. There is always the danger that he may be led by his equipment. His concepts are abstract and most likely of a mechanical nature where he thinks in terms of models he cannot touch, and he can so easily go down a beautiful but wrong path. He must enjoy thinking with his pencil more than with life.

ix

order: in enlargement

Heaven/earth: Nature must be kept within limits, as the earth limits the activities of heaven. . . . This stream of energy must be regulated. . . . It is done by a process of division . . . divide the uniform flow of time into the seasons, according to the succession of natural phenomena. . . . On the other hand, nature must be furthered in her productiveness. This is done by adjusting the products to the right time and the right place, which increases the natural yield.

Fire/water: Fire flares upward, water flows downward; hence there is no completion. Therefore one must separate things in order to unite them. One must put them into their places as carefully as one handles fire and water, so that they do not combat one another. . . .

I Ching, translated by WILHELM/BAYNES

This order has not yet been discovered. It is an assumption based on the belief that there are laws operating among molecules as their masses grow and yield complexity and form. To say that enlargement is a product of the thermodynamic principle that matter will continue to amass with a minimum entropy when it is in a steady-state system or on the other hand to say that enlargement with complexity produces an organism that fits its environment better leaves a shallow ring in our present sense of completeness. There seems to be more order with inner purposes based on

331

unique properties of these organizations of organic matter analogous to that discovered within the cell. In fact it is another assumption that this (or many) order(s) will soon come to our attention because of the confidence the order in composition has given to explaining observations of growth and differentiation. The great problem is that there is no obvious uniformity to these overall phenomena from stage to stage or organism to organism. Explanations on the molecular level are considered most promising. Yet even at this level there is no simple building unit such as the protein or the polynucleotide. The best at present is the Jacob-Monod model of gene regulation, which is obviously incomplete for complex organisms. Much more knowledge about regulation for ever extending layers of matter is necessary. As there were inner circuits of cross regulation within the catalytic-template loop of interdependence, so there must be other circuits of regulation encompassing larger units of matter—organelles, cells, tissues, and organs. This "order in enlargement" chapter actually is a survey of the known systems of regulation operating between these units as they emerge during enlargement and of various strategies used to explain them. The near future promises great activity toward explaining growth and thus adding new insight into just what the organism is, both unicellular and multicellular. As a result this chapter will no doubt be the first to be obsolete and could easily be replaced by several chapters when new orders are revealed.

Most of the material presented will stimulate questions rather than suggest models. To place it into some systematic arrangement, however, we shall consider smaller units of organization and progress toward the more complex. We must consider control involving the spatial and temporal division of labor in enlarged unicells and multicells. We shall also consider the "final" differentiation reaching the adult state of organization of matter. This latter attempts to place a time dimension on the development of a potentially unified molecular basis of adult physiology. Finally the stability of this adult organized state and models of regulation at the various layers of matter will be vaguely thought about. Two extreme models of differentiation are used to attempt some focus. These are the again somewhat arbitrary clock-candle contrast. These apply fairly well as they are natural continuations of the major questions of the experimental embryologists. Clocks follow up the preformation-mosaic explanations, while candles are derived from the epigenic-regulatory models.

SPATIAL CONTROL IN UNICELLS

Spatial descriptions of cells are best derived from electron micrograph and differential diffusion studies. *E. coli* has some spatial distribution of the catalytic and template activities. Most of the anaerobic and synthetic enzymes are found uniformily mixed in the general peripheral regions of the cell. The aerobic enzymes and electron transport system are associated with infoldings of the cell membrane in certain parts of the cell. The template activities exist in the central region of the cell and are probably all attached to one another. That is, the DNA is the backbone upon which the DNA and RNA synthesizing enzymes are found. The synthesized RNA extends from this backbone with ribosomes and new polypeptide chains coming off at given intervals. Thus close or direct interaction bewteen these activities can occur as they are confined to a small volume.

The eucaryotic cell is quite different. A typical one with about 500 times the volume is full of organelles. The aerobic catalysts may still be found rather randomly dispersed throughout the nonconfined regions of the cell. The aerobic and photosynthetic systems are found in complex mitochondria and chloroplasts. The nucleus is a large-pored entity that also has much substructure. There is the complex endoplasmic reticulum that houses the ribosomes and is related with special compartmentation roles not well understood. The Golgi apparatus generates excretion products and membranes. There are many other vesicles and complexes. Of most interest is that the template activity is scattered in organization. The nuclear DNA generates most of the RNA that accumulates in the nucleolus where it is processed; i.e., sRNA is converted to tRNA and the ribosome subunits are constructed. The nuclear membrane is involved in transporting these RNA's to the cytoplasm where most of the protein synthesis occurs on the endoplasmic reticulum. Thus in eucaryots, transcription is spatially isolated from translation allowing that many more sensitive centers for regulation. The membranes of the mitochondria, chloroplasts, and endoplasmic reticulum supply counterpart sites of regulation for the catalytic processes.

Of even more interest, however, is the finding that mitochondria and chloroplasts have their own template centers. Although these centers could use the factors present in the nucleus or cytoplasm, they are peculiar to organelles. For example, the ribosomes of mitochondria are smaller than those of the cytoplasm. The DNA is circular and for mitochondria contains enough nucleotides for about 50 genes, which is far too few for the mitochondria's own needs. It has also been shown that cytochrome c and some structural proteins are generated from nuclear genes. Cytochrome b and a are synthesized by mitochondrial genes. Thus the reason for this particular distribution of roles is far from obvious. A popular hypothesis is that mitochondria and chloroplasts were originally bacteria that formed a synergistic complex with nuclear-type bacteria. Evidence in support of this is that the structure of mitochondria is similar to that of bacteria: they are about the same size; the template factors are similar in many details to those of bacteria; and these factors are sensitive to the same inhibititors as those of bacteria. Although this may be one mode of enlargement of a cell, the question remains, why were the particular gene products distributed as they were? It would seem beneficial to generate one's own structural proteins at least. The nucleus would have to contain those genes whose products must adapt most responsively to cell-centered natural selection pressures—or would they?

There are many other unicells with DNA at more than the one nuclear region. There are many multinucleated unicells about which little is known as far as differential regulation of these nuclei are concerned. There are also other unicells with DNA concentrated into substructures for particular purposes. A typical case is that found in some *Paramecium* where a macronuclei contains about 50 times the diploid amount of DNA. This nuclei is produced by a polyploid synthetic process from the micronucleus and it is involved in general RNA synthesis for the cell's functions. There are apparently many ways to support a larger mass of material but the DNA template must be present in larger numbers if such is the case.

Of especial interest in studies of the spatial distribution of templates is that generated during the maturation of large eggs such as those of the amphibians diagrammed in Figure 9–1. It has been shown that in these latter kinds of eggs DNA sites involved in rRNA synthesis self-replicate themselves, thus producing over 1000 additional rDNA's, twice as much total DNA as found in the chromosome. Each of these form a ring-like structure and produce rRNA at rates over 200,000 times that of an average adult cell during early stages of egg maturation. Over 95% of the nuclear RNA is this rRNA and its DNA and ribosomes make up much of the thousands of nucleoli. This rDNA breaks down by the end of maturation, but the ribosomes so produced may last months to years and are used after the egg divides. Such cases of gene amplification are not found at other stages of differentiation but the fact that it is found at all represents a potential mode of selective gene regulation.

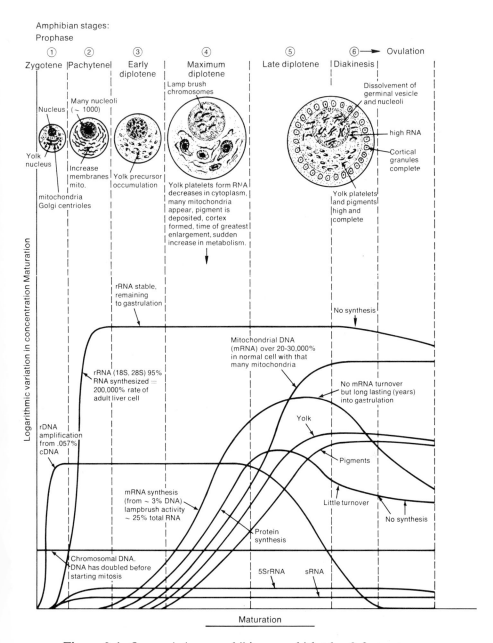

Figure 9-1. Oogenesis (e.g., amphibian egg which takes 2–3 years; most oogenesis occurs during prophase I of meiosis).

It is at about midway through maturation of the amphibian egg that the famous lampbrush configuration of chromosomes appears with the arms generating *s*RNA and some *m*RNA, this latter occupying about 3% of the chromosome template. It is also at this time that the number of mitochondria multiplies so rapidly that their total DNA may be many times that of the nuclear DNA. Over half of the protein synthesis in the egg at this time involves mitochondrial RNA. These mitochondria are involved in the formation of yolk platelets that occupy much of the cytoplasm of the mature egg, located at the vegetal pole by the presence of the enlarged nucleus at the animal pole. In this way there is a polar distribution with the nuclei and much of the cytoplasmic RNA near the animal pole, the mitochondria at the periphery and the yolk more densely found toward the vegetal pole. By maturation protein synthesis has almost ceased altogether; however, most of the *m*RNA as well as the other RNA's are preserved and become especially active by the fertilization trigger. Here is another obvious mode of regulation, stabilized *m*RNA. However, this device of using long-life *m*RNA is common with certain other adult cells.

With most of the spatial organizations above, there does not seem to be mechanisms to maintain exact patterns. Eucaryotic cells, for example, have varying numbers of mitochondria and their location within the cell depends on where within the cell ATP is in most demand. These numbers and movements do not seem to follow set ways but operate in a statistical random fashion. Mitochondria can multiply by budding or condense in number by fusion. Organelle culturing is a most difficult strategy, that could be used to understand their interactions. This difficulty suggests the high dependence on other cell constituents. This will obviously be a valuable tool when perfected. In lieu of this procedure, organelles are injected into or removed from cells. Of special interest here is the finding that added mitochondria are maintained, even those from other strains of unicells. It has even been possible to produce hybrid mitochondria of several strains, as a result of their fusion. Another strategy that should add insight into organelle-organelle interaction is that of induced cell fusion. In this procedure cells of even different organisms can be fused in the presence of certain viruses that disrupt the cell's membrane integrity. These studies are also in their infancy and little can be said except that some hybrid cells can survive and multiply. Perhaps those studies that have revealed most about spatial dependency of the subcellular parts have been those on certain mosaically developing eggs. In these eggs either removal of material from parts of the egg or displacement of materials within the egg by centrifugation can cause abnormal or arrested development. These clearly suggest that material is allocated to set positions within the egg for separation upon cellular cleavage with cellular differentiation resulting. How these materials are segregated and maintained within the egg is not understood.

TEMPORAL CONTROL IN UNICELLS

Temporal patterns of metabolism and template activity are being revealed. Most but not all of these are related to cell division. We are beginning to ask ourselves the rather naive question, just why is there cell division? This seems obvious from the view of increasing numbers of the same kind of cell, but there seems to be much more involved. One might well ask, does division exist for other purposes? Must it exist as some manifestation of the organization of living matter? Was it present with the first forms of life? There is no requirement within the laws of thermodynamics of open systems that they grow and divide. This is important to consider as it may be that the very order we are assuming for this chapter is involved in some way with cell division and/or DNA replication.

Our simple cell system, *E. coli*, shows a continuous replication of DNA with the two loops of DNA produced separating and the cell dividing in absolute constant fashion, the cell dividing when the DNA has been doubled about every 20 minutes. Each "open" *m*RNA and its specific protein are synthesized just after its corresponding DNA has replicated. These processes are all closely coupled as judged by the stopping of them all if any one of them is inhibited. Here is an apparent machine-like clock making its own image over and over again with an internally derived organizer and pacemaker.

In most eucaryotic cells the complicated processes of budding, furrowing, or cell wall formation during cell division are only weakly coupled to chromosome separation or duplication, these processes easily being uncoupled and acting independently of one another with inhibitors. During cleavage, *m*RNA and some proteins generated and stored in the egg are used for this process and may continue for several divisions without the nucleus. All of these cell divisions do require ATP with aerobic metabolism most often a requirement to assure its abundance. In eucaryotic cells there are distinct periods of the cell cycle based on periods when DNA is being replicated and periods of mitosis when the chromosomes are being separated. As shown in Figure 9–2, there are normally four periods

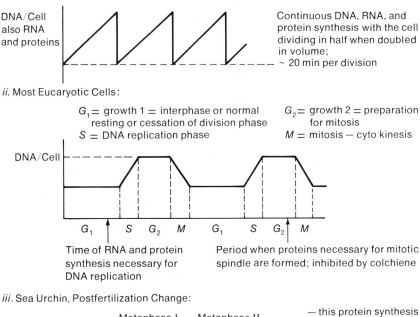

i. E. coli:

DNA/Cell also RNA and proteins

Continuous DNA, RNA, and protein synthesis with the cell dividing in half when doubled in volume; ~ 20 min per division

ii. Most Eucaryotic Cells:

G_1 = growth 1 = interphase or normal resting or cessation of division phase
S = DNA replication phase

G_2 = growth 2 = preparation for mitosis
M = mitosis – cyto kinesis

DNA/Cell

G_1 S G_2 M G_1 S G_2 M

Time of RNA and protein synthesis necessary for DNA replication

Period when proteins necessary for mitotic spindle are formed; inhibited by colchiene

iii. Sea Urchin, Postfertilization Change:

Metaphase I   Metaphase II

Protein synthesis

DNA synthesis

Protein synthesis

DNA synthesis

— this protein synthesis is not dependent on new mRNA synthesis
— uses mRNA produced during oogenes.

min.

80   95   130 140

Fertilization   Cleavage 1   Cleavage 2

iv. Synchrony Possible by Light-Dark Cycle in *Chlorella*-Green Algae:

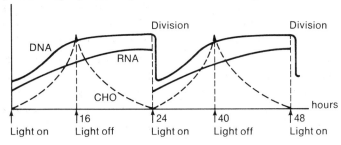

Division   Division

DNA

RNA

CHO

hours

16   24   40   48
Light on   Light off   Light on   Light off   Light on

Figure 9-2. Macromolecular abundance with cell cycles.

of the cell cycle: the G_1 growth phase, the period of preparation just prior to DNA replication; the S period of DNA replication; a G_2 growth period after DNA replication; and the M or mitotic (meiotic) period. G_1 is the most variable depending upon the cell's environment. There may even be a G_0 period preceding it in which there is no activity in terms of the cell division per se. This may be a general resting period for the cell or a period in which certain other specialized processes are occurring. The S period is fairly constant for a given cell type. The G_2 may vary considerably depending on external conditions (and stage of development in a multicellular organism). Mitosis is fairly constant in length except for certain division processes like those of the egg or sperm.

RNA and protein synthesis normally occur only during the G_1 and G_2 periods. This is based upon the notion that in these complex cells RNA synthesis cannot occur when DNA is self-replicating or when the chromosomes are in their compact form during their separation. There are many exceptions to this general picture. It is true that some protein synthesis is required before either the S or the M phase can occur and that these proteins are used in their respective processes. Under fine analysis, however, it can be shown that proteins involved in these same processes are also produced during the early stages of their respective phases. In fact most of the proteins are again produced at a specific time in the cell cycle. For example, proteins found in the spindle are produced during part of the G_2 phase and during early prophase, while proteins necessary for the mechanical movement of the chromosomes are produced later extending into early metaphase. Although overall nuclear protein syntheses decrease or cease during this later part of the M phase, this is not necessarily due to a reduction of RNA synthesis at that time. There is some evidence that protein synthesis decreases because of a regulative factor preventing polyribosome formation and thus mRNA-ribosome attachment at this time. Perhaps this is the same factor that prevents polyribosome formation in the egg cell until fertilization. There are also cases known where protein synthesis, e.g., histones, occurs during the S phase. This may result from the presence of long-lasting mRNA as is the case during cleavage. It is RNA synthesis that is more consistently retarded during the S and M phases, but there are even exceptions to this. rRNA is known to be synthesized in a few species during the M phase of early cleavage. More frequently RNA synthesis occurs beside DNA replication during the S phase. Finally we must remember that mitochondrial RNA synthesis occurs quite independently of these nuclear periods. It is most difficult to claim any hard and fast rules.

One of the most interesting observations regarding cell cycles of these eucaryotes is that the overall cycle tends to have a most natural period suggesting an internal regulatory device. These endogenous rhythms are especially evident when the cycle of a group of cells is synchronized and the synchronizing agent is removed. For example it can be shown that if the single-celled green algae *Chlorella* are exposed to 16 hours of light and 8 hours of dark, cells divide and continue to do so for several generations after the light sequence has been removed and the colony is placed in the dark. There is a natural oscillation as if plucking a string at the right moment. If the pattern of light is changed, the synchrony is lost. As diagrammed in Figure 9–2, in this synchronous colony DNA replication, RNA and protein synthesis begin at about 5 hours after the light is turned on. DNA replication stops by about 16 hours, and RNA and protein synthesis continue to about 24 hours, at which time the cells divide. Nuclear division, the M phase, actually overlaps with the S phase in this organism, occurring from about the twelfth to the eighteenth hour. One of the controlling factors that seems to set the optimum period and length of light and dark is the amount of photosynthesis and carbohydrate synthesis, the dark period using most of the carbohydrates synthesized during the light period.

This particular example has tuned in with a natural day length and its rhythm is classified as a "circadian rhythm." Cell cycles may well be other than day length. Regardless, the question arises as to the source of their regulation. There have been generally two types of explanation for these rhythms, neither of which is entirely satisfactory. The *Chlorella* example is a good one, which suggests that the rhythm is based upon the concentration of some key factor serving as a negative feedback agent. If this substance, which may be a carbohydrate, is abundant, then division is inhibited. When its concentration is below some critical level, then division can proceed. This example is using light to swing the pendulum. In other systems there may be a natural product that serves as an inhibitor to its own synthesis at high concentrations. End product inhibition of enzyme pathways could serve as a model if there was a time delay in the response of the enzyme to product and if the enzyme remained inhibited at fairly low concentrations of end product once affected. One can also think of systems where repressor action blocks a vital gene product that also could act as a corepressor for its regulator gene. The substance would thus block its own synthesis, but this inhibition would stop as product concentration drops and becomes ineffective as a corepressor. One difficulty with all these "candle-like" cycles of regulation is that the periods of most endogenous rhythms are comparatively temperature-insensitive. This would mean that several systems operating in opposition would both have to be affected in the same way by change in temperatures, a rather complicated arrangement.

A counterpart clock-like model has assumed that RNA transcription is cyclical in nature, reading off the DNA in a set direction along the DNA and repeating the read-off in fixed periods. In this way the time of synthesis of key protein products would determine the endogenous rhythm in machine-like precision. *E. coli's* cell cycle would seem to fit this model. If one observes the points of RNA synthesis along the DNA, however, they are not occurring sequentially over the whole chromosome; some RNA is being synthesized continuously, while others occur in bursts. Also, there are huge gaps where no RNA synthesis is occurring. Perhaps the best model for a clock is exhibited by the *Paramecia tetrahymena*. It can be shown that *m*RNA synthesis that uses macronuclear DNA as a template follows temporal regulation in which the synthesis increases and decreases each cycle from one division to the next. But of even greater interest is the "calendar-like" memory of number of cell divisions. It can clearly be shown that specific surface proteins are formed about the second cell division after conjugation of complementary mating types, phosphatases after 10, and esterases after 40 divisions. The mating type of the daughter cells is established shortly after conjugation but does not express itself until the eightieth division. This memory effect can be shown to be attributed to genes, as mutants can be found that arrest mating-type development. Thus some record of the cycles is kept related to the number of times the macronuclei divide. Nothing is really known about the mechanism of this process but its reproducibility demands a closely regulated system; however, such a system would not necessarily depend on the repeated sequential reading along a chromosome. A complicated sequence of regulator genes affecting genes at remote sites could also be operating. Or even the DNA replication process could somehow affect the open and closing of certain genes.

SPATIAL CONTROL IN MULTICELLS

A great deal of knowledge about cell and subcell differences within a multicell cluster has been found because of the larger sizes of material involved and because of the ease of separating cells and carrying out interaction experiments. In many mosaic eggs, differences within the egg are partitioned off by highly regulated cleavage processes. Quantitative if not qualitative differences have been found for these cleavage products in pigmentation, yolk, and organelles such as mitochondria, metabolites, and RNA's. It is often difficult to tell if these differences occurred before or after the cleavage processes. Separating the cells from one another would not necessarily settle the question. It has become a basic assumption that there is no alternation in the nuclear DNA of the product cells, although they may be operating differently. This is based on the finding that the purified DNA of these cells can hybridize completely with one another. Basically one has become accustomed to define cell differentiation on the basis of differences in DNA operation (kinds of RNA being synthesized), kinds of proteins being synthesized or accumulated, and characteristics of the metabolism.

An indication of how extensive cellular differentiation may be, even during early cleavage in highly regulatory eggs, is the example found for organelle differentiation in sea urchin. In the prefertilized sea urchin egg, there are seven isozymes for MDH (malic dehydrogenase) and if the nucleus is removed, two of these disappear, suggesting that they arise from the nucleus. The number of isozymes decreases upon fertilization. No new MDH is produced during this change, and the whole egg is reduced to four isozymes with the anucleated fertilized egg having one less. By the 64-cell stage, there are only three MDH isozymes found and these are all of mitochondrial origin. The interesting thing is that these three are not evenly distributed. Only two of these isozymes are found in the large blastomeres, while all three are found in the small blastomeres.

These early differences most often exist as a gradient. In hydra it can be shown that eight different cell types result depending on their location along the nerve net. This has led to the further discovery that the type of cell resulting depends on the concentration of a single neurosecretory hormone that is emitted from these nerves. Thus a gradient concentration of a single substance can exert many differences in cell operation depending on where the cell is located in the gradient. In frogs, as an example of many vertebrates, gradients of RNA synthesis and carbohydrate metabolism can be found along the dorsal surface reaching a high point at the dorsal lip. Morphogenetic movements resulting from microtubule formation within the cell is a product of this gradient. Thus spatial differences can cause specific changes with new spatial arrangements resulting.

Although cell-cell interaction is obviously at work in simple multicell systems, the agent of interaction is difficult to determine. Success has been best where abundant tissue is involved. In these studies the agent has apparently been combinations of proteins and/or ribonucleic-protein complexes. In some systems of tissues lipids seem to play a role. One of the best studies of how these cellular differentiation inducers operate has been that carried out with the pancreas. Changes in cell structure observed with the electron microscope are correlated with the occurrence of various cell secretion products as 8-day mouse ectoderm destined to become the pancreas is exposed to mesoderm.

It is found that although little change in cell morphology or synthesis of secretory products is evident, the tissues must be in contact continuously from the equivalent of the ninth to the fourteenth days. During the ninth and tenth days rapid cell division and DNA replication occur which is necessary for the induction response. This DNA replication drops off during the eleventh and twelfth days to essentially no division by the fourteenth. The first sign of response is the synthesis of insulin and glucagon by the beta cells, starting about the eleventh day and rapidly leveling off with the production of about 10^6 molecules per cell by the fifteenth day. The acinar cells first show an increase in secretory enzyme synthesis during the twelfth day with the accompaniment of an increase in number of ribosomes and polyribosomes. By the fourteenth day small flattened vesicles appear near the Golgi apparatus containing these enzymes. The fifteenth and sixteenth days see an increase in size and density of these vesicles, now distinguished as zymogen granules. By the seventeenth day these granules can be found in abundance and attached to the cell membrane where their stored enzymes are extruded.

Of special interest is the finding that not all the enzymes are synthesized at the same time. As shown in Figure 9–3, amylase and carboxypeptidase A seem to appear first; then chymotrypsin, ribonuclease, and lipase A; and finally trypsin carboxypeptidase B and lipase B. Insulin production is also increasing several hundredfold during the time the first group of enzymes is being synthesized. The rate of synthesis of these proteins appears to increase in two steps before finally leveling off at the adult rate. Although the spread of these three groups extends over a period of 3 to 4 days, the enzymes within each group are produced at the same time as if there is a sequence of genes being turned on in groups. It has also been possible to inhibit the synthesis of any of these groups of proteins by exposing the pancreas tissue to an RNA synthesis inhibitor a day prior to the normal appearance of the protein. This finding also

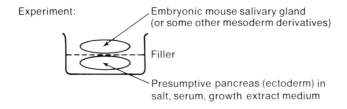

Experiment:

Embryonic mouse salivary gland
(or some other mesoderm derivatives)

Filler

Presumptive pancreas (ectoderm) in
salt, serum, growth extract medium

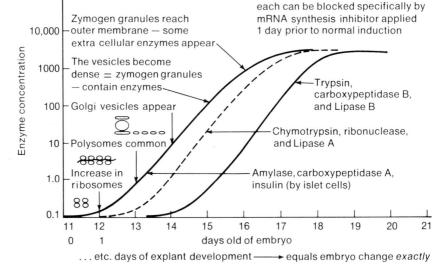

Observation by EM of acinar cells:

Sequential appearance in groups —
each can be blocked specifically by
mRNA synthesis inhibitor applied
1 day prior to normal induction

Zymogen granules reach
outer membrane — some
extra cellular enzymes appear

The vesicles become
dense = zymogen granules
— contain enzymes

Golgi vesicles appear

Polysomes common

Increase in
ribosomes

Trypsin,
carboxypeptidase B,
and Lipase B

Chymotrypsin, ribonuclease,
and Lipase A

Amylase, carboxypeptidase A,
insulin (by islet cells)

Enzyme concentration

10,000
1000
100
10
1.0
0.1

11   12   13   14   15   16   17   18   19   20   21
0   1   days old of embryo

. . . etc. days of explant development ——→ equals embryo change *exactly*

Figure 9-3. Induction process: an ideal example in mouse pancreas.

suggests that there is a sequential expression of genes. If these inhibitors are added after the enzymes appear, there is no effect on further enzyme production or the formation of the vesicles. This would suggest that the *m*RNA responsible for the enzymes is of a long-lasting type.

TEMPORAL CONTROL IN MULTICELLS

In many ways the example of induction above cannot be separated from any discussion on temporal regulation. DNA replication was required and a long sequence of events transpired. There are still other temporal changes that accompany all known induction processes worthy of exampling. The capacity for the invaginating mesoderm of amphibians to induce neural derivatives in the overlying ectoderm develops just prior to invagination and drops off again by the tail bud stage. Likewise the ectoderm has a period of competence during which it can be induced by the mesoderm. If the age of either of these tissues is off, the process is lost and cannot be recovered. This may be complicated, as in the case of amphibian lens induction, by the necessity of more than one inducing tissue. As shown in Figure 9–4, the initial induction involves endoderm,

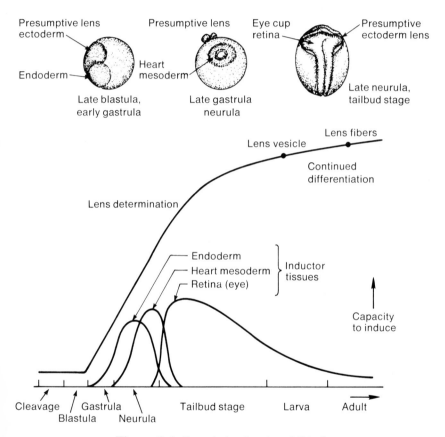

Figure 9-4. Lens induction (amphibian).

then heart mesoderm, and finally retinal ectoderm. Each has its period of capacity. What is interesting in this example is that it can be shown by transplanting tissue of different ages that any one of these three tissues can serve in the place of the others if its capacity (age) to induce is high at the time. This would mean that each of the tissues is producing the same inducer substance(s). It can also be found that tissues not normally involved in an induction process can serve as an inducer. These cases remind us that at least in many instances the inducer does not direct the response—this is rather set and prearranged in the responding tissue.

Some insight into the patterns of response that may occur to these induced tissues has been assumed by analogy to the responses of insect salivary glands to hormones during metamorphosis. Metamorphosis is considered to be sudden developmental changes. Some insects have giant chromosomes in their salivary glands that display differential "puffing" patterns depending on the environmental conditions surrounding these cells. These puffs are due to the unfolding and extending of the DNA strands in the chromosome. As much as 10% of the chromosome may be in a puffed condition at the same time. These puffs contract if inhibitors of RNA synthesis are added, and RNA synthesis can be shown to occur in these regions by tracer experiments. As shown in Figure 9–5, puffing

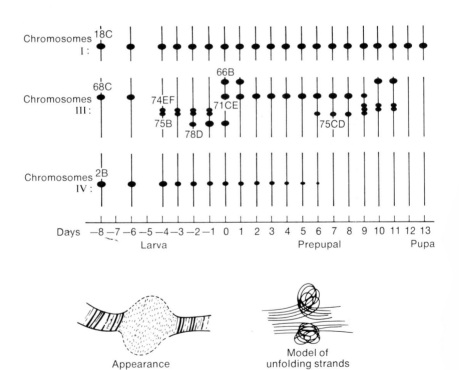

Figure 9-5. Puffing patterns in chromosomes of salivary gland.

patterns follow distinct changes during the metamorphosis. These are chords played along the keyboard of DNA. If one injects the molting hormone ecdysone (a steroid) into the larva at intermolt, molting is induced and corresponding patterns of puffing can be followed. This pattern is identical with the natural pattern. Even isolated chromosomes respond to ecdysone treatment in a natural manner. Puff 18C on chromosome I rises within a few minutes of exposure followed in 30 minutes with puff 2B on chromosome IV. This can happen with as few as 100 molecules of hormone per chromosome. During normal development the 18C puff lasts throughout all the molting periods, while the 2B is gone during the larva-pupa molt. It would be interesting to see if this decrease in 2B puffing is due to the presence of juvenile hormone. If inhibitors of RNA synthesis are added, these molting puffs are delayed until the inhibitor is removed. Upon the removal of inhibitors and in the presence of ecdysone, the puffing patterns continue with the same sequential time pattern as before indicating an interdependence of gene expression. This suggests that the products of one gene may act as an internal inducer for the next, etc. It is important to know more about this kind of sequencing and the details of its mechanisms. Insects may help here as they did for genetics.

That patterns of differential RNA synthesis occur during development between different tissues and within the same tissue can be demonstrated by DNA-RNA hybrid experiments. This is most easily shown by using the pure single chain of DNA and exposing it to *m*RNA extracted from two different tissues or the same at different ages. The concentration of one of the kinds of *m*RNA is kept constant, while the other concentration is varied. In this way competition for common DNA sites of complementation are indicated and also the percentage of the DNA active in the RNA synthesis is revealed. These studies definitely show a continuous change in types of *m*RNA generated with no more than about 15% of the DNA ever active at the same time. Many more kinds of *m*RNA are synthesized in early embryos, fewer kinds in later embryos. On the other hand more kinds of the *m*RNA are conserved (long life) in the cytoplasm of older embryonic cells than in their younger counterpart.

Another closely related temporal question is that regarding cell division and differentiation. This in short means the relationship between DNA replication and *m*RNA synthesis. It used to be assumed that cell division and DNA replication inhibit *m*RNA synthesis and therefore differentiation. As indicated in the pancreas example above, there is a need for DNA replication for the pancreas cell to synthesize the new proteins. It has also been found that *m*RNA synthesis occurs during the G_2 and S period in frog embryos. The length of G_2 is dependent on the rate of cell division, which does vary with age. In older embryos the rate of division is slower, the G_2 period is elongated, and more *m*RNA is synthesized. Some investigators believe that the rate and number of cell divisions play a direct role on differentiation by affecting the amounts of *m*RNA being synthesized. They feel that for later differentiation there must be an accumulation of many *m*RNA's of the same kind to produce enough of certain kinds of proteins for the cell to assume a differentiated condition. There would also be more *r*RNA synthesized making the *m*RNA more stable. There would still be a decrease in number of kinds of *m*RNA synthesized with age and there would be some new *m*RNA sites opened on the DNA but these alone could not bring about the later stages of differentiation. Evidence in support of this hypothesis has been the finding that the addition of LiCl, which increases the G_2 period and RNA synthesis per cell division, also increases the amount of meso-endodermal tissue compared to ectoderm. The addition of $NaHCO_3$ has just the opposite effect. Whether this hypothesis is correct or not, it is clear that the cell cycle periods can influence the quantity and perhaps quality of *m*RNA synthesized, and DNA replication could play a more direct role in differentiation than had ever been previously imagined.

The data above are telling us that embryonic induction is far more complicated than the enzyme induction processes of *E. coli* or other microorganisms. Genes are being turned on and others off and probably some kind of regulator-operator gene systems are operating, but there are distinct differences that make any models for embryonic induction difficult to design and support. For one thing this latter process takes days, while enzyme induction takes seconds to minutes. There are apparently many complex and sequential steps with presumably several operating genes involved. There may well be more than one inducer required for some embryonic induction processes. Also, at least in some cases, the chromosomes must become competent or sensitive to the inducer with this ability coming and going with age. The embryonic induction is an irreversible process normally where the synthesis of the *m*RNA remains after the inducer is removed. Finally embryonic induction appears to depend in some way upon DNA replication and *m*RNA synthesis seems to take place at only certain times of the life cycle. It may be that the inducer is only effective at a particular part of this cycle. Obviously the two processes are very different.

Models for progressive differentiation involving induction are more often endowed with clock-like designs than candles. The sequential patterns of capacity to induce and competence to be induced seem to suggest a structured time regulator. Also inhibitors of RNA synthesis can delay early stages of development. When these inhibitors are removed, the processes pick up again in their normal sequence. Also the obligatory sequencing of puffing in the salivary gland chromosome implies a definite built-in temporal sequence of gene regulation. The dependence on DNA replication and RNA synthesis being related to a particular part of the cell cycle also suggests a dependent regularity of developmental processes on some necessary rate and number of DNA replications. We cannot think of this pattern depending on the direct arrangement of genes along the DNA, as considered for unicells, but rather that each DNA replication produces some environment around the chromosome that could effect the pattern of genetic expression during the following replication.

Once having stated this view, however, it has been pointed out that each of these clock-like processes could be accounted for by a regulator-operator gene system in which the internal inducer or corepressor is a metabolic intermediate or energy carrier that exists as a pool and the concentration of this pool depends on the general operation of many metabolic steps and the abundance of related factors including meta-bolites from the environment of the cell. It has been shown that differen-tiation decisions can clearly be influenced by metabolite concentration and/or starvation, which causes metabolite competition and the shifting to certain preferred metabolic operations. This type of control would be of the dynamic cyclical nature as a flame of the candle.

THE FINAL DIFFERENTIATED STATE— THE ADULT

There is reason to believe that in some tissues of plants and most tissues of animals there is a final adult state of differentiation. These are states at which no further "progressive" differentiation results and yet in many cases the protein synthesis and functions of the cell can still be modified considerably depending on their environment, the difference being that these changes are fully reversible and thus physiological in nature. There are cases in which there is no apparent difference in the final step of differentiation from early steps except that the nuclear operation of these cells becomes minimal. For example when lens, skin, feather, or red blood cells are finally formed, they each tend to produce only one or a few proteins from long-lasting mRNA. This mRNA is apparently stabilized by the fact that it is long and combines with many ribosomes forming a less vulnerable compact unit. In these kinds of cells the nuclei usually abort and the cell fills up with the protein product. This last step is obviously irreversible and stable as long as the cell exists. Some of these cells have a fixed life-span once the nucleus has been extruded, for example being about 107 days for red blood cells.

That a final differentiation step exists may be implied by the finding that some cells separated from the multicellular organism at different ages and placed in growth media continue to divide only a fixed number of times depending on their age. If cells are taken from the embryos of a small animal, they may divide about 10 to 20 times and then they die. If they are taken from an older equivalent animal, the cells divide correspondingly fewer times before they stop and die. Similar studies have been done with human cells in culture with the finding that many cells can live no longer than 50 divisions. This would suggest that after some final division no further growth can occur.

This is more directly supported in the case of the differentiation of chondrocytes. DNA synthesis is required for the final adult step and a special kind of DNA appears to be produced at this time. At least the synthesis of this DNA is sensitive to inhibitors that previous syntheses are not. It is even likely that this DNA is formed at a time corresponding to the G_1 period of the more typical cell cycle. This DNA can also be shown to be the only one that in its native state can produce adult-type RNA. After this final synthesis, no further DNA synthesis occurs. Finally one of the necessary conditions for the synthesis of this DNA is cell-cell contact, this acting in someway as an inducer. Cell contact is also necessary for the final differentiation of the end plate and skeletal muscle in many vertebrate nerve-muscle connections.

Another system that has been studied not only for itself but as a means to gain insight into final differentiation steps is that of molecular antibody formation, one of several immune reactions found in mammals. Foreign molecules, molecular antigens, eventually induce the formation of proteins specific for reacting with the antigen; thus specific gene action is induced. This process also has a memory effect and thus is at least analogous to other developmental processes. This analogy is made complicated by the fact that antigen-responding cells, stem cells, macrophages, and monocytes themselves must develop under the action of thymus gland hormone.

The ability for antibody reactions to occur does not arise at the same age from one species to the next nor at the same age within the same species for different specific antigens. In man they normally do not occur until after birth, but in sheep it may be but a third of the way through prenatal development. Also sheep may respond to a virus at this age but not to a kind of bacterium until much later. The entire responding apparatus appears to develop over a long period with each different specific reaction rapidly appearing as a mature complete reaction at different times.

Once formed it is assumed that macrophages found in the circulation and lymph systems are the antigen-sensing cells. It has been shown that antigens enter the macrophage, move to the nucleus, and induce *m*RNA synthesis. This *m*RNA is passed to the stem cells by cytoplasmic bridges where it somehow induces these cells to divide rapidly, at least once every 10 hours, and differentiate into antibody-producing cells. The stem cells must divide with DNA replication before they are able to produce antibodies. This developmental process is precisely timed so that nine divisions occur during which the cells enlarge and increase their amount of endoplasmic recticulum and ribosomes, with RNA and protein synthesis increasing exponentially. Over 90% of the proteins synthesized by these developed stem cells, which are now called *plasma cells*, are specific antibodies. These macromolecular antibodies are secreted from these gland-like plasma cells and complex with the antigens.

Since theoretically antibodies are produced for every kind of antigen protein conceivable, this would be a very large number of kinds of antibodies and if only one plasma cell produces only one antibody, there would have to be a correspondingly enormous number of potential plasma cells and thus an enormous variety of differentiation products. It has been shown that some plasma cells can produce more than one antibody at the same time, but this is a rare occurrence and could not reduce the problem significantly. It might be assumed that the antigen is broken down by, say, the macrophage cells with antibodies being produced for each class of reactive groups, being reasonably small in number. This does not seem to be the case, however, since only a few if that many antibodies are produced for an average-sized protein antigen and specific antibodies are produced for large structures such as viruses. Such a viral antibody cannot react with the individual protein units of the virus coat. Thus the problem remains rather formidable.

It is true that not every antigen can evoke antibody synthesis in any one organism and thus there may be a limit to the number of kinds of antigen-specific plasma cells. Also the fact that part of the antibody chains is constant from one kind of antibody to the next suggests that perhaps this part of the antibody is made on one gene and the variable part made on other gene sites, one corresponding to each possible combination of amino acids necessary in the active site region. This would not reduce the number of genes but these genes would be significantly smaller than most and probably enough DNA is present to account for them. However, there is still the problem of how the constant and variable parts of each chain are put together.

Another way in which many kinds of specific antibody structure could be produced is to have any two of a number of chains that could be produced by separate genes able to combine in random combinations. In this way if there were, say, 100 different genes for both light and heavy chains, there could be 2^{100} different antibodies produced, a significantly large number and a big help.

Some do not believe that even this could account for the large number of antibodies and suggest that genes responsible for the various regions of the chains may be able to cross over with one another easily and thus increase the possible number of combinations. An even more drastic model is to suggest that a gene site is constantly mutating in rapid fashion and that the antigen somehow stabilizes the gene into some corresponding state. Immunoglobin synthesis begins in the G_1 and extends into the S periods of the cell cycle. It is conjectured that perhaps the final differentiated plasm cell stops in the G_1 phase set for antibody production. The ceasing of the cell cycle may somehow stabilize unstable genes or gene crossover combinations into one pattern. We are a long way from understanding how the antibodies are constructed but when this problem is clarified, another regulatory device at the chromosome will be uncovered and a possible final differentiation step mechanism revealed.

That there are distinct differences between the transcription processes of adult tissue and the final differentiation step producing that adult state is most clearly shown where hormones are involved. It has been found, for example, that NADP dehydrogenase and glucose-6-phosphatase, as well as tyrosine aminotransferase and serine dehydrase which appear immediately after birth, can be induced in the liver earlier than normal by prenatal injections of glucagon, epinephrine (adrenalin from the adrenals), and perhaps by insulin. There is a competency factor involved; i.e., if the hormones are injected too early, more than 3 days prior to the time they appear in normal development, then there is no effect. Likewise, it is interesting that responses of the tissues to the hormones are different as to the age at which competency begins. The order of appearance of competency corresponds with the order that the enzymes appear in normal development. Also, since these enzymes do not appear in the liver if all of these hormones are absent during the course of development, it is concluded that these hormones are necessary to initiate enzyme synthesis and that this constitutes a developmental step.

This is to be contrasted with the response of adult liver to hormones. In this case, the enzymes above are induced to be rapidly synthesized when the liver is exposed to the hormones above. In the case of at least tyrosine aminotransferase, the enzyme most studied, it is induced in adult liver to higher concentrations by not only glucagon, epinephrine, and insulin but also by steroid hormones such as hydrocortisone. There are also other differences that must be noted between the responding mechanisms of the developmental versus the physiological systems. The most obvious difference is that if the artificially applied hormone acts only briefly upon the enzyme induction system of one of these enzymes prenatally and prior to its natural appearance, then enzyme synthesis returns to a negligible amount in a manner quite similar to that experienced with enzyme induction in microorganisms. In adult liver, however, there is always a significant residual enzyme synthesis present, the hormones enhancing this rate of synthesis. The kinetics of the two reactions are also quite different. This is rather amazing since both response systems seem to involve RNA synthesis. Adult liver enzyme synthesis is also less inhibited by RNA synthesizing inhibitors than is the developmental step. Finally the results of the above and other liver enzyme studies suggest that where glucagon or epinephrine are necessary for some first developmental change, cortical hormones are required soon afterward for a further change. This latter somehow alters the mechanism into an adult state of operation.

Another important finding was that both response systems may be induced by the addition of cyclic AMP (see Figure 9-6), even sooner than the hormones themselves. In adult liver, this amounts to a more rapid response. In the developing liver, competence of the responding system to cyclic AMP appears even earlier than the competence to respond to hormones. Cyclic AMP substitutes for many nonsteroid hormones and its mode of action on adult liver (and other tissues) has been elucidated to some extent. Hormones are known to act at many different sites. Even the same hormone may act directly on an enzyme to modify its catalytic

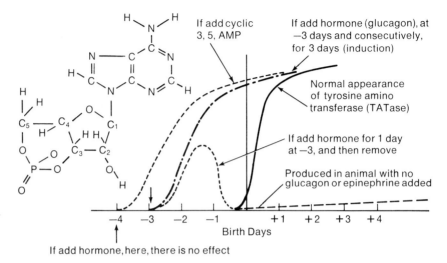

Figure 9-6. Hormone effect on development (in rats) of tyrosine amino-transferase response in rat liver.

activity. It may increase a membrane's permeability to certain ions. It may also act directly upon the ribosomes to influence the rate of protein synthesis. It may even act as an uncoupling agent in the electron transport-phosphorylation system. But of primary interest to us here is the finding that tissues are specifically affected by certain hormones by the presence of an enzyme in their outer membranes that has a complementary site of response to the hormone. For example, liver cells have an enzyme, adenyl cyclase, that is specific for glucagon. When glucagon combines with this enzyme, it changes allosterically and a catalytic site for the conversion of ATP to cyclic AMP is created. A different adenyl cyclase may be present that is activated specifically by, say, epinephrine, etc. In this way cyclic AMP is generated inside the cell and acts as a "second messenger" to affect possibly several different cellular responses. This is apparently why cyclic AMP can substitute for the hormones. Cyclic AMP may influence several activities but most of them seem to involve the activation of phosphorylating kinases, enzymes that transfer phosphate groups onto or off other proteins. These kinases may in turn have multiple effects, as diagrammed in Figure 9-7. One of these effects found in liver cells is the conversion of phosphorylase *b* to the active form phosphorylase *a* that is used in glycogen degradation as mentioned in the last chapter.

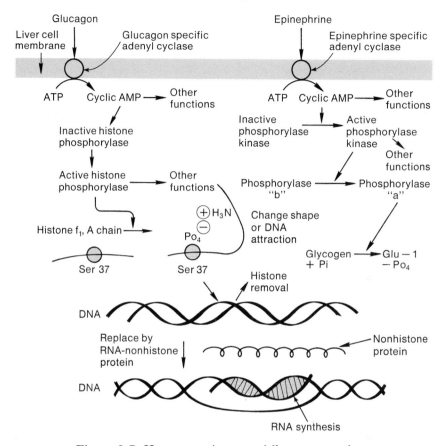

Figure 9-7. Hormone action: especially on gene action.

Our particular interest here is that there is a cyclic AMP activated kinase in liver that adds phosphate to a particular serine of a specific protein closely associated with the liver's nuclear DNA, a basic protein known as f_1 histone.

Histones are found in all eucaryotic cells and have been implicated in the regulation of RNA synthesis in these cells. This is based on many facts. The histone to DNA ratio by weight from many kinds of chromosomes is close to 1.0. Even more important, when DNA replicates, histones are synthesized simultaneously. When chromatin, native chromosomal DNA-protein, is separated from cells and used in in vitro experiments with RNA polymerase, the percentage of the total DNA that is template active in RNA synethesis is inversely related to the amount of histones present. The histone content is replaced by other protein. There are different kinds of histones based on relative proportions of basic amino acids. It has also been found that the relative abundance of these different fractions of histones varies from species to species and from young to old within the same species. All this is very suggestive of their role as repressors of RNA synthesis and that they are selectively removed and replaced during the development of these eucaryotes. There are many difficulties, however. One is that in many cases the same relative proportion of the various kinds of histones is found within the different kinds of tissues, hardly expected. Even more damaging is the finding that if the histones are removed and replaced on the DNA, the type of RNA synthesized after replacement is different from that synthesized beforehand. There is the problem of histone specificity. There does not seem to be that many different kinds of histone. Also histone from one species can inhibit RNA synthesis from another species' DNA. One other point is that when even the smallest amount of histone is removed, as much as 3 to 7% of the DNA has been derepressed indicating that a large number of genes would be effected at the same time.

In many ways this lack of specificity of histones is in accordance with expected response on RNA synthesis as mediated by hormones. It appears that many hormones affect the rate of protein synthesis not only by influencing mRNA for these proteins but also by affecting the synthesis of tRNA and rRNA, thus releasing any pool restrictions on the overall rate of protein synthesis. Histones are bound to DNA by ionic bonds, presumably the basic amino groups are attracted to the dissassociated phosphate groups of the DNA. These histones are also believed to be hydrogen bonded to the smaller amount of non-histone protein, which in turn is believed to be bound to the DNA through base pairing of an RNA moiety which is covalently bound to this non-histone protein. This chromosomal RNA is small (only 40 to 60 nucleotides) and could serve as the more specific "derepressor" factors by holding open specific gene sites. With this in mind, a model of how the glucagon-cyclic AMP-kinase phosphorylation of histone action can be understood as diagramed in Figure 9–7.

Hormones may also have an effect upon the rate of cell division and thus differentiation where temporal control is important. Cases have been found where hormones increase and others where they decrease the rate of division. Regardless, RNA and protein synthesis must occur before this effect is felt. Another hormone example that indicates that hormones must act at certain parts of the life cycle of the developing tissue is that found for mammary glands. In order for this tissue to produce and secrete the milk protein casein, it must be exposed to insulin, hydrocortisone, and prolactin in that order for specific periods of growth. Insulin must also be present during the exposure to prolactin. This occurs in immature or adult animals whether they are pregnant or not. The early insulin exposure is necessary to enhance DNA synthesis and cell division that must occur for further development. This proliferation must extend for about 2 days and during the time that the other hormones act. Hydrocortisone cannot initiate DNA synthesis but must act at the time that the cells are proliferating and probably during the G_2 period between DNA synthesis and mitosis. Prolactin and insulin on the other hand act after mitosis, during the next G_1. Thus as outlined in Figure 9–8, at least

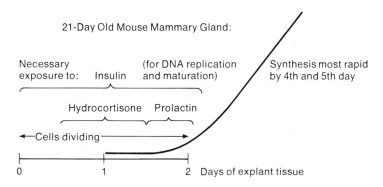

Figure 9-8. Development of mammary gland synthesis of casein.

three hormones are necessary for this development and probably act at different times of the life cycle of dividing cells. Notice that protein hormones act before steroid hormones and that hormones that influence casein synthesis by the gland are also involved in the gland's development, both conditions similar to that found for liver enzyme system development.

We need not discuss the matter but remind ourselves that the above also indicates the manner in which homeostatic systems emerge. Their separate parts are even aided in the last steps of development by those hormones that shall serve as homeostatic messengers and yet the whole system represents a new interplay on the organismic level.

STABILITY OF THE DIFFERENTIATED STATE

One of the characteristics of the process of development is its memory effect, as the change at each step is not normally reversed. But just how stable is this new state and what is it that regulates this stability? Any understanding of development and the adult state of a multicellular organism must include explanations of their stability. Again we know so little and must resort to some detailed examples rather than mechanisms.

In general it can be said that the more developed the organism, the more stable the state of differentiation of most of its cells. The ability of some cells in all organisms to replace lost cells is an example of how the whole organism influences the stability of the final functioning state. Some cells, like those of liver, can begin to multiply when their neighbors disappear below some critical mass of cells. Otherwise cell division is held in check. Regeneration of complex limbs and organs in some animals is an even more dramatic indication of a natural stabilizing influence that one tissue has on another. No more dramatic example can be found than that of the salamander lens. When this is removed, the iris cells dedifferentiate characteristic structure and protein synthesis, begin to multiply, and redifferentiate with the production of both new iris and lens cells. This also demonstrates the remaining totipotential of the chromosomes of these iris cells. There must have been a drastic alteration of the chromosome state of gene expression. If lens is removed, ground up, and placed inside the eye cup, the iris does not dedifferentiate. This indicates that some lens product holds the iris cells into its adult state and the removal of this substance releases that state. That all regenerations may not depend on the same processes could be surmised by studies on the regeneration of amphibian limbs, which apparently depend on a blastema covering and some exudate from nerves in the stump region. In this case substance(s) seems to initiate the dedifferentiation process.

An artificial but informative method of studying the intercellular effects of preserving a differentiated state is that of cell culturing. For most studies the cells are diluted from their normal tissue concentration about a thousandfold. Under these conditions at least 20 to 30 growth factors must be added, depending on the cell type, to even maintain 0.1% of the cells. These include amino acids, vitamins, salts, metabolites, nucleotides, and even unknown substances collected from the serum of almost any animal. Besides, the media must be buffered against pH change and often exposed to a CO_2 atmosphere. Obviously these cells are quite leaky and all sorts of materials must be able to diffuse from one cell to another while in a tissue state. Embryonic cells are considerably easier to culture than adult ones, suggesting that more factors are involved in maintaining the adult state. The interesting thing is that in spite of these additives, the cells are not the same as those in the tissue. Either cell-cell contact is necessary or some other regulatory substance is not available to the cells. One clear indication of this is the finding that animal cells in culture divide once every 16 to 30 hours versus 30 to 90 days while in the tissue.

The cells also lose many of their characteristic structures as less of the proteins involved are being synthesized. They are said to be degenerate and usually lose any of their specialized functions they had as a tissue. This can be checked by the return of structure and function if the cells reaggregate into a compact mass. Contact and number of cells reverse this degenerate condition in many cases. If the cells are kept separate over extended periods, they lose the ability to reaggregate and there is a question of whether the degeneration has gone too far and/or if some critical regulatory factors have been lost. Degeneracy is not necessarily dedifferentiation in the sense of the cell losing its determined state. For example, single skeletal muscle cells have been grown for a number of generations and yet when they reaggregate only skeletal muscle is found and its function returns although the isolated cells were quite degenerate.

One the other hand, it has been possible to grow complete plants from single root phloem or pith cells. This is another obvious case of totipotency being retained in these differentiated cells. It also suggests that whatever regulator substance was active in maintaining a single determined state was lost with these particular cells.

There are other changes worthy of note. Isolated cells from most organisms assume ameboid-like shapes and move about with pseudopods. This movement permits the reaggregation process to occur. Also it is not uncommon to find certain metabolic pathways blocked and subject to activation by substrate induction. On the other hand many metabolic intermediates or products are found in excess. Obviously the modes of internal regulation are altered, presumably by the regulating agents leaking into the medium. Although many specialized proteins are not being synthesized, there is still general RNA and protein synthesis occurring.

Cancer is considered to be a loss of some internal regulatory process. Some kinds of cancers can be caused by mutagens and viruses and in these cases are most often hereditary-consistent thereafter. A common theory is that all cancers are caused by the activation (or inhibition) of a controlling agent, a virus-like "C particle," which somehow maintains the cancerous state and replicates with cell division, thus retaining its effectiveness. According to this hypothesis all cancer-producing agents affect this endogenous particle present in all cells. Although this may be the case for some cancers, there must remain the open question of whether all cancers could possibly result from the loss of a single factor in the complex interdependent regulatory processes of the cell. There appear to be many kinds of cancers even for the same cell type. A cancerous state seems to be one of many altered states of differentiation within which the cell may exist. It may be misleading that there is one common property of cancerous cells, mainly that they divide rapidly. This would obviously affect the number of specialized proteins that they could synthesize, but this may not be all that is involved. It may mean that this is the most sensitive regulatory process that is upset. It may not be the only one.

There is good evidence that some cancers do not result from mutations of genes per se. Cells can become cancerous quite easily when growing separately in cultures. They essentially flip out and are no longer restricted to a fixed number of cell divisions before their death. They also often lose the ability to aggregate, but if they can, they continue to divide rapidly and remain degenerate. These cancerous cells never seem to return to normal as long as they remain in culture media. They may change to what appears to be another cancerous state.

Also unlike mutagens and viruses, which tend to produce a single state of differentiation, other inducers may produce different "levels" of cancer depending on the conditions of their exposure. This might best be exampled by the cancerous growth in plants, crown gall. This results from a combination of substances emitted from wounded tissues and certain bacteria, this latter being the inducer. When the inducer is added a day after the wound, the gall grows slowly. If applied about 2 days after wounding, then the growth of the gall is moderate. And if it is applied at least 3 days after the wound, then the rate of growth of the crown gall is very rapid. The gall looks the same, only the rate of growth varies. What is most amazing is that the rate of growth of these three differently treated tissues remains constant over many generations (years) even if the tissue is transplanted to other culture dishes.

It was also found that crown galls and other plant cancers synthesize considerably more auxin and kinetins than normal tissues do. As it is known that these hormones stimulate growth, both enlargement and cell division, it was natural to assume that all three of the induced crown galls grow faster than normal tissue because of the hormones. In fact, each type of crown gall produces amounts of hormone that correspond with their rate of growth. It was also found that the fastest growing crown gall could maintain its growth on only a simple medium containing salts, three vitamins, and a sugar energy source. Also the moderately growing crown gall could grow at the same rate as the fast one if besides the additions above, auxin, glutamine, and another vitamin is added. Slow gall can grow at the fast rate if in addition to all these substances aspartic acid and the precursors of the nucleotides are added to the growth media. Finally most interesting of all is the finding that even normal cells can be transformed into gall and grow at the fast rate if they are exposed to media containing all the additions above plus kinetin. As outlined in Figure 9–9, it would seem that these crown galls represent tissues with different metabolic disorders that could somehow be systematically increased in kind by varying treatments of a wounding substance and an inducing material from the bacteria. The simplest factor that could be upset was apparently the regulation of the cells synthesis of kinetin. Finally, it can be found that if embryonic cells of different ages are exposed to wounding and inducer substances, different regulatory processes appear to be upset depending on the age.

What is most important is the finding that such cancers as crown gall can be reversed under certain conditions. If for example cancerous buds are grafted onto normal tobacco stems, they will take and grow with their resulting shoot being less abnormal. Also if the tips of these shoots are serial transferred onto other normal stems, by the third such transfer, the growing shoot is completely normal. Apparently some regulatory agent present in the normal cells enters the cancerous ones and restores the missing function. Analogous findings have been reported with the injection of embryonic cancerous mouse cells injected into a normal embryo. A mass of cells results that still contains some cancerous cells but also has several different kinds of normal adult cells. In a like manner, cancerous cells near a regenerating limb in an amphibian may be converted back to normal. Apparently wounding substance and some substance released from regenerating tissues "loosen up" the regulatory processes of the cancerous cell and allow some of them to redifferentiate to normals. Embryonic cells, cancerous or otherwise, are more easily "loosened up" than adult cells.

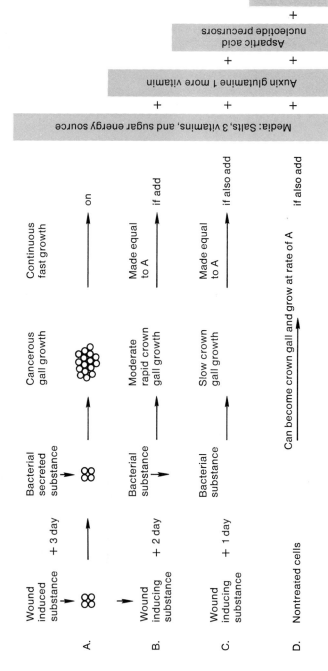

Figure 9-9. Crown gall (plant cancer).

An indication of the relative stability of different determined cell states has been revealed by serially culturing the cells from the imaginal disks of fruit flies into the abdomen of adult flies. In this latter environment, the cells can grow but not complete their determined fate. At any point along the serial transfers, some of these cells can be tested by placing them back into a larva and noting their resulting differentiated structure. The abdomen culturing process can thus be used to retain the imaginal disk cells in their embryonic condition as long as desired. Thousands of divisions of such cells, requiring hundreds of serial transplants, have been followed in this way. The interesting finding has been made that some of these embryonic cells experience a "transdetermination" in certain directions. The directions of change possible are summarized in Figure 9-10. There are distinct frequencies of transdetermination with the

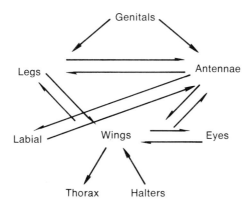

Figure 9-10. Transdeterminations in fruit fly.

frequency in one direction usually differing from that in the opposite direction. These are not mutational changes but apparently changes in internal regulation. Results such as these may some day be used to "map" the characteristics of different regulated states analogous to the way fruit flies allowed us to discover the gene.

That regulation of cell state involves the cytoplasm as well as the nucleus has been shown most effectively by nuclear transplant studies with toads. Nuclei from either the lining of the adult intestine or from the adult brain were injected into an anucleated egg cell and in some cases these eggs gave rise to normal adults. This is another indication of the retained totipotential of the chromosomes of these mature cells. But even more can be shown. When these nuclei that had been engaged in producing *m*RNA for specialized proteins were placed in the egg cytoplasm, this kind of *m*RNA synthesis stopped immediately and assumed the mode of operation of an egg nucleus. With cleavage and further development, new kinds of RNA were synthesized exactly as would the normal egg: *m*RNA by midblastula, *t*RNA by late blastula, and *r*RNA by gastrula. Thus the cytoplasm has a strong regulatory control over the operation of the nucleus, presumably containing whatever internal repressors and inducers required. On the other hand, it was quite remarkable that only about 1.5% of the injected eggs developed into adults and very few more beyond cleavage. Upon examination it was found that the reason for the fatalities was not normally the operation but rather due to the rapid rate of division of the egg during cleavage (several hours). Incomplete DNA replication was very common. Apparently those regulating factors controlling the rate of division were still operating at the adult mode (once every 2 days or slower) and the DNA replication process was retarded accordingly. Thus it would seem that factors controlling DNA replication rate are for the most part confined to the nucleus. It is no wonder that one thinks of a clock when one thinks of cell cycle regulation and a candle when the overall cell mode of regulation is considered. The previous discussion on final differentiation state yelled clock, the above on stability cried candle. With these nuclear studies one begins to wonder if both may not be operating with equal but different effectivenesses.

THOUGHTS

We have done little but bombard the reader with facts and observations. We have omitted the history of the timing of these facts as we do not know where they will fit into some overall painting. Also as indicated before, we know not the order except that the mass of molecules is enlarging. Therefore we have not a clear idea of what questions to ask or what we are to explain. Obviously there is the question of mechanisms of development and adult maintenance, and the seeking of these is assumed to yield an order. What we must do is think in terms of spatial and temporal regulations. As suggested above the clock-like models concentrate upon template centers, while the candlemakers deal with environmental control of these templates including the catalytic centers of metabolism. When we think of the whole mass of a multicellular eucaryotic creature with different operating templates and regulatory states of cytoplasm around them, the whole becomes too complex to cut through. There seem to be ever increasing circles of regulation extending through the mass, but from where do they begin and what are their pathways? This picture is made additionally complicated by finding ubiquitous acting agents like cyclic AMP. This "second messenger" not only is involved with control of metabolism and histone phosphorylation but along with Ca^{2+} and microtubules is implicated in ameboid movement, secretion, contraction, and probably active transport in a variety of cells and various functions. Taken in their most basic expression these constitute every process of the adult organism. This demands that specificity is involved in the localization or receptor sites of this common messenger. Somehow all this has to be sorted out, but it may suggest another way in which regulation is handled on a multicellular level. Again we must wonder about the inherent necessity of such regulation being determined by the physical characteristics of the messenger per se. The similarity with cyclic AMP and other internal regulators (cofactors, template building block pool, enzyme activators and inhibitors) and energy carriers as all being derivatives of nucleotides is striking. Could this be a clue to the need for the whirlpool of life matter to grow?

In any respect the strategies used by searchers are those of the molecular biologist building on models of regulation whether candle or clock, and these have so far involved the same kinds of mental and experimental tools. Development brings us back up into the realm of the visible (by eye) and thus there could be value in attempting to synthesize styles of thinking with those of the physiologist and morphologist who are seeking to explain the order in form. These are the great challenges ahead in the near future.

X

order: in awareness

There is no question of the existence of this order. Every response to
the environment involving the nervous system, every behavior, every
thought, and every product of these thoughts including the words you are
reading are testimony to the ordering and sorting process of the vast
amount of information about us. The brain is the instrument of this order
in awareness and any explanation of the process is an explanation of the
brain. And this is where the difficulty lies. No organ has been thought
about or written about more than has the human brain. This order is not
new to us as some of the others just discussed. The history of explanations
of this order has been too extensive and full of too many diffuse ideas
for this writer to manage. And in spite of any effort, there is no explaining

of the brain. As yet it has remained the most mysterious of all living processes. None is more beautiful to wonder about and more nonsensical to attempt to define. This order must fittingly come at the end of the book as any explanations as to how the brain manages it are destined to be products of the distant future if at all. I must even apologize to assume that anything said here has anything to do with that explanation. But the order exists and its magnificence must be mentioned.

The two quotations given above indicate a small spattering of impressions of the workings of the brain. As with the last few chapters one can arbitrarily imagine the brain as clock or candle. The ancient Greeks and many since have thought of the brain as a vessel for spirits, humors, or even gases. These often flowed from the outside, embodying the man with vitality. In any respect these agents flowed through the body analogous to the flowing of a flame. More recent candle models use flexible and adjustable nerve tracts in circuits to store memories and modify input-output information. They also claim that the brain depends on a constant input of impulses to develop and maintain itself, analogous to the flow of matter and energy of the flame. Clock models of the brain were launched by Descartes who considered that a "rational soul" sitting in the pineal regulated the flow of liquid stored in the brain into nerve tubes that caused all the motions of the body by filling bellows attached to swing parts or as a water clock is turned. These mechanical-hydraulic models have given way to electrical ones analogous to computers with predetermined stable parts that often have cyclical patterns of impulses effecting inputs and outputs. These clock brains are built to respond to but act independently of the environment. These extreme models will again play only a clarifying role and could both be very far from actual.

There will be a need for a great deal of synthesis of styles of thought from the various approaches that have been used to define the brain. The molecular biologist must share and synthesize with the nerve cell physiologist. The invertebrate neurophysiologist must synthesize his findings with those of the vertebrate neurophysiologist. The results of general nerve chemophysiology must be integrated with those of the electrophysiologist. More neuroanatomy and brain physiology must come together. The pathologists and brain surgeons must work and think closer with the research neurophysiologist. All of these must do the difficult task of integrating their results with those of the physiological psychologist. Moreover the ethologist and clinical psychologist must meet on common grounds. And all of these must get to know well the brain evolutionists and the social psychologists. This is a great deal of synthesis to ask for, and this does not necessarily mean it will solve the question of defining the brain. It might help.

With this understanding we give but a few examples of observations made from different points of view. There is no attempt here to present most of the views of the psychologists or ethologists. The major emphasis is on the brain per se. After a summary of some of the information that enters the sense organs, there are a few examples of the stable wiring used to handle this input data, modify and control output, and of centers of the brain used in these inherent control processes. From these we move on to the less well-understood modifiers or shapers of output from the brain involving consciousness, learning, memory fixation, and speech and thought generation. This will bring up questions about that inner universe, the mind. Finally we must be concerned about thoughts for thoughts' sake and their evolution.

INPUT INFORMATION

Our awareness is not unlimited. We tend to think that we know about all that happens about us. We know light, heat, chemical, and mechanical environmental cues. But we, unlike some fish, cannot detect changes in electric fields—some as little as one-millionth of a volt per foot change. We are also very limited in the amount of heat change we can detect compared to some snakes that can detect one-thousandth of a degree difference in temperature in a nearby object from its surroundings. Likewise we cannot detect polarized or ultraviolet light like some insects. And our chemoreceptors are tuned in on specific molecules such as Na^+ or H^+. When one thinks about it, we are extremely limited to the world about us and seem to be in a box peering out of a few narrow slits.

Although our range is narrow, our sensitivity to the physical environment that is perceived is about as sensitive as possible within the limitations of sense organ design. For example, the sensory cells of the inner ear are so sensitive to the bending of their hair stimulators and the amplification of mechanical displacement across the ear is so great that displacements of the outer ear drum in the order of magnitude of a hydrogen atom diameter can cause a sensory nerve impulse. Also pressures as low as 0.0002 dynes per square centimeter can be detected. Such small values are possible because they represent net effects above a random background noise. Also, the eye is as sensitive to light as physically possible. This is made possible by the light-sensitive pigments being in an unstable state, held so by weak bonds. Light easily excites these pigments to the point where they can alter configurations into a stable state and in so doing somehow generate the nerve impulse. This requires a single quantum of light, a theoretical minimum.

What is also amazing is the high content of information that is carried by the ear and eye. Humans can hear about 10 octaves and can distinguish up to about 12 tones per octave, which would represent $\log_2$ 120 or 7 bits per tone of information. Also the human range of intensity of sound can be divided into 250 distinguishable levels or about $\log_2$ 250 or 8 bits. Thus each tone could contain about 15 bits of information and man's ear can accommodate around 10 tones per second or a total of 150 bits per second. On the other hand a person can distinguish about 100 different intensities of light falling upon a single receptor of the eye. This would be $\log_2$ 100 or 7 bits of information. Since an eye sees a new picture 10 times a second, there could be 70 bits per second per receptor cell of information. There are about 10^7 receptor cells but these integrate with one another and eventually affect about 7×10^4 optic nerves. Assuming that integration reduces the amount of information transfer for the sake of pattern recognition, then there would be about 5×10^6 bits per second information reaching the brain from the eyes. Thus there would be about 30,000 times as much information entering the brain by way of the eye than by the ear.

There are also many sensory cells located about the body sending impulses up the spinal cord. Although little information would be traveling in any one of these tracts, there are about 3 million such tracts and if these impulses are added to all those from the sensory organs, there could be a total of about 10^7 bits of information impinging upon the brain each second. The brain of a man could be actively receiving inputs every 10^{10} seconds of life, and thus 10^{17} bits of information could enter and act upon a man's brain. This is a great deal higher than the estimate of 10^{10} bits of information stored in man's DNA and thus available for his structures and functions. An absolute clock could only order this latter figure, a candle somewhat more.

INPUT WIRING

The nerve tracts that leave sensory cells do eventually enter the brain and represent a structural organization of neurons that is believed to be common even within the brain itself. By tracing some of these 3 million input wires, the assumption is that one can see into the organization of the 10 to 20 billion neurons of the human brain. It has been clearly established that these tracts go to the brain stem and then project out to the cortex into rather discrete regions. Along this path the neurons interact producing an integrative action at a series of levels from the sensory cells to the cortex. Although these interactions may seem large, involving tens to hundreds of neurons on any one level, the complexity of this integration is small compared with that of neuron-neuron interaction of most of the brain systems where most neurons synapse with about 1000 others. Nevertheless, we can imagine we understand the former and can only hope it says something about the latter.

The sensory input wiring about which most is known is that coming from the eye. Cross inhibition between neighboring retinal cells is found in most animals. This means that when a retinal cell is stimulated by light, it can decrease the effect of light stimulating neighboring retinal cells. Nerve endings from each of these receptor cells either synapses directly with the neighboring cells or acts on the cell body of a ganglia cell on the next level. In vertebrates there are at least three levels between retina and optic nerve leaving the eye. Recording from these optic nerves and shining microbeams of light onto vertebrate retinas have shown that these receptors integrate in set patterns. So-called receptor fields of retinal cells being as large as a millimeter in diameter integrate so that when their centers are stimulated, impulses can be recorded from a single optic nerve. Some receptive fields cause impulses when the light is turned on the center, others when light that has been shining onto the center is turned off. Each of these "on" or "off" centered fields displays the opposite effect when retinal cells in their peripheral regions are stimulated. Apparently the "on" regions of either of these kinds of fields have less cross inhibition by adjacent receptors than do the "off" regions. It has been assumed that the purpose in this organization is to increase the acuity of the contrast at the edge of an image. If the boundary of a light-dark image crosses a receptor field, the pattern of stimulation from its integrated action will be distinct from one entirely in the light part of the image.

This integration process continues at different levels going "into" the brain depending on the animal. These integrations result from nerve endings arising from specifically spaced ganglia of the last peripheral level synapsing with a single postganglion cell that is only stimulated when all the nerve endings are active at the same time. At one extreme is the frog where most of the integration takes place before the optic nerve level.

In this case many of the receptive fields are integrated so that an image with a curved boundary, light on the concave and dark on the convex side, moving across the retina stimulates single optic nerves. The size of this image is equivalent to that a fly would make and presumably for sensitive recognition of this prey. A frog is built to see mostly fly-sized objects moving across his vision. Mammals, on the other hand, do their integration in the cortex, as oversimply diagrammed in Figure 10-1. The

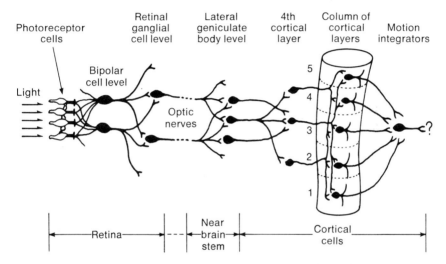

Figure 10-1. Levels of integration of the optic track in the cat (a simplified model).

tracts from the eye enter the fourth layer down of the cortex and from there synapse up and down the cortical layers and back down to the brainstem. The cortex has five distinct layers of neuron cell bodies with their axons running up and down perpendicular to the surface of the cortex, extending very little to the side. Recordings from these cortical layers have revealed that these cells are stimulated when images of straight edges or slits fall across a particular arrangement of receptor fields. A particular cortical cell is only activated when the edge is oriented at a fixed angle. The cortical cells are organized in columns about 1 millimeter in diameter and these extend through the five layers. Each column seems to contain only those cells that are stimulated by some straight-edged image of the same orientation, whether the image is that of a slit, bar, or edge. In addition some of the cells of this column are only activated when the straight-edged image at the fixed orientation is moved laterally across the retina. It is believed that these cells integrate combinations of cells at other layers of the same column when the straight image falls sequentially on these different regions of the retina. Regardless, the cortex responds to lines and one must imagine the complexity of stimulations when one is observing an irregularly shaped object that would be seen as a composite of many small straight lines in a special spatial pattern to each other.

The other sense organs must have analogous spatial projections of their receptor cell stimulation organized by special integrated designs. These seem to remain in simple geometric patterns: lines, circles, etc., when originating from sense organs. When the source of sensory input is from the body, the cortical projections are spatially arranged similarly to the body itself; i.e., there is a sensory "homunculus." These homunculi can be found at least on the cortex and cerebellum. These cortical projections are not permanent traces and must be considered as temporary input data for additional integration with the multitude of other activated neurons throughout the brain.

WIRED RESPONSE

Somehow those 10 to 20 billion cells of the brain take this input and generate a mere 150,000-odd motoneurons that determine our overt behavior. Some of this behavior, especially obvious in less complex animals, is complicated in appearance but may be the sum total of co-ordinated predetermined synapses. In the blowfly, for example, the hungry animal is stimulated to move by excitation of its leg muscles via stretch receptors in its midgut. By random movements, eventually olfactory receptors in its antennae are stimulated by aromatic compounds causing the fly to orient toward the food. A taste receptor on the legs, if stimulated, causes the extension of the proboscis. Stimulation of taste receptors on this proboscis causes sucking by movements of the pharyngeal muscles. Fluid in the esophagus stimulates peristaltic contractions of its muscles driving food into the midgut and crop. When the midgut is full, negative feedback inhibits sucking, extension of the proboscis, etc. Actually each one of these steps may be quite complicated, involving regulatory feedback loops to assure smooth operation and coordination of each muscle. Also a group of nerves may act as a unit. For example flight patterns in insects are controlled by one ganglia in which certain "command" neurons, once stimulated, cause the same motor response output regardless of the intensity or rate of stimulation. The assumption is that the stimulation of any one of these command cells above some critical level sets into motion a sequence of firings, closely regulated by feedback control. These coordinated units may well be responsible for complicated responses such as those of song in birds, feeding, grooming, fright and attack reactions, and courtship behavior in other complex animals.

One can also show that a response depends on the elaborate coordination of inputs from several different sensory organs or brain centers. The amount of turning of a praying mantis' neck toward a prey has been shown to depend on the difference in stimulation of the two eyes. This difference sets a bias into the stretch receptors of the neck muscles equivalent to that required for the turn. This requires an innate calibration of photoreceptor unbalance to stretch receptor contraction. It is thought that a similar process causes the eye of any mammal to turn when an image is presented more to one side of the retina than the other. Probably the most complex of these goal-directing systems is that coordinated by the cerebellum. Every movement of a limb is guided by the barrage from many stretch receptors of the muscles of that limb and sensory organs modifying the motoneuron signals directing the movement to assure smooth and efficient motion with minimal overshoot. Also the celebellum must somehow calculate estimated positions of the body and its limbs for some future position. For example, if one is running to catch a ball, his movements must move toward some goal ahead of the concurrent position of the ball. Anticipation and remote guidance are common in tracking stars or satellites and our models of the celebellum are influenced accordingly, as diagrammed in Figure 10-2. These systems require an

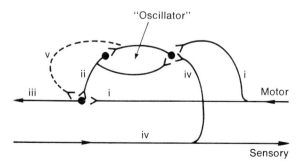

Figure 10-2. Models of cerebellum circuits.

oscillator as a source of bias, with variations in frequency from this standard created by a difference in requested position and actual position of some remote unit (limb for the organism) used as a feedback to correct the outgoing command moving the unit. Clock builders have been encouraged to find that electrical oscillations due to varying nervous activity with frequencies of 2 to 400 cycles per second have been measured in the cerebellum.

There are also those responses that we think of as basic drives, instincts, or emotions. With invertebrates, it is difficult to draw a line between these and reflex-like responses discussed above. Most of the actions above are all or none. When the sucking takes place, it is at a fixed rate. When the insect flies, the wings beat at a fixed rate, etc. There is seldom the sense of urgency or motivation indicated. Some claim these drives exist but are seldom described as we are asking the wrong questions, not able to measure them, or simply prejudiced. One can find good equivalents of motivation in several known cases. For example, with increasing starvation, a neurosceretory gland, the corpus centrale, excites the subesophageal ganglia, which in turn increase the leg movement. The amount of movement is related to the amount of hormone and thus starvation— there is an increase in intensity of effort or motivation with degree of starvation. Moreover, although the blowfly sucks his food at a given rate, his tolerance for salt in sugar water increases with length of food deprivation. One can well ask the question whether these tiny creatures feel emotions including pain. Here is a question that we are very likely never to answer.

With mammals these emotions become more obvious but their source is unclear. It seems pretty clear that these drives are instinctual and thus inherent in most cases and presumably are related to specially designed systems of nerves. It is probably not remarkable in hindsight that centers related to drives were discovered in the hypothalamus, a subunit of the brain adjacent to the brainstem extending toward the forebrain. The hypothalamus is the switchboard or direct sensing element for sensory input from the autonomic nervous system. It has opposing centers for controlling temperature, heartbeat rate, breathing rate, blood pressure, salinity, glucose, and all pituitary function. In this latter respect the hypothalamus has centers which differentiate under the influence of sex hormones into either a male or female dominating center which controls the cycle and those kinds of hormones related to sexuality secreted by the pituitary. In addition, opposing centers have been discovered by electrical stimulation of specific regions of the hypothalamus. These include appetite thirst, sexual activity, and (of especial interest) pleasure-pain. If any of these are stimulated, the corresponding drive continues unsatiably. The animal will literally gorge himself to choking if his positive appetite center is stimulated or starve to death if its opposite is stimulated. If a lever is set in such a way that pressing it causes a stimulation of the pleasure center, the animal will press it at such a rate endlessly that he collapses in exhaustion. On the other hand, if the pain center is stimulated, the animal shows fear, pain, and anger and if this is continued for a few hours it will even kill the animal. These centers have been found in man and the pleasure-pain ones have been described in terms of heaven and hell.

Although these centers are usually strongest in the hypothalamus, equivalent centers can also be found in other regions of the brain, especially in the brainstem. Pleasure centers have even been found in the cortex; however, these are weak in effect. As the different pleasure centers affect one another during stimulation or if one is excited, they are assumed to be interconnected by tracks. Similar loops or tracks of the other centers are found, most of which are related to what has become known as the limbic system. This system is a figure eight-shaped loop with about six synapses, as shown in Figure 10–3. This system was traced by injecting

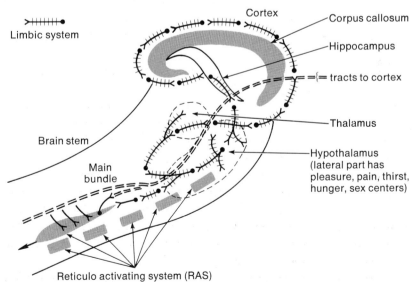

Figure 10-3. Limbic system, pleasure centers, and reticulo-activating system (brainstem and adjacent regions).

either acetylcholine or noradrenalin (not epinephrine) into regions of the brainstem and noting similar responses, for example the urge to drink with acetylcholine. Apparently certain nerve endings secrete specific neurohormones and post synaptic receptor sites are specific for these neurotransmitters. This has turned out to be a powerful tool for tracing tracks. Other tracks leading through and beyond the limbic system have been detected by injecting and tracing by fluorescent-marking dopamine and serotonin.

The finding that different amines are used to discriminate tracks as well as anatomical connections has suggested that these amine tracks may also be influenced by general level of the amine concentration in areas of the brain. These concentrations could be determined by the amine level in the circulation and cerebral fluid systems. Thus there could be parallel or specialized regulation of the emotions by the neural and humoral inputs. The neural inputs come from other regions of the brainstem and cortex, the humoral from body fluids. There has been the discovery that enhancement of moods is related to high concentrations of general amine, especially in the hypothalamus. Tranquilization and sometimes depression are associated with a low or ineffective concentration of these amines.

In addition, specific effects have been found by the injection of separate amines into specific regions of the hypothalamus and brainstem. In addition to thirst being stimulated by acetylcholine, it also stimulates pleasure in some spots and pain in others. It also enhances sleep. One of the most curious new findings is that it can stimulate a kill or aggressive behavior, giving some support to the contention that psychopathic killers may have a mental disorder. Generally noradrenalin is antagonistic to acetylcholine at the same injection spots. It is also effective at stimulating appetite, arousal from sleep, and increasing sexual acivity. Dopamine, a precursor of noradrenalin, is also effective at most of these spots, especially the last two. Serotonin generally increases passivity, decreases sexual activity, and can prevent sleep. Serotonin appears to be somewhat specifically antagonized by the hallucogen LSD. Obviously there is need to know a great deal more about the complex effects and interactions of these amines on the emotions. Finally it should be mentioned that testosterone induces aggressive sexual activity, while estrogen causes sexual receptivity when injected into specific regions of the hypothalamus. There could well be other humoral agents acting in specific ways.

There are other activities present in the brain that have some kind of reciprocal interaction with this limbic system, the combination affecting all responses. One of these is the reticular activating system (RAS) which is a long cylindrical mass of less differentiated cells which runs down the brainstem below the limbic system. This mass has many tracts entering and leaving, including branches from all sensory and motoneurons, tracks on down the spinal cord, tracks to the thalamic and hyperthalamic regions, and some tracks projecting into the cortex. Recordings from its neurons indicate that every neuron receives some stimulation from every sensory nerve but that each of these neurons receives different relative amounts of the various stimulations. For example, one neuron may respond more from light landing on the retina than from a stretch receptor in the right leg. A different neuron would respond in just an opposite fashion. The RAS seems to act as a selecting and balancing device for tuning up certain nervous activity and down others, acting on both sensory input and motoneuron output. Presumably the total input pattern, from limbic cortex, sensory, motoneuron, etc., influences the manner in which it selects which sense or response to emphasize. It apparently works when we tune out TV commercials, for example. One of its selections is sleep or arousal, its track coinciding with those of the limbic centers where sleep can be affected by humoral injection. It is interesting how electrical rhythms stimulating these regions at the frequency found by recording from the sleeping brain induces sleep. Vice versa, a waking-type electrical stimulation can arouse, suggesting that the frequency of rhythm is the important mode of operation deciding this particular form of selection. Dreaming cycles during sleep also seem to be fixed important operations of the brain. Obviously very little is known about this complex system. It is hard to study a part in order to understand the whole.

Finally there is the mysterious frontal lobe that probably interacts in some close manner with the limbic system in man. All that is known is that this huge cortical area is related to personality where responsibility and initiative are involved. It can induce rage, anger, silliness, guilt, and ethical judgments. One may wonder if ethical good and bad judgments are derived from centers in this region of the brain. This large region is most significant to an explanation of the brain and yet it is essentially an unexplored continent.

INFORMATION STORAGE

Where above we were concerned with complex wired systems that affect response and operate in the most complex of organisms, there are those systems that must be involved in memory. These former are trans-scribing or even translating signals and interacting these with one another in ever changing temporal patterns. Memory must involve a developmental process with something changing with the inflow of new data in an accumulative irreversible fashion. It is here where we must somehow integrate the findings of the psychologist, the neurophysiolgist, the molecular biologist, and the theoretical computer designers. Where wires complex, however, they may be hooked up to fit clock models; learning involves a plasticity that has a natural appeal for the candlemakers.

Studies have revealed that there are probably three kinds of learning; habituation, response conditioning, and operant conditioning. These are distinct on the basis of procedure used during the learning and probably involve distinct systems of neurons. Habituation is found when there is a diminishing response to some repeating stimulus. In mammals this negative form of learning involves the RAS, the limbic, and cortex as indicated by the drop off of electrical signals when the animal is exposed to repeated stimulations of clicks, light, or touch. As diagrammed in Figure 10–4, it is assumed that habituation requires some form of negative inhibition.

Response conditioning occurs when one and only one nonresponsive stimulus is given slightly ahead of a responsive stimulus. This is the classical conditioning Pavlov discovered in which the response stimulus, the image or smell of food, which causes salivation, is given just after the nonresponse stimulus, the buzzer or bell. A neural model of how this may occur is also shown in Figure 10–4. Several response-conditioning stimuli can be built upon one another. The sight of food did not cause salivation until many tries of the sight of the food just prior to the food being placed in the dog's mouth, which causes the salivation by taste buds. Also the range of response conditioning is broad. Even painful electric shocks or cuts could become a conditioning stimulus to produce salivation with no apparent pain being experienced by the dog if the intensity of the stimulus was gradually increased to the painful level. This type of learning is thought to work primarily through the autonomic system although it has been possible to condition a skeletal motor response by artificially stimulating the motor region of the cortex while also electrically stimulating a nonmotor responsive region. Even so this learning could not be produced by external stimulations. Response conditioning may be operating in subtle ways—a child's fear of the dark when some painful experience occurs with no lights, the effect of advertising, and the directed use of hypnosis and yoga. There is however a limit to the amount of learning possible by response conditioning. If only the nonresponsive

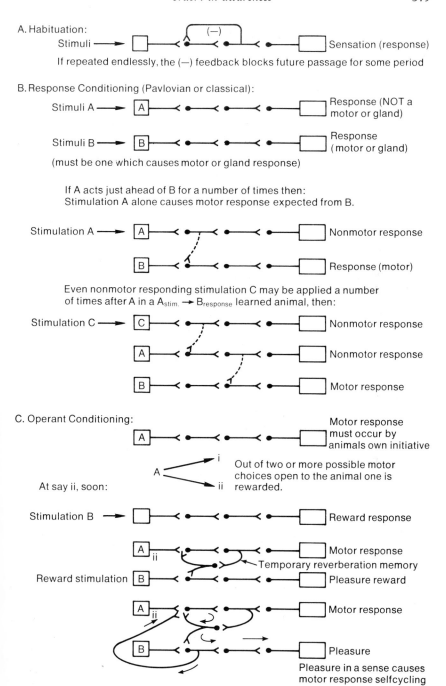

A. Habituation:

Stimuli → Sensation (response)

If repeated endlessly, the (−) feedback blocks future passage for some period

B. Response Conditioning (Pavlovian or classical):

Stimuli A → A — Response (NOT a motor or gland)

Stimuli B → B — Response (motor or gland)

(must be one which causes motor or gland response)

If A acts just ahead of B for a number of times then:
Stimulation A alone causes motor response expected from B.

Stimulation A → A — Nonmotor response

B — Response (motor)

Even nonmotor responding stimulation C may be applied a number of times after A in a $A_{stim.}$ → $B_{response}$ learned animal, then:

Stimulation C → C — Nonmotor response

A — Nonmotor response

B — Motor response

C. Operant Conditioning:

A — Motor response must occur by animals own initiative

A → i, ii — Out of two or more possible motor choices open to the animal one is rewarded.

At say ii, soon:

Stimulation B → Reward response

A ii — Motor response

Temporary reverberation memory

Reward stimulation B — Pleasure reward

A ii — Motor response

B — Pleasure

Pleasure in a sense causes motor response selfcycling

Figure 10-4. Diagrammatic representations of learning models.

signal is given over an extended period of time without some reinforcement of a response signal, then the response drops off. For this reason plus the fact that it seems only to involve the autonomic system, response conditioning is believed to have little direct effect upon our thinking processes.

The most familiar and understood form of learning is that of operant conditioning in which some action of the organisms is rewarded or punished. This operates naturally with some voluntary muscular process. If a pleasure center is electrically stimulated when an autonomic (often brainstem or limbic) electrical pattern is detected, however, then this pattern is encouraged and becomes increasingly displayed. Again, this appears to be an artificial condition only. We normally think of Skinner boxes or T-mazes when we think of operant conditioning. Animals as simple as the ant have been taught in this way, ants having only 256 neurons in their brains. Perhaps most remarkable is the finding that regardless of the animal roughly the same number of trials, about 150, are required to learn the T-maze. This agrees with the number of times a 6 months old baby has had to have its hand stopped from reaching toward a flame before it learned not to do so. The neural model in Figure 10–4 is a highly simplified version of many models that have been conceived for this process. We must elaborate on the facts known about this highly studied form of learning. Any further reference to learning implies operant conditioning-type learning.

Practical experience, learning experiments, and experiments involving an excision of parts of the brain all suggest that there is an area of the brain involved in learning and short memory and a separate region for long-term memory. Memory fixation is a long process involving at least minutes and it can be interrupted by electrical stimulations, insulin shock, or hibernation in hibernating animals. The clearest results with anatomical location of these processes have been done with the octopus, but analogous findings have been found for mammals where the experiments can be performed. In this latter group if the cortex is entirely removed or deactivated with potassium chloride, a new task can be learned equally well but it will not be retained over an extended period of time. Short-term memory apparently occurs in the brainstem regions and is transferred to the cortex. The curious thing is that long-term memory of specific kind, say involving vision, can still exist to a lesser extent if the visual area of the cortex is removed. The whole area seems to be involved although those areas most correlated with the kind of stimulus are of greater importance. Curiously enough if most areas of the cortex are electrically stimulated, no sense of memory is felt. There is the notable exception of regions around the temporal lobe where such stimulations have induced the experience of "time strips" of past events. Discovered by the brain surgeon, if a patient is stimulated in a specific spot, he may experience a beautiful concert with all its details, visual, auditory, and tactile, and at the same time realize that he is in the operating room. These are events that he may have forgotten completely but have been verified by others in enough cases to know that they are real past events. If the electrical stimulation is stopped, the time strip stops. If started again at the same position in the brain, the same strip is usually experienced starting from the same place in time. Although stimulated at only one point, these time strips are influenced by the existence of other cortical regions and presumably some circuits of neurons are involved in this storage. Another common belief is that all memories once fixed remain in the brain even though most of these are consciously forgotten—that is, unless physically removed or damaged.

It has also been discovered that both hemispheres record the same memories. If the optic chiasma is severed in such a way that the optic track runs back from each eye to only the hemisphere on its own side, tasks taught through one eye can still be carried out by stimuli presented only to the other eye. It can be shown by "split brain" experiments that information is transferred between the two hemispheres by the corpus calossum—a bundle of about 10 million tracks that runs between the two hemispheres. If this is severed and the optic chiasma is cut as described above, different tasks may be learned and stored in the separate hemispheres if done so through the different eyes with each completely independent of the other as far as stimulation and response is concerned. However, a complex task involving both separated hemispheres, a visual stimulus to one, a tactile to the other, can be taught and stimulated. This indicates that the stored information is coordinated by some regions subcortical and does not involve the corpus calossum. By further splitting of the brain it can be shown that this coordination is lost if the hippocampus, a region just beneath the corpus calossum and included within the limbic system loop, is severed. Even if this region is damaged alone, no new tasks can be learned and stored as memory. Any new task learned is forgotten within hours. This suggests that the hippocampus is central for even short-term memory as well as memory fixation for long periods and the coordination of these fixed memories. A brief model of this role is diagrammed in Figure 10–5.

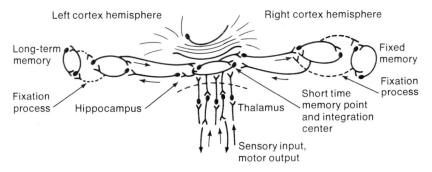

Figure 10-5. Hippocampal role in memory fixation and hemisphere coordination.

Electrical recordings have also supported these findings. It was found that an animal normally shows an erratic pattern of oscillations ranging from about 4 to about 7 per second, which is recorded from either the hippocampus or temporal lobes. As soon as the animal begins to seek for food in a T-maze, however, the oscillations settle down to a more steady rhythm of about 6 cycles per second. The interesting thing is that as the task is learned the rhythms become more regular and intense, first in the hippocampal region and spread into the cortex as fixed.

Models of learning and memory fixation have tended to be clock-candle types. The clock kinds rely upon some molecular or morphological change to set neurons, singly or in circuits, the particular change and pattern depending on the task in some predetermined manner. The potential is set; the stimulation evokes the change. This model and variations of it are called the *persistent synapses hypothesis*. The candle counterpart relies on reverberation circuits of neurons, set into reverberation by the stimulations, and once set they are influential on further stimulations. These circuits form at random depending on the pattern of stimuli and thus there is a minimal amount of predetermined organization. Variations and hybrids of these two extremes have been proposed, as summarized in Figure 10–6. It is not inconceivable that a reverberation-type process is involved in short-term memory and the persistent change for long-term. The least certain of either of these models is the predetermination aspect.

Most experimental results favor that memory fixation involves a

(a) Persistence Synapse Hypothesis:

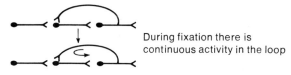

or

During fixation
there is a change
in the synapse

(b) Reverberation Circuit Hypothesis:

During fixation there is
continuous activity in the loop

(c) Synthesis between (a) and (b):

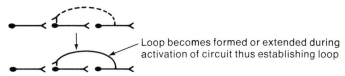

Loop becomes formed or extended during
activation of circuit thus establishing loop

Figure 10-6. Memory fixation (long-term memory) circuit models.

permanent change to at least some neurons. Reducing the temperature around the brain eliminates most of the electrical activity that can be recorded; yet memory that had previously been fixed returns upon returning the temperature to normal. If the temperature is lowered within a half hour after a task is learned, it is forgotten. Memory fixation takes at least a half hour in most animals. Also, repeated stimulation of neurons in tissue culture causes their nerve endings to grow. Moreover, the size and extent of nerves in the brains can be increased with specific sensory stimulation or task learning. Protein and RNA concentrations increase under these same conditions, being as great as 60% in goldfish brain. These changes actually follow the track of sensory input, for example along the optic track. The RNA increase is found mainly in the axonal regions and its G/C base ratios can be found to be specific for the neurons affected. The glial cells, of which there may be about 10 around each neuron in the brain, multiply, and their protein and RNA content may drop 45% per cell with their RNA G/C ratios dropping upon such stimulation.

Correspondingly the neuron RNA and protein level increase about 25%
with an increase in RNA G/C ratio. Thus perhaps some transfer is
involved in memory fixation. It has also been shown by the use of inhibitors
that protein synthesis is required for memory fixation but only for early
stages of learning or for response once the memory has been fixed.
Whether RNA synthesis is required is less certain—it being necessary in
goldfish but not mammal memory fixation. It has also been found that
some drugs increase the rate of memory fixation but have no effect on what
is learned. Also RNA content per neuron increases with age along with
learning and seems to drop off again in old age. Perhaps the most interesting finding has been that a specific protein has been formed in certain
cells of the interior regions of the hippocampus when a specific task is
learned. For example, if the animal leans to turn left in a maze, it appears,
but not if it turns right. There is no question that with memory fixation
some permanent changes occur. The question remains as to whether this
is in terms of specific neurons, tracks of neurons, or what, and how
predetermined these are. The ability to transfer memory would depend
on a very specific set of task-associated RNA as well as devices to prolong
the life of the RNA. However, the success at memory transfer by the
transfer of some substance from a trained to an untrained animal has
been far from convincing.

COMPUTERS AND THE BRAIN

Information stored affects responses to stimuli, so we believe. The
processes discussed above must somehow as a mass influence the wired
responses of the section preceding it. Our mind staggers at this task and
we are caught in our need for explanations within the limitations of
analogous systems about us. The computer stands out as the only system
available. We have no idea whether it is helping us or leading us down
into a logical trap. Digital computers with their simple yes-no unit
operation seem to be the simplest to store information (as bits) and to
allow decisions to be made. The all or none effect of nerve stimulation
fits. The 10^7 odd neurons of the brain could easily be found in enough
combinations of yes-no states to store the 10^{17} odd bits of information
that is thrust upon it. Even this figure is very extreme in amount by many
orders of magnitude as so little of the information that enters the brain
is actually retained as fixed memory. This is because of competition with

other input signals and especially redundance. The computer and information theory have reminded us of how little of the new messages has to be stored as they can be built on past learned data. Every time one looks at a house he could use a house generalization memory plus combinations of specifics he has most likely already been exposed to, and the house need not be memorized at all. Some estimates are as low as 10^7 bits of new information that is fixed—this being remarkably close to the number of neurons found. But let us remember the methods of estimating these figures are extremely inaccurate—we know too little. Some support for the digital system of neuronal operation is given by learning processes that occur in the eye of an octopus in which there are a limited number of nerve cells and simple learning. Models have been developed that most easily explain this learning by assuming that there are two choices for visual stimuli, attack or retreat; these choices are represented by two nerve endings on the photoreceptors that each act upon hypothesized memory calls whose operation is determined by reward or punishment feedback stimulation arising from corresponding body receptors, depending on whether food or pain has been the result of the attack. A negative response causes cross inhibition from these memory cells to affect the operation of each other and prevent further attack stimulation. A diagram of this hypothesis is shown in Figure 10–7. Also, digital computers have

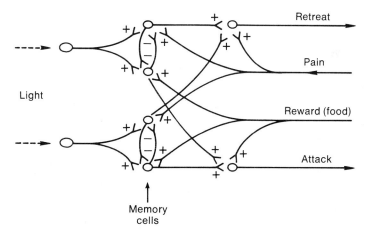

Figure 10-7. Dual switching model as proposed for learning in octopus' eye.

been built that can solve problems by making decisions based on all stored information causing a yes-no-type output. Such machines can play chess with amazing skill and presumably the ability to make decisions of the most complex type can be performed if enough information can be put into and stored by the computer.

There are of course the problems that neurons in the brain of complex animals do not seem to rely on simple on-off-type operation or with simple dual nerve endings for switching operations during learning. Moreover there may not be the specificity of operation where a single switch could influence the operation of all other follow-up circuits. It seems a bit too simple to fit brain function. Potentially the more successful computers are the analogue types that can be designed to produce the characteristics of the brain regardless of how complicated it might be if these characteristics are known and electrical circuit analogues of this operation can be built. This latter, although possibly requiring a great deal of space, is very probable even for nerve function. The problem is the former. It is very difficult and perhaps impossible to measure all the characteristics of the many parts of the brain. It may well be physically impossible to build and energetically run such a computer unless scientists happen to learn how to build the brain itself. The problem with all computers is that they are just as good as the assumptions that go into their basis of operation. Even if they may do some of the things brains do, there is no saying that they are operating as a brain. It may well be wise to ask, in spite of all that has just been said, whether if it was possible to build an electrical analogue computer with all the characteristics of a human brain, could that computer think and be a self-conscious being with an inner universe of its own?

IDEAS AND THE INNER UNIVERSE

The above indicates from an overall model view how it might be that decisions of action may be formed and enacted. Where our brain becomes even more obscure, but related to the above, is in the formation of ideas or thoughts that may or may not be acted upon. Our present fussy concept is that ideas exist in our mind when particular neurons that relate to one another in some common way(s) are activated and interconnected at the same time. Presumably there is an integration of memory units that have something in common. We use the concept of classification frequently to imagine how commonalities are grouped. We imagine also that brain-stem areas combine with a barrage of inner projections from the cortex to make up "words" of even a most specific meaning. Generalizations may well include further groupings of these, etc.

The closest view we seem to have of this mysterious ability is that exposed by speech. It must be quickly added that speech is not the only form of communication of inner thoughts. Gestures and other forms of motor action are well-known and the main mode of communication in other animals. It has been interesting to find that two-way communication can be set up between humans and chimpanzees when sign language is used in place of words. Over 50 nouns, verbs, and adverbs have been used in this way. Apparently the limitation with speech communication is the limitation in the vocal cords of the chimps. In any respect speech in humans should give us more insight into idea formation.

Speech cannot be induced by electrical stimulation of the cortex or any other region. Rather, electrical stimulation may inhibit the formation of certain words. If this happens, the person will try to say the word in an indirect manner, e.g., for shoe, "you put your foot in it." If the person knows more than one language, he cannot say the word in any of these languages. Sometimes the electrical stimulation will cause the person to use another word that may sound like the correct one, but not always. The speaking of words is clearly associated with related expressions. For example, an electrical stimulation may also block the ability to write the word or any gestures such as nodding or smiling if appropriate with the word. The person knows the word he is being asked to say but cannot do it. The point at which a given word is located according to its ability to be blocked in any one person is quite different from the corresponding point in another person. Also the regions are flexible in that if a certain region is damaged, a different region of the cortex will take over. The speech areas, of which there are three, are also flexible in that they are normally found in the dominant hemisphere of the person, but if that hemisphere is damaged, they will appear in the opposite hemisphere. Thus we can see there is close association of all associated memories but that the point of association on the cortex is extremely flexible and can re-form in new places if damaged. This is a rather discouraging finding for those who feel that associations have a predestined pattern. There must also be an integration of rather distant memory traces.

With this flexibility the problem becomes one of trying to imagine how any associations are formed around a given idea, and of even greater interest and speculation is how new ideas or combinations of ideas could come about. It is this question that is most often ignored or rationalized away in any discussion of brain mechanisms from a clock point of view. The question demands some statement about free will. Clock builders most often claim that although it may not be obvious there are specific predetermined associated routes available to form. Sweeping waves of excitation would cause the associations to come to mind depending on some modulation, depending on which sensory inputs were present and which previous idea was stimulated. Original thoughts would be chance associations of sequentially stimulated old ideas. Free will would appear to exist because there are so many combinations possible and so many ideas that quite different associations can form in any one person. Candle builders would keep the associations random and flexible and original ideas and free will would be the rule rather than a chance association. Sweep generators may also be operating to highlight the ideas that happen to be in a certain region, this being the current thought of a type not induced by some sensory stimulation.

The dependence on sweep generators exciting regions of the cortex as they sweep has been a popular notion since EEG (electroencephalograms) waves were discovered. Also there are so many rhythms found among the brain parts. We have mentioned those found in the cerebellum, the RAS, the hippocampus, etc. Dream cycles and sleep cycles are obvious cases, the latter even having a molecular change in neuronal enzyme concentration—being high during sleep and low during wakefulness. There are also endogenous rhythms in the concentration of amines in the hypothalamus and pineal, the former affecting pituitary secretion of hormones that in turn induce those liver enzymes described in the last chapter. Endogenous rhythms are found associated with the nervous systems of every animal studied. These rhythms are used apparently as reference systems for keeping time for navigation and other behaviors as exampled in the well-known bee and bird stories. The brain does have sweeping voltage changes and the question must be asked, what are they for? They just could be used some way in forming our ideas.

The most fascinating thing about ideas is that they can be felt from both sides of the brain, the outside and the inside. The ideas may be expressed as words and gestures but they are also known from the inside. With eyes closed or open we can revisit much of our past and discuss it with ourselves at any length we wish. We each have this mysterious and wonderful inner universe, the mind. To say that it is merely brain function does not do away with this amazing self-consciousness. Since the mind is still a very complicated organization of matter, we must recognize it as a form of matter, aware that it is matter and has organization and is capable of knowing it. But if we can each be internally aware, why are not other animals? The answer we seem to come out with is that they must be. There are animals who can abstract by their ability to count. Animals must also be able to generalize; a dog knows that all trees are a class—he does not have to learn each tree. Also we know mammals dream, as we all know who have had pets wimper or cry in their sleep. Then we might say that any animal with a cortex can think his inner thoughts. But we must ask again whether indeed any animal that learns must have several impressions upon which to make discriminations. And say if such is the case, does this animal not feel within? Could not the ant with his few hundred neurons sense within? If we agree the answer is yes, we must of course say that each animal would have his own limitations of impressions depending on the number of sensory impressions and kinds of memory possible. He would think within these. There would still be a kind of self-consciousness although we may wonder if there were ideas—associations of memory traces. There would have to be some cutoff point below which an "idea" or "word" would not exist, but rather a few selections of persistent neurons or reverberating circuits during learning because of the limited number or organization of neurons. We must know more about learning to be anything but speculative, but we have good reasons to ask these questions. In so doing we face the possibility of a self-conscious computer, do we not?

The beauty and problem in attempting to answer these questions are that the mind is a completely private thing. We can never experience the mind of another organism. We may be able to someday attach electrodes to some part of our brain and see a thought register as a particular electrical pattern. A second person may look at this same pattern and even think until an identical electrical pattern could be obtained from recordings of his own brain and still not know if what he feels in his mind is what we feel in ours. Also this unique property of viewing a living process from within could never be explained in terms of the molecular or neuronal activity of the brain. Perhaps someday we may just accept this dual explanation of the mind and not expect to resolve it.

THOUGHTS ON THOUGHTS

We have in a sense gone about as far as we can in making our rash generalizations and ill-founded judgments of what the brain is and we have finally reached where we started at the begining of this book, talking about thoughts. We cannot claim to explain the ways of thought in terms of brain activities and yet there are interesting relationships that seem worthwhile considering. Perhaps we are made prejudiced by our methods of analysis, but one can find similarities between the learning and decision-making processes and science itself.

Science is a way of thought in which hypotheses are offered, experiments tried, and on the basis of results these hypotheses are supported or denied. Also hypotheses build on one another resulting in generalizations that in the process include explanations for more and more phenomena of the world. The process is never complete; there is an endless readjustment to what is the best explanation possible at the time. There is in this process the tendency for greater abstraction with increasing generalizations, but there is to be no absolute or final explanation expected.

One cannot help but note that learning theory is based upon the same general procedure. Some random activity of the animal causes some sensory input which stimulates a pleasure or pain response which in turn influences the animal's future behavior—whether to repeat the activity again or not. There is reinforcement or discouragement. With many such inputs we imagine that circuits of ideas are integrated and formed, a generalization process. Abstract thinking, a relative term, increases with the extention and development of generalization. Of especial importance to both science and learning, with generalization comes a greater amount of order. The random events about one are placed into an explanatory framework in science and into integrated memory traces in the brain. In both, order comes from disorder by a self-perfecting and self-directing process.

The question that remains is how similar are the sources of judgment during the perfection process of the two systems? Science depends on the reliability of prediction versus an unreliability to predict. The basis of perfection during learning is not known for certain but may depend on opposite reference centers such as those for pleasure and pain. In either case these sources would influence the direction of choices with no anticipation of the goals to be achieved.

Scientists of course are human and capable of as many misjudgments as anyone else and will continue to blunder as well as occasionally find new order. They will even quibble as will anyone about what science is and whether one form of science or one explanation is better than another. One of the purposes of this treatise is to emphasize that there are many ways to find order and many orders and no one way or kind should be considered superior to any other. No kind of abstraction whether number or correlation should be considered superior to another. Likewise an order found for one level of matter may not be expected to apply to other levels. We need every approach and idea man is able to conceive.

Perhaps the most important thing that can be said is that even if the mind does not work as science does, they both bring order and science serves as an extension of the mind's processes, funneling many diverse impressions down to fewer kinds of abstractions that can be learned. This is not to mean that scientific ways of thought or ordering of facts are the only mental decisions made. Any personal decision by man depends on ethical and aesthetic considerations as well as the ordered facts or predictable outcome of some action. There must be other reference levels for these other kinds of decisions. They may be similar to or quite different from the pleasure-pain centers. We know so little about these matters. There are even a few who feel that some day enough generalizations about man's behavior will have accumulated that clear references can be used to decide moral issues. Some have even attempted to define goodness as the most ordered result produced by a personal action. There does not seem to be any justification for such statements at this time. Any attempt to put poles of goodness and badness in terms of defined order will be as subjective as any other decision and cannot be used to make further value judgments with any claim of exactness. Similar arguments apply to aesthetic decisions. At least for the time being it would seem wise to keep an open mind and refrain from subordinating any one way of thought to another.

There must be some relationship among the products of science, knowledge, and man's rate of evolution. There are some evolutionists who are convinced that man has stopped evolving physically. Their argument is that speciation can no longer take place since there are no geographic barriers remaining on earth. They claim that the gene pool of the present human population is stabilizing since our altruistic nature has overcome the natural selection against hereditary diseases and less fit individuals. Also our cultural "plastic bubble" has permitted us to become so adaptable or able to control our own environment that we are no longer under pressure to change.

I believe that this is a rather shortsighted view of a quite different trend. I maintain that not only will man continue to evolve but that he will speed up his own rate of evolution considerably and alter that of other organisms faster than it has been in the past. This is based upon the assumption that cultural evolution contains more factors to change man than it does to preserve his present nature. The source of all these progressive factors is education. Man accumulates and passes along knowledge by means of education. The rate of this accumulation in knowledge seems to be increasing with no upper limit. Knowledge is contributed by all ways of thought that are searching for explanations. As this knowledge is accumulated and acquired by others over the generations, it also becomes more common to all men. Improved means of communication ensure this and this knowledge should lead to an increase in the amount and uniformity of actual physical changes to the brains of men. This physical form of knowledge is not restricted by natural selection pressures of the type that operate on the gene pool. The amount and diversity of physical change seem to be increasing much more rapidly than have the number of genes themselves and thus there should be that many more possible kinds and combinations of ideas. The potential for future change seems immense, especially if one thinks in terms of potential variations in macromolecules and neuron interconnections that are involved but not yet expressed. This should easily permit man to understand the molecular basis of his and all other organism's hereditary development, and even the workings of his brain. With this potential all that is necessary is the drive to acquire that knowledge. It takes a brief look at the history of man and especially the history of science to realize the inherent force of that drive. In spite of the haphazard methods of fluctuating civilizations, man has continued to learn more and more about his environment and life itself. There is an inherent need to understand what is about us. In fact, physically we must learn or die. If our brains are not stimulated, they deteriorate. This drive has to continue to push the evolution of man's brain at rates that are inconceivable at present.

The result of this drive has affected and will affect all about man. For example, the affect of man on the evolution of other organisms is just beginning but is already immense. There is a gradual loss of natural species and an increasing proportion of cultivated ones that have been generated by man, e.g., corn, tomatoes, cows, horses, and many more. We have altered species more rapidly by selective breeding than nature could possibly have done. Man can select very specific characteristics and eliminate others altogether. Although he is also eliminating many species by his lack of concern or present knowledge, this could well reverse. He now knows how to manufacture DNA and thus he could add more species than have been lost. We are now terribly concerned that he will upset ecological balances and destroy all wildlife and then himself. This is doubtful. Although he will destroy much of the wildlife as we know it now, he will not destroy all forms of life. He is too worried to do that and fortunately he knows too much. He will make his environment over more and more to suit himself for food, for recreation, and for beauty. He will design and control the whole earth as he now does gardens about himself. In this sense his plastic bubble will grow and grow to include all the spaces about himself. He will do this for the most efficient use of other life processes to serve himself and for the predictable and yet ever pleasing aspect of a nature he will learn to fabricate. And in order to make his ecosystem more efficient, he will have to create more organisms than have existed.

In order to assure the maintenance of his created environment, he must also conquer the elements. There must be an increase in order, a decrease in entropy, and an improvement in the efficient use of energy as well as a greater flow of energy into our plastic bubble. Man will harness more and more of the energy available to improve himself and other forms of life and control the nonliving environment about himself. He will control fire, water, sunlight, soil, and minerals all to his end. More and more of the mass and energy of the universe will move into an ordered state dictated by the organization of matter that makes up the brains of man. This will not violate any laws of energy because an excess of energy can continue to flow through the system suffering an even greater increase in entropy to keep the whirlpool of life and its plastic bubble tossing back upon itself. He will do all this for himself because he can and he will judge it good, for it will help the maximum number of persons.

Even more important, he will change himself. He will direct his own physical evolution. He has already started this. So far, most of this has been of a negative nature. There are programs of eugenics wherein individuals are prevented from having their detrimental gene succeed to other generations. But positive eugenics have also started wherein individuals choose sperm from donors of their choice because of some advantage, perhaps the lack of a disease or some improved potentials of the brain. When more is understood about the DNA of man, he will use this knowledge to eliminate disease and to improve himself, whatever that means at the time. Probably included among his self-improvements will be that of altering his brain structure to improve his knowledge potential.

As he has discovered that an increase in efficiency, regulation, and order requires an increase in number of specialized integrated parts, so he must produce an even greater diversity in his environment and society. There will no doubt be found to be an advantage in increasing the ways of thought and modes of thinking. The few represented in this book will be miniscule. Hopefully, each person will think pluralistically with even more specialized modes and also be able to integrate his thoughts with others forming a common knowledge analogous to a food-web. The belief-web must continue to become increasingly complex. This could be a utopia of sorts; however, there would still have to be opposite reference points in order to make personal decisions. There will still be beauty and ugliness but the endless search into their deeper meaning and expression will probably never cease. The modes of expressing beauty will most likely diversify immensely. And there shall still be heaven and hell, pain and punishment, discontentment, fear, and hatred as well as satisfaction, peace, and love. But again, there will no doubt be new diverse ways of sensing and expressing these ethical constructs.

No one has expressed with more acceptability the future of our cultural evolution as has Pierre Teilhard de Chardin (1881–1954), a Jesuit priest who is considered one of our greatest paleontologists and spiritual leaders of all time. Where Darwin dealt with change of life and thus introduced direction and Oparin was concerned with the beginning of life from matter and energy, de Chardin explored the culmination of life. He sees the ending as an inward folding combining matter, living and non-living, into a new reciprocal relationship with the force of cultural evolution bringing this about. As shown in Figure 10–8 de Chardin feels that this infolding results when the brains of all men acquire a maximum amount of knowledge and merge into one immense complete belief-web with common generalizations and abstractions. There would still be individual styles of logic but all could understand one another. In this condition there would be a common thread of communication within one system of thought and love. There would be one super mind. This ultimate end point, the omega point as he calls it, is the vision of the universal truth and ultimate of all religious prophets and mystics who have been able to sense this future mental state. They each may have described it differently and prescribed different paths to be used to reach there, but there is but one. This omega point is somewhat analogous to infinite space folding back upon itself. These impressions are still vague as they must be, for they are but the seed of a growing concept that someday may well be explained. By this conception no ending to evolution is foreseen but rather an endless refluxing about itself with no further goals or purposes.

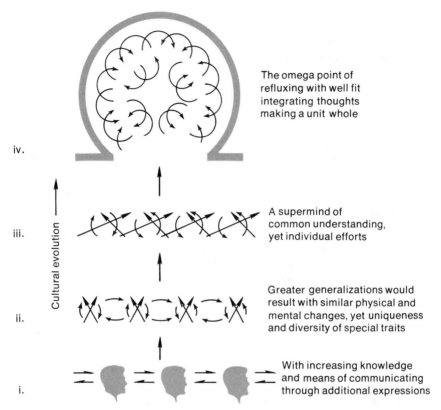

Figure 10-8. De Chardin's model of the culmination (Ω) of life.

A PERSONAL THOUGHT

Finally I would like to impose myself upon you, fair reader, and tell you what I have gained in value judgments from my own particular views of life. I see that each of you must be unique. Each of you has different genes (unless you have an identical twin) and different acquired learning patterns. You could not have experienced the same impressions in the same sequence. You are physically and mentally unique. You cannot be like anyone else and therefore you are in your own mind alone. You must always travel alone with no one else really ever sensing what is in your mind. You can do no other than your own thing. You will resist institutions because their efficient organizations treat all as one. Educational institutes will have standards that tend to make you compare yourself with others. The standards are necessary, however, and valuable if you can only let yourself see them in a different light. These standards serve as a reference point to compare your present self with an older self. Since you are alone, you need this reference point, not as a point or place you should be, this you could never do. But rather, to treat this point as a point some distance off about which you can move. In this way you can tell which way you are moving. This does not mean you necessarily need to move toward or away from the point. You will then know how far and in what direction you have moved and this is the important thing. The exact location has no meaning, all is relative and you will be able to be comfortable with change per se if you can but measure the amount of change. We also need educational institutions because they serve as the storehouse of knowledge.

And when you have mastered the relative sense, then you know that you are neither better nor worse than any other living things, especially other men whom you can more easily imagine. Life is too complex and relative to think that one form of life is better than any other. You must also remember that as a person you have one of the most beautiful entities in the world, a human brain and mind. There is nothing so complex and so beautiful as that brain of yours. No matter how inferior you may feel, and we all do many times, think about your brain and how miraculous it is that this structure could have arisen from a single cell and think—imagine that!

You are and must be alone, but to say this is not enough. Many in fact spend much of their life thinking about the question "Who am I?" This question can never be answered alone. As there are special properties for each different kind of atom to allow it to interact with only certain other atoms to form molecules, so unique sets of different kinds of macro-molecules interact to form the cell, and different kinds of cells interact in specific ways to form the organism. There must also be different kinds of individuals to form societies and in so doing preserve each individual. As pieces in a jigsaw puzzle, if they are all square in shape, they will fall apart—they must be different to stay together. It is our differences that allow us to complement each other and keep us together. We need each other. And the search for who we are is answered by what we see in others we try to fit. Through life, we must all continue to adjust and match our differences with one another. Psychologists will call this process many things: I will call it love. Thank you.

Only a beginning.

name index

Adamson (1727-1806), 29
Aristotle (384-322 B.C.), 4-5, 13, 16, 24, 28, 125-8, 130, 132-4, 144, 158, 165, 168-9, 171, 187, 235, 237
Arrhenius, Svante (1859-1927), 250
Ashbury, W. T., 291, 293, 309
Avery, O., 308

Bacon, Francis (1561-1626), 100, 111
Barrow, Isaac, 2
Beadle, George, 307
Benzer, Seymour, 313
Bernard, Claude (1813-78), 143, 161, 165, 254-5, 285-6
Berzelius, Jons (1779-1848), 243-5
Bichat, Xavier (1771-1802), 188-9
Blyth, Edward (1810-73), 109-11
Bonner, John T., 177-8, 226-7
Bonnet, 150, 169
Botero, Giovanni (1540-1617), 40, 43
Boveri, Theodor (1862-1915), 194-5, 197
Boyle, Robert (1627-92), 138-40
Bridges, Calvin (1889-1938), 211-2, 233
Brown, Robert (1773-1858), 190
Brucke, Ernst (1819-92), 193
Buchner, Edward (1860-1917), 257, 281-2
Buckland, William (1784-1856), 82-3, 85-6, 111

Caesalpinus, Andreas (1519-1603), 134
Caspersson, T., 309
Chambers, Robert (1802-71), 111, 115
Chase, Martha, 308
Christian, John, 53
Clements, Frederic, 68-9
Columbus, Resaldus (1510-59), 133
Correns, Karl (1864-1933), 201, 209
Cowles, Henry, 65
Crick, Francis, 309-10
Cuvier, Georges (1769-1832), 81-3, 94, 104, 111, 166

Dalton, John (1766-1844), 242, 245
Darwin, Charles (1809-82), 74-5, 84-95, 98-102, 104-11, 113-20, 122, 145-6, 179-80, 193, 196, 201, 219, 225, 231-2, 249, 258, 272, 394
Darwin, Erasmus (1731-1802), 101, 104-5, 107, 109, 111
da Vinci, Leonardo (1452-1519), 124, 147-8, 161, 168, 236, 285, 322
de Buffon, Georges L. L. (1707-88), 41-2, 44-5, 78, 100-2, 111, 169-70, 238

de Chardin, Pierre Teilhard (1881-1945), 366, 394
Delbruck, Max, 307-8
Democritus (460-362 B.C.?), 11, 168, 187
Descartes, Rene (1596-1650), 100, 111, 143, 161, 169, 237, 367
de Tournefort, Joseph (1656-1708), 23
de Vries, Hugo (1848-1935), 201, 218, 220
Dobzhansky, Theodore, 223-5, 232
Dreisch, Hans (1867-1941), 174
Dujardin, Felix (1801-60), 191
Dumas (1800-84), 191
Durer, Albrecht (1471-1528), 148, 161
Dutrochet, R. J. H. (1776-1847), 189-90

Einstein, Albert (1879-1955), 8, 13
Elton, Charles, 39, 47
Embden, Gustav, 284-5, 297, 316, 325

Fabricius, Hieronymus (1537-1619), 134, 137, 169, 178
Fisher, Emil (1852-1919), 252, 281, 291
Fisher, Ronald, 222
FitzRoy, Robert, 75, 78, 83-4, 86
Flemming, Walter (1843-1905), 194
Fox, Sidney, 268, 270
Franklin, Rosalind, 309

Galen (131-201), 128-34, 147, 158, 165
Galilei, Galileo (1564-1642), 3-9, 12-16, 20, 43, 72, 122, 135, 149-50, 152, 154
Galton, Francis (1812-1911), 201
Gause, G. F., 40, 49-52, 56, 71-2
Gay-Lussac, L. (1778-1850), 241, 245, 249
Gibbs, Willard (1832-1903), 259
Gilbert, W., 321
Goethe, John Wolfgang von (1749-1832), 124, 145, 150, 166-7, 171-2, 179
Gray, Asa (1810-1888), 111, 117
Grew, Nehemiah (1628-1711), 144, 169, 187
Griffith, F., 308

Haeckel, Ernst (1834-1919), 181-3, 193, 225, 259
Hales, Stephen (1677-1761), 145
Harvey, William (1578-1657), 134-9, 165, 169, 171, 179, 187
Helmholtz, Herman von (1821-94), 253
Henderson, L. J., (1860-1917), 257-9, 261
Henslow, John, 77-8
Hershey, Alfred, 308
Hooke, Robert (1635-1703), 10, 138, 188

Hooker, Sir W. J. (1785-1865), 28, 111, 115, 117
Hutton, James (1726-97), 80-1, 85-6, 111
Huxley, Julian, 155, 177
Huxley, Thomas (1825-95), 117-8, 182

Ingen-Housz, John (1730-99), 145

Jacob, F., 321, 332
Joule, James (1818-89), 253

Kalckar, Herman, 287, 289
Karpechenko, 227
Kekule, Friedrich (1829-96), 251
Kendrew, John, 294
Kolliker, Rudolph A. von (1817-1905), 191 194
Koshland, D., 299
Krebs, H. A., 284-5, 287-9

Lamarck, Jean de (1744-1829), 99, 101-5, 109, 111, 115, 118-9, 166, 179, 189
Lavoisier, Antoin-Laurent (1743-94), 140-2, 145, 240-2, 245, 252-3, 255, 286
Lawrence, William (1783-1867), 109, 111, 115
Lederberg, Joshua, 307
Leeuwenhoek, Anthony van (1632-1723), 138, 169-70, 187, 238
Lehninger, Albert, 289
Leibnitz, Gottfried W. von (1646-1717), 79, 100, 111, 169
Liebig, Justus (1802-73), 44, 55, 62, 244-8, 250-1, 281, 291
Linnaeus, Carolus (1707-78), 25-9, 33, 41, 79, 93, 100, 107, 111
Lipmann, Fritz, 287, 289
Lotka, Alfred, 45-9, 52, 56, 71
Lower, Richard (1632-91), 139
Lyell, Charles (1797-1875), 83-6, 89-90, 92-5, 98-9, 104, 107-8, 110-11, 115-7, 259

MacArthur, Robert, 54
Maillet, Benoit de (1656-1738), 79-111
Malpighi, Marcello (1628-94), 138, 169, 188
Malthus, Thomas Robert (1766-1834), 41-2, 44, 107, 111
Margalef, Raman, 69
Matthew, Patrick, 109
Maupertius, Pierre Louis M. de (1698-1759), 100-1, 111, 187
Mayer, Julius (1814-87), 253
Mayow, John (1643-79) 139, 145
Meckel, J. F. (1761-1833), 179
Mendel, Gregor (1822-84), 201-7, 209, 211, 213-4, 219, 221, 233, 307
Mendeleev, Dmitri (1834-1907), 37, 252
Meyen (1804-40), 190-1
Meyerhof, Otto, 284-5, 287, 316, 325
Miller, Stanley, 267, 270-1
Mirbel, Brisseau de (1776-1854), 189
Mobius, Karl, 60, 68
Mohl, Hugo von, 190
Monad, J., 321, 332

Morgan, Thomas H. (1866-1948), 211-8, 223, 233
Muller, Herman, 218

Nafis, Ibn, 133
Nageli, Carl von (1817-91), 192-3, 202, 219
Naudin, Charles (1815-99), 109
Needham, J. T. (1713-81), 239-41, 247
Newport, George (1803-54), 191
Newton, Isaac (1642-1727), 1-3, 8-13, 16, 20, 138, 154, 171, 233, 237, 253
Nicholson, A. J., 48
Nirenberg, Marshal, 313

Odum, Eugene, 64, 69
Oparin, A. I., 234, 261-72, 275, 325, 394

Paraceleus (1493-1541), 21
Pasteur, Louis (1822-95), 246-50, 255, 257, 259, 281
Pauling, Linus, 292-4, 309-10
Payen, A. (1795-1871), 243
Peltier, Pierre (1788-1842), 191
Persoz, J. F. (1805-68), 243
Perutz, Max, 294
Pfluger (1829-1910), 255-6
Phillips, David, 297-8
Plato (437-347 B.C.), 7, 20, 40, 144, 147
Pouchet, F. (1800-72), 247-8
Prevost (1751-1839), 191
Priestley, Joseph (1733-1804), 140, 145
Pritchard, James (1786-1848), 109

Quetelet (1796-1874), 42, 201

Ray, John (1628-1704), 22-5, 29, 34, 100, 111
Redi, Francesco (1626-97), 238
Ross, Sir Ronald (1857-1932), 45
Roux, Wilhelm (1850-1924), 173-5
Rumford, Count (1753-1814), 253

Saint-Hilaire, Geoffrey (1772-1844), 104, 111
Sanger, Frederick, 292, 307, 325
Saussure, N. T. de (1767-1848), 145
Schleiden, Malthias (1804-81), 190-4
Schneider, Anton, 193
Schrodinger, Edwin, 266, 308, 311
Schultze, Max (1825-74), 192-3, 209, 247
Schwann, Theodor (1810-82), 190-2, 245-7
Sedgwick, Adam (1785-1873), 77-8, 82-4, 86, 89, 100, 111, 115
Serres, E. R. A. (1787-1868), 179
Servetus, Michael (1509-1553), 133
Smith, William (1769-1839), 81-2, 111
Spallanzani, Lazare (1729-99), 170, 185, 188, 191, 239-41
Spemann, Hans (1869-1941), 175
Strasburger, Edward, 194, 196-7
Sumner, James, 284
Sutton, Walter (1876-1916), 205-9, 219-20

Tatum, E., 307
Theophrastus (370-287 B.C.), 21

Thompson, D'Arcy (1860-1948), 152-5, 157-67, 176-7, 185
Tinbergen, Niko, 53
Tschermak, Erick, 201
Turpin, J. P. F. (1775-1840), 190

Urey, Harold, 266-7

Valentin, Gabriel (1810-83), 190
van Beneden (1846-1910), 192, 194-5
Vesalius, Andreas (1514-67), 133
Virchow, Rudolf (1821-1902), 192
von Baer, Karl (1792-1876), 172, 179, 191

Wallace, Russell (1823-1913), 111, 116-7
Warburg, Otto, 283-4, 287
Watson, James, 310
Weinberg, 221-2, 225
Weismann, August (1834-1914), 196-201, 206-7, 219, 226, 259
Wells, William (1757-1817), 109
Werner, Abraham (1749-1807), 80-1, 111
Wilkins, Maurice, 309
Wilson, Edmund (1856-1939), 209-10
Wohler, Frederick (1800-82), 244
Wolff, Caspar (1738-94), 171-2, 188
Wright, Sewall, 222, 224-5

subject index

abiosynthesis (abiogenesis), 263, 267-8
active pump, 305-6, 318
active site, 298-9, 318
adaptation, ecological, 94-5, 103-4, 106, 108, 112, 118
adaptive radiation, 113-4, 122, 227
adenyl cyclase, 354
adult state, 332, 349-52
age of earth, 40, 79, 83, 262, 269
age of life, 269
Age of Reason, 2
albuminous matter or glutinous component, 243, 248
allele frequency (*p* and *q*), 221-3, 225, 227-8, 232
allometry, 155-7, 177-8
alpha helix, 293-4, 309
anabolism concept, 255-6, 260, 268, 285-7, 315-6
anastomoses, 130-1
"Anatomie Generale," 188
animal spirits (pneuma), 128, 132, 142, 144
animiculists, 170, 185, 188
antibody formation, 350-2
aperiodic crystal, 266, 309
apple, the fall of the, 3, 6, 9, 10, 12
Aristotle's gravity, 4-5, 13
arterial vein, 131
atmospheres of planets (primitive earth), 262-3, 265-7, 269-71, 278
atom abundance on earth, 275
atom (element) concept, 242-3, 275
ATP, 287-90, 306, 314-19, 325, 336-7, 354

"Beagle, H. M. S.," 76
belief-webs, 5, 7-8, 12, 17, 20, 110, 185, 394
binomial nomenclature, 27
biogenic amines, 376
biogenic law, 180-3
biogens, 256
biomass, 64-5, 271
biophores, 198-9
blastema, 358
building block pools (e.g., amino acids, etc.), 318, 349, 365

calendar-like cell division, 341
calorimetry, 141
cancer and tumor growth (including crown gall), 359-62
candle models, 234, 236-7, 239-40, 255, 257, 260-1, 264-5, 270, 285, 290, 322, 332, 341, 349, 364-5, 367, 369, 378, 382, 388
carbon properties, 245, 251
cartesian coordinates, 148, 161-2, 167
catabolism concept, 255-6, 260, 268, 285-7, 316
catalytic center, 316-7, 365
catalytic (enzyme) reactions, 243-4, 252, 254, 257, 268, 281-3, 285, 297
catastrophism (cataclysm), 82-3, 85, 93, 99, 119
cell-cell contact, 350, 359
cell concepts, 188-9, 247, 249, 256-7, 264-5, 275, 280-1, 285, 301-2, 306, 311, 314-7, 322, 324-5
cell culturing, 358-60

cell (cycle) division, 190-1, 194, 207-8, 210, 212, 337-41, 347-50, 352, 356-8, 360, 363-4
cellular catalysts, 245-9, 251-2, 255
central dogma, 309-10, 313
centriole, 195
centrosome, 195
cerebellum, 372-4, 388
chromatography, 285, 291
chromosome concepts, 194-6, 198, 209-17
classification:
    artificial system, 23-7
    criterion, 19-25, 27-30, 33
    natural system, 22-3, 29-33
climax and seres, 65, 67-9, 72
clock - models, 234, 236-7, 244, 252, 256-7, 265-6, 270, 291, 306, 313, 322, 332, 337, 341, 349, 364-5, 367, 369, 378, 382, 388
coacervates and colloids, 264-5, 268
coefficient of association, 58
command neurons, 373
community concept, 40, 60-70, 72, 155, 185
competition, ecological, 42, 48-51, 69
computer use and significance, 33, 296, 312, 384-6
coordinate transformation, 161-5, 167
coral island formation, 86-9
corepressor, 321, 340, 349
corpus calossum, 381
cortical (cerebral) column, 371-2
cosmic egg, 275
cosmic evolution, 235, 259, 274-5
c-particle, 360
creationism, 79, 83, 93
crossover (and frequency of), 213-8
crystals and form, 160-1, 167, 191
cultural evolution (plastic bubble), 231, 391-5
cyclic AMP, 353-6, 365
cytoblast, 191

Darwin's early life and character, 76-8, 86-7
"Das keimplasma" (germ plasm), 196, 198-9
degeneracy, 359
dendrogram, 32
dephlogiston (vital air), 140-1
derepressors, 356
"The Descent of Man," 118
determinancy, 176, 185, 359, 363
diagram of forces, 151, 161
diastase and zymase, 243, 247, 257
"Die Thierchemie," 246
differential centrifugation, 301
differentiating characters, 202, 219
dihybrid cross, 205-6, 216
"Discours sur les Revolutions de la Surface du Globe," 82
distribution (ecological) studies, 56-8
DNA, 308-14, 321, 328-9, 333-4, 339-41, 343-4, 347-52, 355-7, 364, 369, 382, 393
Doctrine of Elements, 20
Doctrine of Signatures, 21
dominant characteristics, 203-7, 221
dominant species, 58, 65-7
dream cycles, 377, 388

ecosystem concept, 40, 62-4, 67, 69, 72, 393
efficiency of production, 64
Einstein's theories, 13
electroencephalogram (EEG), 388
electron microscopy, 301-2, 322
electron transport (cytochrome) system, 284-5, 288-9, 303-6, 318
elementary organs, 190
Embden-Meyerhof pathway, 284-5, 287, 316, 325
embryonic induction, 343-6, 348-9, 352-3
endemicity, 95-7, 229-30
endogenous (circadian) rhythm, 340-1, 388
endoplasmic reticulum, 302, 333
endproduct inhibition, 319-20, 340
energy coupling, 256-7, 260-1, 267-8, 289-90
energy diagram, 276-8
energy sinks, 260-1, 272, 324
entropy (entropic doom) and life, 260, 272-90, 312, 323, 331, 393
enzyme concepts, 257, 281-4, 297-9
enzyme induction, 321, 352-3
epigenesis, 168-9, 173, 175, 332
"Escherichia coli," 286, 300, 306-7, 312-3, 319, 321, 323, 333, 337, 341, 348
"Essay on Population," 107
"An Essay on the Principle of Population as It Affects the Future Improvement of Society," 41
eugenetics, 232, 393
evolutionary clock, 326
evolution, meaning and significance, 111, 120, 122, 230,-1, 271-4, 328, 391-2, 394-5
extinction, 90, 92-5, 99, 122
extra-terrestrial life, 271

feedback inhibition in behavior, 372-3, 378, 385
fermentation, 142, 238, 240-1, 245-9, 255, 257, 263, 281, 284, 287
fertilization concept, 169-70, 191, 193-4, 196, 198-200, 206, 342
Fibonacci series, 151
"The Fitness of the Environment," 258
foci of creation, 93
food chain and web, 39, 48-9, 52, 55, 62, 64, 67, 69, 122, 394
force of reaction, 244
fossil horse study, 121, 163-4
fossils, interpretation, 78-9, 81-2, 89-92
free will, 388
frequency profile, 57-8
frontal lobe, 377
fruit fly studies, 210-14, 216, 223, 226, 307, 363
function-form concept, 126, 128-9, 132, 134-5

Galapagos Islands, 95-8, 100, 102, 105-7, 112, 120, 229-31
Galileo's thoughts on gravity, 3-9, 12-5
gastraea, 181
Gause's law (principle), 40, 51, 54, 72
gemmules, 193
gene concept, 216-8, 233, 307, 310-13

gene pool, 228, 232-3
"General Plantarum," 23
gene regulation, 221-2, 332, 334
"Genesis," 78-9, 81, 83
genetic code, 311, 313, 327, 329
genetic drift, 222
genetic load, 232
genotype, 221-2, 232
geographic isolation, 229-31
germ cells (gametes), 196-7, 206-9, 212, 216-7, 220, 222, 226
gill slits, 181-2
glial cells, 383
Goethe's cycle of growth, 172
golden mean, 151-3
gradients of differentiation, 177, 342
growth resistance, 41, 43, 54

habituation, 378-9
hand and glove model of evolution, 272-3
Hardy-Weinberg law, 222
heart concepts, 126-8, 130, 132-5
hemoglobin, 284-6, 289, 300, 319, 325-6
herbalists, 21
hereditary atoms, 217
hereditary factors, 203-5, 207, 214-6
heterochrony, 183, 226
hippocampus, 381-2, 384, 388
histones, 355-6, 365
"Historie Naturelle," 41
homeostasis, 143-4, 146, 357
homologues, 167
homunculi, 169
hormone action, 352-7, 361, 374
hybrid cells, 336
hydrogen-bonds, 293, 304, 310, 356
hypothalamus, 374-6

idant, 198-9
ideal angle, 151
ideoplasm, 194, 198-200
ids, 198-9, 206
immediate principles, 240
independent segregation, 204-7
inducers (for protein synthesis, etc.), 321, 346-7, 349-50, 353, 360-1, 364, 388
information, bits, and life, 309, 311-2, 369, 377-8, 384-5
inhibition of metabolism studies, 284
inserted adaptations, 182
inversion, 223-4
irreductionism, 324
isomers, 245, 251
isozymes, 300, 342

Jacob-Monod model, 332

K_M, Michaelis-Menten constant and relationship, 283
Krebs citric acid cycle, 284-5, 287-9

ladder of life, 21
law of frequencies, 58, 72
law of parallelism, 179-82
laws of von Baer, 172, 179-82

"Lectures on Physiology, Zoology, and Natural History of Man," 109
life cycle concept, 225-8
limbic system, 375-8
living particles, 256
logistic equation, 43, 45, 49, 54
long life mRNA, 336, 339, 344, 350, 384
long term memory, 380-3
lung concepts, 129, 131, 138-9, 141

macronuclei and micronuclei, 334, 341
"Magazine of Natural History," 109
manometry, 283
meiosis, 195, 199, 206, 214-7, 220, 339
membranes, 302-6, 311, 316, 325, 333, 336
memory fixation (protein synthesis, etc.), 383
metamorphosis, 179, 345-7
meteorite composition, 262
micelles, 193
microspheres, 268
microtubules, 365
migration of alleles, 222
mind, 388, 390-1, 394-6
mitochondrial DNA, 333-4
mitosis, 194, 339
molecular ecology, 328
molecular fossil, 327, 329
mosaic development, 173-7, 332, 342
motivation, 374
mRNA, 313-4, 320, 336-7, 341, 344, 347-8, 351, 356, 364
mutability of species, 100-5
mutationists, 221, 227
mutations, 218, 220, 222-3, 227-8, 232, 310, 313-4, 328, 360

natural spirits (pneuma), 124, 130-2, 142, 144
near collision hypothesis, 262
neo-Darwinism, 200
neo-Lamarckism, 200
neotony, 183
neptunism, 78, 80-1, 85
neurosecretory hormones, 376
nitro-saline and nitrous spirits, 139-40, 145
nondisjunction, 212
nuclear rods, 194
nuclei abundance in universe, 275
nucleus concept, 191, 193

omega point, 394-5
one gene-one protein hypothesis, 307
"On Generation in Animals," 169
"On Growth and Form," 152
"On the Tendency of Varieties to Depart Indefinitely from the Original Type," 116
operant conditioning, 379-80
operation taxonomic unit (OTU), 30-3
operon, 321
orbit of the moon, 9-11
organic particles, 238
organic soup, 271
organic substance concepts, 238-40, 244-6, 251-2, 256

origin of cell models, 264-5, 269-71, 281
"The Origin of Life," 262
"Origin of Species" (version 4), 117, 188, 219, 250
oscillatory circuits of neurons, 374, 377, 382, 388
ovists, 169-70, 188

pangenesis, 187, 193, 196, 200
panmerism, 192-8, 207
persistent organs, 193
persistent synapse hypothesis, 382
"Personal Narrative," 77
"The Phenomenon of Man," 394
phenon, 32
phenotype, 218, 221-2
"Philosophia Zoologique," 103, 189
phlogiston, 140-1
phosphorylation, 287-9, 303-6, 318, 335, 354, 356, 365
phyllotaxy, 147, 150-2
physiological units, 193
plant juices (auxin and kinetins), 144-6, 360-1
plateau's surfaces, 158
pleasure-pain centers, 375, 380
pleated plane structure, 294
polyribosome, 339, 343
population cage, 224-5
population growth (S-shape curves), 40-2, 44-5, 50
preformationism, 168-9, 171-5, 187-8, 332
primitive metabolism, 263, 269-70
principle of minimal work, 158
principle of similitude, 150, 152, 154
"Principles of Geology," 83-6, 93, 104
production-respiration ratio, P/R, 69
profile analysis, 147-8
progressionism, 81, 83, 99, 220
protein concepts (including primary, secondary, tertiary, and quaternary structure), 245-6, 264, 266, 268, 282, 290-301, 327
protoplasmic theory, 193
puffing (chromosome), 346-7
push-pull enzyme mechanisms, 297-8

radioisotope methodology, 284-6, 308, 322
Raunkaier profile, 58
rDNA, 334
reaggregation, 359
receptor fields, 370-1
recessive characteristics, 203-7, 221
"Recherche sur l'Organisation des Corps Vivants," 202
reductionism, 324
regeneration (dedifferentiation and redifferentiation), 358-9
regulative development, 173-7, 332
regulatory genes, 321-2, 340, 348-9
replication (of DNA), 309, 337-9, 343-4, 349, 364
repressor, 321, 340, 364
reproduction concepts, 170, 186-7, 201-7, 211, 219-23, 225-6
response conditioning, 378-9

reticule-activating system (RAV), 375-8, 388
reverberation hypothesis, 382
ribosome, 313, 320, 333-4, 343, 350
RNA, 309-11, 313, 333-4, 339, 341-3, 346-51, 353, 355-6, 359, 383-4

science and values, 15-6, 390-2
second messenger, 354, 365
sedimentation, geological, 80, 85
selection:
    artificial, 106, 109, 222
    natural, 106-9, 116-9, 122, 167, 181, 186, 200, 219-22, 224-7, 230-3, 273, 328, 334, 392
self-assembly (reassembly, reconstitution), 301, 303-4, 313
self-consciousness, 386, 388
sexual dimorphism, 33
sex chromosome, 209-13
sex determination, 209-11
sex-linked, 211-4
short term memory, 380-1
"Size and Cycle," 226
sleep, 377, 388-9
soap bubbles, 158
sooty vapours, 135
speciation, 227-32
species definition, 20, 34-5
species diversity, 40, 59-62, 72
spectrophotometry, 283
speech, 387
spiral form, 151-3
split brain, 381-2
spontaneous (auto-) generation, 235, 238, 241, 247-8
star temperatures, 262
statistical analysis of heredity, 201
steady state, ecological, 63, 69
steady state, enzyme kinetics, 283
stem (plasma) cells, 350-2
stratification, geological, 80-1, 89
succession, 40, 65-70, 72
synthesis of analoques, 292
"Systema Natura," 25
"Systeme de la Nature," 100
sweep generators (neuron circuits), 388

template centers, 316-7, 333-4, 364
template concept, 309-10, 315, 321, 328, 337
"Theoria Generationis," 171
"Theorie de la Terre," 79
"The Theory of the Earth," 80
thermodynamics (open and nonequilibrium systems) and the cell, 323-5, 337
time strips, 381
tissue staining significance, 188-9
tissue transplants, 345
T-maze studies, 380, 382
tolerance, ecological, 67
totipotency, 174, 358-9, 364
transcription, 311, 333, 352, 378
transdetermination, 363
transformation, inanimate to animate, 235-9
transforming principle, 308

translation, 311, 333, 378
transmutation of species, 101-6, 108-10, 113, 119
tRNA, 313, 329, 356, 364
trophic levels, 62-4, 69
turnover, *P/B*, 64, 69

uncertainty principle in biology, 324
uncoupling agents, 289, 304, 354
uniformitarionism, 81, 83, 86
utricles, 188

variation (in offspring), 107, 109-10, 113, 115-6, 118-9, 219-20, 226
venous artery, 131-4
"Vestiges of Creation," 115
virus, 308, 313, 336, 351, 359-60

vital (heat) forces, 239-40, 252
vitalism, 175
vital principle, 238-9
vital spirits (pneuma), 127-8, 131-5, 138, 142, 144
vulcanism, 78, 80-1

water cage, 296
water, properties for life, 258
Watson-Crick model, 310
"What is Life?" 266, 308
Winds of Orpheus, 127

X-ray analysis, 291-4, 297, 299, 301, 309-10, 322

"Zoonomia", 102, 105